The Library, National University of Ireland,
Maynooth

This item is due back on the date stamped below.
A fine will be charged for each day it is overdue.

ECOLOGICAL ENGINEERING FOR PEST MANAGEMENT

To our partners: Donna Read, Claire Wratten and Clara I. Nicholls.

ECOLOGICAL ENGINEERING FOR PEST MANAGEMENT

Advances in Habitat Manipulation for Arthropods

Editors

Geoff M. Gurr
Pest Biology and Management Group,
Faculty of Rural Management, University of Sydney,
PO Box 883, Orange, NSW, Australia

Steve D. Wratten
National Centre for Advanced Bio-Protection Technologies,
PO Box 84, Lincoln University,
Canterbury, New Zealand

Miguel A. Altieri
Division of Insect Biology,
University of California,
Berkeley, CA, USA

CSIRO PUBLISHING

CABI Publishing

National Library of Australia Cataloguing-in-Publication entry

Ecological engineering for pest management : advances in habitat manipulation for arthropods.

     Includes index.
     ISBN 0 643 09022 3.

     1. Environmental engineering. 2. Arthropod pests – Control.
     3. Agricultural ecology. I. Gurr, G. M. (Geoff M.). II.
     Wratten, Stephen D. III. Altieri, Miguel A.

     628.96

Published exclusively in Australia and New Zealand, and non-exclusively in other territories of the world (excluding Europe, Africa, the Middle East and the Americas), by:
CSIRO PUBLISHING
PO Box 1139 (150 Oxford St)
Collingwood VIC 3066
Australia

Tel: (03) 9662 7666 Int: +(613) 9662 7666
Fax: (03) 9662 7555 Int: +(613) 9662 7555
Email: publishing.sales@csiro.au
Website: www.publish.csiro.au

Published exclusively in Europe, Africa and the Middle East and non-exclusively in other territories of the world (excluding Australia, New Zealand and the Americas) by CABI Publishing, a Division of CAB International, with the ISBN 0 85199 903 4

CABI Publishing
CAB International
Wallingford
Oxon OX10 8DE
United Kingdom

Tel: +44 (0) 1491 832 111
Fax: +44 (0) 1491 833 508
Email: cabi@cabi.org
Website: www.cabi-publishing.org

Published exclusively in the Americas by Cornell University Press
www.cornellpress.cornell.edu

Set in Minion 10/12
Cover and text design by James Kelly

Front cover (clockwise from top): Strip cut lucerne in Australia provides continuity of habitat for natural enemies (Photopgraph: Z Hossain); A carabid beetle (Photograph: Mike Mead-Briggs); A strip of phacelia around a wheat field in England (Photograph: Steve Wratten); An Australian hoverfly feeding on a flower (Photograph: M. Bowie)

Back cover (from left to right): *Helicoverpa punctigera* larva (Photograph: CSIRO Entomology); Predatory coccinellid feeding at a lucerne flower (Photograph: Z. Hossain); Adult *Helicoverpa punctigera* (Photograph: CSIRO Entomology); Sown wildflower strip (Photograph: G.M. Gurr)

Printed in Australia by Ligare

Foreword

Professors Geoff Gurr, Steve Wratten and Miguel Altieri and colleagues present readers with a thoughtful outlook and perspective in *Ecological Engineering for Pest Management: Advances in Habitat Manipulation for Arthropods*. The stimulating chapters provide timely strategies concerning pest control in crops worldwide.

There is currently a critical need for food production, as the world population is rapidly growing. At present, the World Health Organization (WHO) reports that more than 3 billion humans are malnourished. People are dying from shortages of calories, protein, vitamins A, B, C, D and E, plus iron and iodine. Half of the world population is malnourished, the largest number recorded in history. People who are malnourished are also more susceptible to a wide array of diseases, including malaria, tuberculosis, schistosomiasis, flu, AIDS and numerous other diseases. The WHO reports that human diseases worldwide are increasing.

In part, the number of malnourished is increasing because the world population is increasing faster than the growth in the food supply. The world population now numbers more than 6.3 billion. More than a quarter-million people who need to be fed are added to the world population each day.

Unfortunately, more than 40% of all world food production is being lost to insect pests, plant pathogens and weeds, despite the application of more than 3 billion kilograms of pesticides to crops. Insect pests destroy an estimated 15%, plant pathogens 13% and weeds 12%. These estimated losses vary based on the 'cosmetic' standards that exist in each nation. For example, many fruits and vegetables sold on the Guatemalan or Indian markets would not be saleable in the USA or Australia. Large quantities of pesticides are applied in the USA, Australia and other developed nations to achieve the 'perfect-looking' apple or cabbage.

An excessive amount of pesticide is being recommended to replace the sound habitat manipulations previously employed in crop production. For example, since 1945 the amount of insecticide applied to US crops has increased more than 10-fold, yet crop losses to insects nearly doubled from 7% in 1945 to about 13% today. The reasons for the doubling of crop losses to insects, despite the 10-fold increase in insecticide use, include the reduction of crop rotations, the planting of some crop varieties that are more susceptible to insects, the destruction of natural enemies, the elimination of hedgerow and shelterbelts, an increase in monocultures, reduced crop diversity, reduction in sanitation, the practice of leaving crop residues on the surface of the land and the use of herbicides that increase crop susceptibility to insect attack.

The authors and editors are not opposed to the judicious use of pesticides; their concern is the neglect of various environmentally sound pest controls. The wide array of habitat manipulations currently include agroforestry, biological control, crop rotations, crop diversity, flower strips, natural enemy refuges, trap crops and other technologies. Each of these technologies, and combinations of these pest suppression technologies, offers opportunities to reduce crop losses to pests while at the same time reducing the use of pesticides. The authors of this book are leaders in the development of such approaches. The various chapters present valuable, up-to-date information on how ecological engineering approaches to pest management can be developed and applied, as well as pointing out technical and practical frontiers for future research.

The editors and authors are to be commended for producing an outstanding book that provides many opportunities to help reduce the 40% food-crop losses to pests and increase the food supply to the malnourished billions of the world. We in pest management owe a debt of gratitude to the authors for this timely book on habitat manipulations for pest suppression.

David Pimentel
Cornell University, Ithaca, USA

Contents

Preface

The future for pest management

The ecological engineering discussed in this book involves manipulating farm habitats, making them less favourable for pests and more attractive to beneficial insects. Although they have received far less research attention and funding, ecological approaches may be safer and more sustainable than genetic engineering of crops. This book brings together contributions from international workers at the forefront of the fast-moving field of habitat manipulation. Chapters explore methodological frontiers of ecological engineering ranging from molecular approaches to high-tech marking methods and remote sensing, as well as reviewing theoretical aspects and how ecological engineering may interact with its controversial cousin, genetic engineering. Examples from recent and current research, combined with liberal use of figures and tables, illustrate the elegance and utility of ecological engineering for pest management, showing that it is much more than so-called 'chocolate-box ecology', where the practices are aesthetically pleasing but lacking in rigour and efficacy.

With contributions from Australia, Germany, Israel, Kenya, New Zealand, Switzerland, the USA and the UK, this book provides comprehensive coverage of international progress towards sustainable pest management. We are grateful to many of the contributing authors who have acted as referees for other chapters, but are especially indebted to the following people for also acting as referees: Pedro Barbosa (University of Maryland, USA), Robert Bugg (University of California, USA), Paul De Barro (CSIRO Entomology, Australia), Martin Dillon (CSIRO Entomology, Australia), Les Firbank (Lancaster Environment Centre, UK), Shelby J. Fleischer (Penn State University, USA), David Goldney (University of Sydney, Australia), Matt Greenstone (USDA-ARA,USA), Dieter Hochuli (University of Sydney, Australia), Robert Holt (University of Kansas, USA), Wolfgang Nentwig (University of Bern, Switzerland), David Pimentel (Cornell University, USA), Wilf Powell (Rothamsted Research, UK), Peter Price (Northern Arizona University, USA), Craig Phillips (AgResearch, New Zealand) and Nancy Schellhorn (Adelaide University, Australia). The editors thank also Maureen Mackinney and Dianne Fyfe for wordprocessing support and Fiona Wylie for tireless help accessing even the most obscure references.

We hope this book will raise awareness of the value and potential of international efforts to develop ecologically sound pest management approaches that confer 'ecosystem service' benefits including conservation of wildlife. The ecological engineering approaches described here combine in a dynamic way a knowledge of ecology, behaviour, agronomy, molecular biology and communication in arthropods to reduce pest numbers in low-input farming – surely the essence of a new integrated pest management approach for the 21st century.

Geoff Gurr, Orange, Australia
Steve Wratten, Lincoln, New Zealand
Miguel Altieri, Berkeley, USA
January 2004

Contributors

M.A. Altieri, University of California, Berkeley, USA
Division of Insect Biology, University of California, Berkeley, CA, USA
Email: agroeco3@nature.berkeley.edu

J.M. Alvarez, University of Idaho, USA
University of Idaho, Department of Plant Soil and Entomological Sciences,
Aberdeen R & E Center, 1693 S. 2700 W. Aberdeen, ID 83210, USA

L. Berndt, University of Canterbury, New Zealand
Forest Research, University of Canterbury, PO Box 29 237, Fendalton, Christchurch, New Zealand

M. Coll, Hebrew University of Jerusalem, Israel
Department of Entomology, Hebrew University of Jerusalem, PO Box 12, Rehovot 76100, Israel
Email: coll@agri.huji.ac.il

G.M. Gurr, University of Sydney, Australia
Pest Biology & Management Group, Faculty of Rural Management, University of Sydney,
PO Box 883, Orange, NSW 2800, Australia
Email: ggurr@orange.usyd.edu.au

J. Hagler, Western Cotton Research Laboratory, Arizona, USA
USDA-ARS, Western Cotton Research Laboratory, 4135 E. Broadway Road, Phoenix, AZ 85040, USA

G.E. Heimpel, University of Minnesota, USA
Department of Entomology, University of Minnesota, St Paul, MN 55108, USA

N. Irvin, University of California, Riverside, USA
University of California, 3401 Watkins Drive, Riverside, CA 92521, USA

M.A. Jervis, Cardiff University, UK
Cardiff School of Biosciences, Cardiff University, Cardiff CF10 3TL, UK
Email: jervis@cardiff.ac.uk

Z.R. Khan, International Centre of Insect Physiology and Ecology, Kenya
International Centre of Insect Physiology and Ecology (ICIPE), PO Box 30772, Nairobi, Kenya
Email: zkhan@mbita.mimcom.net

C. Kinross, University of Sydney, Australia
Faculty of Rural Management, University of Sydney, Orange, NSW 2800, Australia
Email: ckinross@orange.usyd.edu.au

D.A. Landis, Michigan State University, USA
Department of Entomology and Center for Integrated Plant Systems,
204 Center for Integrated Plant Systems, Michigan State University, E. Lansing, MI 48824, USA

B. Lavandero, Lincoln University, New Zealand
National Centre for Advanced Bio-Protection Technologies, PO Box 84, Lincoln University,
Canterbury, New Zealand
Email: lavandbl@lincoln.ac.nz

J.C. Lee, University of Minnesota, USA
Department of Entomology, University of Minnesota, St Paul, MN 55108, USA

F.D. Menalled, Montana State University, USA
Department of Land Resources and Environmental Sciences, Leon Johnson Hall,
PO Box 173120, Montana State University, Bozeman, MT 59717-3120, USA
Email: menalled@montana.edu

R.K. Mensah, Australian Cotton Research Institute, Australia
NSW Agriculture, Australian Cotton Research Institute, Locked Bag 1000, Narrabri, NSW 2390, Australia
Email: robertm@csiro.au

C.I. Nicholls, University of California, Berkeley, USA
Division of Insect Biology, University of California, Berkeley, CA, USA

L. Pfiffner, Research Institute of Organic Agriculture, Switzerland
Research Institute of Organic Agriculture (FiBL), Postfach, CH-5070 Frick, Switzerland
Email: lukas.pfiffner@fibl.org

J.A. Pickett, Rothamsted Research, UK
Rothamsted Research, Harpenden, Hertfordshire, AL5 2JQ, UK

S.L. Scarratt, Lincoln University, New Zealand
National Centre for Advanced Bio-Protection Technologies, PO Box 84, Lincoln University,
Canterbury, New Zealand

M.H. Schmidt, Georg-August University, Germany
Department of Agroecology, Georg-August University, Waldweg 26, D-37073 Göttingen, Germany
Email: m.schmidt@ns1.uaoe.gwdg.de

R.V. Sequeira, Queensland Department of Primary Industries, Australia
Farming Systems Institute, Agency for Food and Fibre Sciences,
Queensland Department of Primary Industries, Locked Bag 6, Emerald, Qld 4720, Australia

C. Thies, Georg-August University, Germany
Department of Agroecology, Georg-August University, Waldweg 26, D-37073 Göttingen, Germany

T. Tscharntke, Georg-August University, Germany
Department of Agroecology, Georg-August University, Waldweg 26, D-37073 Göttingen, Germany

J. Tylianakis, Georg-August University, Germany
Department of Agroecology, Georg-August University, Waldweg 26, D-37073 Göttingen, Germany

S.D. Wratten, Lincoln University, New Zealand
National Centre for Advanced Bio-Protection Technologies, PO Box 84, Lincoln University,
Canterbury, New Zealand
Email: wrattens@lincoln.ac.nz

E. Wyss, Research Institute of Organic Agriculture, Switzerland
Research Institute of Organic Agriculture (FiBL), Postfach, CH-5070 Frick, Switzerland

Chapter 1

Ecological engineering, habitat manipulation and pest management

G.M. Gurr, S.L. Scarratt, S.D. Wratten, L. Berndt and N. Irvin

The management of nature is ecological engineering (ODUM 1971).

Introduction: paradigms and terminology

This book is essentially about the management of arthropod pests, though at least some of the principles described will have relevance to other pests, weeds and pathogens. Over recent decades, integrated pest management (IPM) – the combined use of multiple pest-control methods, informed by monitoring of pest densities – has emerged as the dominant paradigm. Each of the specific methodological approaches used in IPM (mechanical, physical and cultural control; host plant resistance; biological control etc; see Figure 1.1) has tended to become a specialised area of research with sometimes only limited communication between researchers across areas. Even sub-areas, such as the four forms of biological control (conservation, classical, inoculation and inundation) recognised by Eilenberg et al. (2001) (Figure 1.1), have tended to become the domain of specialists. This has led to calls for greater cooperation and exchange of ideas between different sub-disciplines. In the case of biological control, for example, Gurr and

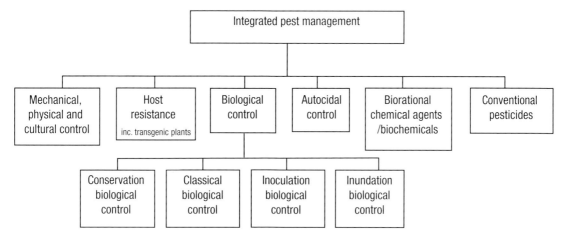

Figure 1.1: Biological control approaches in relation to other tactics available to integrated pest management.

Wratten (1999) proposed the concept of 'integrated biological control', which uses conservation biological control techniques to support classical, inoculation and inundation biological control.

Conservation biological control (CBC) has been defined as 'modification of the environment or existing practices to protect and enhance specific natural enemies of other organisms to reduce the effect of pests' (Eilenberg et al. 2001). In practice, CBC is effected by either (1) reducing the pesticide-induced mortality of natural enemies through better targeting in time and space, reducing rates of application or using compounds with a narrower spectrum efficacy, or (2) by habitat manipulation to improve natural enemy fitness and effectiveness. The second approach often involves increasing the species diversity and structural complexity of agro-ecosystems.

In the context of CBC, habitat manipulation aims to provide natural enemies with resources such as nectar (Baggen and Gurr 1998), pollen (Hickman and Wratten 1996), physical refugia (Halaji et al. 2000), alternative prey (Abou-Awad 1998), alternative hosts (Viggiani 2003) and lekking sites (Sutherland et al. 2001). Habitat manipulation approaches, such as those pictured in Figure 1.2, provide these resources and operate to reduce pest densities via an enhancement of natural enemies. For example, 'beetle banks' (Figure 1.2b) are raised earth ridges that typically run through the centre of arable fields and are sown to perennial tussock-forming grasses.

Figure 1.2: Examples of ecological engineering for pest management: (a) buckwheat strip in the margin of an Australian potato crop providing nectar to the potato moth parasitoid, *Copidosoma koehleri* (Hymenoptera: Encyrtidae) (Photograph: G.M. Gurr); (b) 'beetle bank' in British arable field providing shelter to predators of cereal pests (Photograph: G.M. Gurr); c, strip cutting of a lucerne hay stand in Australia provides shelter to within-field community of natural enemies (Photograph: Z. Hossain); (d) New Zealand vineyard with buckwheat ground cover for enhancement of leafroller parasitoids (Photograph: Connie Schratz).

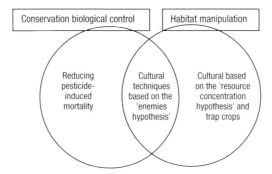

Figure 1.3: Comparing and contrasting habitat manipulation and conservation biological control approaches to pest management. Resource concentration and enemies hypotheses are as defined by Root (1973), see text for detail.

© Kluwer Academic Publishers. Adapted from and originally published in Gurr, G.M., Wratten, S.D. and Barbosa, P. (2000). Success in conservation biological control. In *Biological Control: Measures of Success* (G.M. Gurr and S.D. Wratten, eds), p.107, Figure 1. Reproduced with kind permission of Kluwer Academic Publishers.

During the winter, far higher densities of predatory arthropods shelter on the well-drained, insulated sites than in the open field. In the spring, beetles and other natural enemies emerge from the beetle bank to colonise the growing crop and prevent pest aphid outbreaks (Thomas et al. 1991). When herbivores (the second trophic level) are suppressed by natural enemies (third trophic level) in this manner, control is said to be 'top-down'. Root (1973) referred to pest suppression resulting from this effect as supporting the 'enemies hypothesis'. Importantly, however, within-crop habitat manipulation strategies such as cover crops and green mulches (components of the first trophic level, as is the crop) can also act on pests directly, providing 'bottom-up' control. Root (1973) termed pest suppression resulting from such non-natural enemy effects as the 'resource concentration hypothesis', reflecting the fact that the resource (crop) was effectively 'diluted' by cues from other plant species. These mechanisms are explored in detail in chapter 3, 'The agroecological bases of ecological engineering for pest management', by Nicholls and Altieri.

Though considerable attention has been devoted to testing the relative importance of bottom-up and top-down effects, they are not mutually exclusive and in many systems both are likely to operate (Gurr et al. 1998). Thus habitat manipulation, though it makes a major contribution to CBC, includes a wider series of approaches that may operate independently of natural enemies (Figure 1.3) and, as discussed below, constitute a form of ecological engineering. Examples of ecological engineering for pest management that operate largely by top-down effects are detailed by Pfiffner and Wyss in chapter 11, 'Use of sown wildflower strips'. Natural enemies use such strips for resources such as nectar and pollen in ways explored by Jervis et al. (ch. 5, 'Use of behavioural and life-history studies'). The push–pull and intercropping approaches described in the two chapters by Khan and Pickett (ch. 10) and Mensah and Sequeira (ch. 12) employ top-down effects, but the operation of bottom-up effects is also clearly evident.

Ecological engineering

Odum (1962) was among the first to use the term 'ecological engineering', which was viewed as 'environmental manipulation by man using small amounts of supplementary energy to control systems in which the main energy drives are still coming from natural sources'. In more recent years, Mitsch and Jorgensen (1989) have defined ecological engineering as 'the design of human society with its natural environment for the benefit of both'. Among the characteristics of this

Table 1.1: Applications and examples of ecological engineering.

Application	Examples
Ecosystems used to reduce or solve a pollution problem	Wastewater recycling in wetlands, sludge recycling
Ecosystems imitated to reduce or solve a problem	Integrated fishponds
Recovery of an ecosystem after disturbance is supported	Mine restoration
Existing ecosystems modified in an ecologically-sound manner to reduce an environmental problem	Enhancement of natural pest mortality

Adapted from and reproduced with permission from Mitsch, W.J. and Jørgensen, S.E. (2004). *Ecological Engineering and Ecosystem Restoration.* Wiley, New York.

form of engineering are the use of quantitative approaches and ecological theory as well as the view of humans as part of, rather than apart from, nature. Ecological engineering is a conscious human activity and should not be confused with the more recently developed term 'ecosystem engineering'. This refers to the way in which other species shape habitats via their intrinsic biology rather than by conscious design. For example, termites alter the structural characteristic of soils (Dangerfield et al. 1998), and such ecosystem engineers thereby moderate the availability of resources to other organisms (Thomas et al. 1999).

Recently, Parrott (2002) has discussed the ecological engineering field as having evolved to incorporate a growing number of practitioners whose endeavour is the 'design, operation,

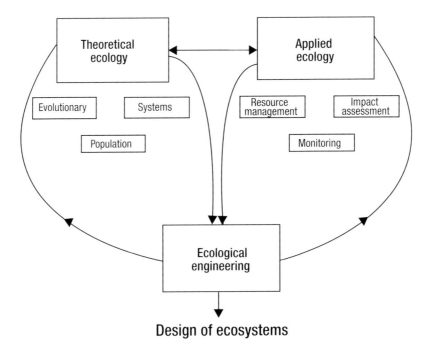

Figure 1.4: The relationship between ecological engineering, and theoretical and applied ecology.

Adapted from and reproduced with permission from Mitsch, W.J. and Jørgensen, S.E. (2004). *Ecological Engineering and Ecosystem Restoration.* Wiley, New York.

management and repair of sustainable living systems in a manner consistent with ecological principles, for the benefit of both human society and the natural environment'. Possibly, however, the most elegant definition of ecological engineering comes from Chinese approaches where a long history of complex land use systems was, in the closing decades of the 20th century, formalised into a 'design with nature' philosophy (Ma 1985). The existence of the well-established periodical *Ecological Engineering: The Journal of Ecotechnology* is evidence of the level of activity in this research field. This title reflects the synonym for ecological engineering, 'ecotechnology'.

Various disciplines are allied to ecological engineering: restoration ecology, sustainable agroecology, habitat reconstruction, ecosystem rehabilitation, river and wetland restoration and reclamation ecology (Mitsch 1991). These sub-sets indicate the range of areas in which ecological engineering has been applied (Table 1.1), including the restoration of wetlands, treatment and utilisation of wastewater, integrated fish culture systems and mining technology (Mitsch and Jorgensen 1989) as well as wildlife conservation (Morris et al. 1994).

The contrast between ecological engineering and other fields, such as theoretical and applied ecology, has been explored by Mitsch (1991). In its role of supporting the design of ecosystems it draws from both theoretical and applied branches of ecology (Figure 1.4) and, as shown by the feedback loops in that figure, can contribute to the knowledge base in these domains.

The last of the types of applications listed in Table 1.1, 'Existing ecosystems modified in an ecologically sound manner to reduce an environmental problem', has particular relevance to agroecosystems. Pimentel (1989) identified several 'ecotechnological principles' that underpin productive, sustainable agricultural systems:

- adapting and designing the agricultural system to the environment of the region (e.g. choice of appropriate crop species and cultivars);
- optimising the use of biological resources in the agroecosystem (e.g. the use of biological control);
- developing strategies that induce minimal changes to the natural ecosystem to protect the environment and minimise use of non-renewable resources (e.g. appropriate fertiliser formulations and application patterns).

Reflecting the utility of the ecological engineering paradigm to agriculture, the term 'agro-ecological engineering' has developed currency (e.g. Hengsdijk and van Ittersum 2003) and this has been viewed explicitly as a way towards sustainable agriculture in China, where it is said to be thriving (Liu and Fu 2000). These authors hold that agroecological engineering produces agricultural systems with multi-components and multi-storey vegetation giving higher vegetative cover than is typical of monocultures. As explored by many authors in the present volume, vegetational diversity plays a central role in habitat manipulation.

It could be argued that all pest management approaches (Figure 1.1) are forms of ecological engineering, irrespective of whether they act on the physical environment (e.g. via tillage), chemical environment (e.g. via pesticide use) or biotic environment (e.g. via the use of novel crop varieties). It is, however, the use of cultural techniques to effect habitat manipulation and enhance biological control (Figure 1.3) that most readily fit the philosophy of ecological engineering. These cultural techniques typically:

- involve relatively low inputs of energy or materials;
- rely on natural processes (e.g. natural enemies or the response of herbivores to vegetational diversity);
- have developed to be consistent with ecological principles;
- are refined by applied ecological experimentation;
- contribute to knowledge of theoretical and applied ecology (Figure 1.4).

The development of habitat manipulation

Contemporary habitat manipulation has its genesis in practices that have been used to promote generalist predators in agricultural systems for centuries (Sweetman 1958). An example of an early habitat manipulation technique, used by Chinese farmers for over 2000 years and still in use today, is the use of straw shelters to provide temporary spider refugia and overwintering sites during cyclic farming disturbances (Dong and Xu 1984). Another technique, developed in Burma in the 1770s, used connecting bamboo canes between citrus trees to enable predatory ants to move between the trees to control caterpillar pests (van Emden 1989).

An analysis of the habitat manipulation literature

There have been a number of important reviews detailing the effects of environmental manipulation on natural enemies (Sweetman 1958; van den Bosch and Telford 1964; Risch et al. 1983; Landis et al. 2000) as well as two major edited volumes on the subject (Barbosa 1998; Pickett and Bugg 1998). However, despite statements that CBC is the least studied form of biological control (Dent 1995), there have been no comprehensive assessments of the historical development of literature for CBC and the allied discipline of habitat manipulation. The only such temporal survey published (Naranjo 2001) was concerned solely with the whitefly, *Bemisia tabaci*. That review supported the notion that the numbers of studies concerned with CBC of this pest had risen in both absolute terms and as a proportion of the literature on *B. tabaci* (Figure 1.5). To address the gap in knowledge about broader research interest in ecological engineering for arthropod management, a survey was undertaken.

A computer-based search of the CAB abstracts database was made for the years 1973 to 2002 to identify articles relating to ecological engineering for pest management for arthropods. References from 2003 were not included, as that year's record was incomplete. The initial search used the terms 'arthropod' and 'biological control' or 'biocontrol', as well as one of the following terms: supplementary food$, companion plant$, cover crop$, enhanc$, habitat manipulat$, flower$, nectar, pollen, shelter, overwinter$, wildflower$, wild flower$, habitat manag$, conservation biological control, landscape ecology, uncultivated corridor$, trap crop$, resource concentration, associat$ resistance. The $ symbol functions as a wildcard truncation to find either no or any ending to the term. The fields 'abstract', 'title' 'original title' and 'heading words' were used for the search.

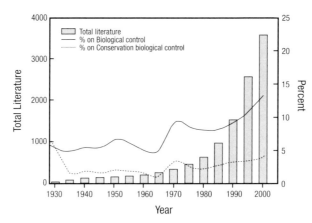

Figure 1.5: Historical summary of the total published research on *Bemisia tabaci/argentifolii* and percentage of the total concerned with biological control and conservation biological control.

Reprinted from *Crop Protection* 20, Naranjo, S.E. Conservation and evaluation of natural enemies in IPM systems for *Bemisia tabaci*. pp. 835–852, 2001, with permission from Elsevier.

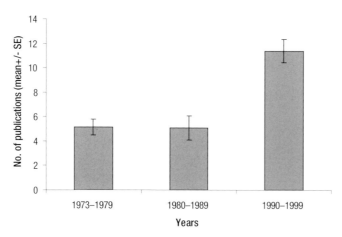

Figure 1.6: The mean number of ecological engineering for pest management publications published annually in each decade from 1973 to 1999.

An article was deemed relevant if the study involved primary research on the management of a habitat to enhance the effectiveness of natural enemies against an arthropod pest. The selected research could not simply involve an observation of a condition that enhances natural enemies, such as landscape complexity, but had to describe or suggest the manipulation of the habitat for that purpose. Literature discussing inundative release of natural enemies was not included, unless habitat manipulation techniques were used to enhance the effectiveness of released natural enemies. The use of bacterial, food or kairomone sprays were excluded because they are essentially input-substitutes for pesticides rather than ecological engineering approaches based on habitat manipulation. Literature that described habitat manipulation to aid in breeding natural enemies for release was also excluded.

This procedure yielded articles concerned with CBC. However, other habitat manipulation studies that dealt with mechanisms such as resource concentration hypothesis effects, companion planting or trap crops in which natural enemies were not involved would not have been captured. To address this, additional searches were done using only some of the search terms previously used: companion plant, trap crop, resource concentration and associat$ resistance. The resulting articles were included if they described habitat manipulation studies that used any of the mechanisms described above. Articles were not included if they used trap crops to attract and/or retain an arthropod pest which was then treated with an insecticide.

Six or fewer ecological engineering for pest management papers were published each year during the 1970s. Apart from 1984, 1986 and 1988, when more than seven papers were published in each year, fewer than six such publications were published in each year in the 1980s. In the early 1990s (1990–92) there were fewer than 10 papers published per year. Following this, the number increased to 11 or more per year. Fewer papers were published in 2001 and 2002 than the annual totals since 1998. However, this does not necessarily indicate a decline in habitat manipulation work in recent years. The apparent decline is more likely due to delays in entering work into CAB abstracts, making data for these years incomplete. When the mean annual number of ecological engineering for pest management publications is considered for each decade, there was a marked increase in the number published in the 1990s (Figure 1.6).

The numbers of ecological engineering for pest management papers published over the surveyed period is modest compared with the total number of CBC studies reported for *B. tabaci* (Naranjo 2001). This may be accounted for by the relatively broad interpretation of CBC in that study. It included articles concerned with reducing pesticide-induced mortality of natural

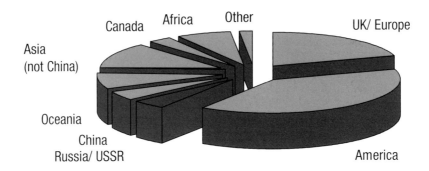

Figure 1.7: The percentage of the total ecological engineering for pest management papers published from each geographical region from 1973 to 2002.

enemies as well as work of a survey nature that would lead to the identification of potentially important natural enemy species or of strictly biological studies (e.g. life table studies). Attempting a similarly broad review of all aspects of habitat manipulation for all species of arthropod would have been unmanageable, so the scope of the present review was deliberately set to identify only those studies in which a habitat manipulation strategy was implemented and tested. Accordingly, the absolute numbers of studies in the present work and that of Naranjo are not directly comparable. Both, however, show a distinct increase in the level of research activity in habitat manipulation and CBC, respectively, from the 1970s to the turn of the millennium. In the case of *B. tabaci*, this appears to have been a continuation of an increasing level of activity that has occurred fairly gradually since the mid-1930s.

The collection of data to examine the historical development of habitat manipulation provided the opportunity to examine other trends in research within this field. In terms of the geographical setting of work, the USA, the UK and Europe together accounted for 60% of the 3529 ecological engineering for pest management papers published since 1973 (Figure 1.7). Almost half were from the USA. The UK and Europe accounted for 20% of the total habitat manipulation papers published; Russia/USSR, China, Oceania and Africa each accounted for 4–8% of the total papers published on this topic.

Another trend apparent in the ecological engineering for pest management literature is that orchards are a common choice of agricultural system. Examples of such studies include Collyer et al. (1975), Yan and Duan (1988), Halley and Hogue (1990), Bugg et al. (1991), Wyss (1995), Stephens et al. (1998) and Kinkorova and Kocourek (2000). One reason may be that such perennial systems which normally include more than one plant species (i.e. some type of groundcover) are more amenable to habitat manipulation than are annual, row-crop monocultures (Altieri 1991). This is borne out experimentally. For example, the response to adding floral diversity to orchards has shown positive results in suppressing pest (specifically aphid) abundances (Halley and Hogue 1990; Wyss 1995). Wyss (1995) examined the effects of 21 weed species on aphids and their natural enemies in an apple orchard in Switzerland. By sowing weed strips between the tree rows and at borders of the orchard, more aphidophagous predators were observed on apple trees within the strip sown area, compared with the control areas. Wyss (1995) also demonstrated that aphid abundances could be reduced in these weed strip areas during the vegetative period. In a two-year study by Halley and Hogue (1990), the effects of different groundcovers on the apple pest *Aphis pomi* and its associated predators were examined in an apple orchard in British Columbia. Although results were mixed across the two years, in 1987 the total season aphid and predator densities were four times lower in the clover-grass ground-cover treatments than in the other treatments, which included rye, treated with herbicide, herbicide strips and grassed alleys, and woven black plastic strips and grassed alleys. In 1988, aphid and predator densities were variable and therefore no differences in arthropod abundances could be detected between the different groundcover treatments for that year. Ecological engineering using sown wildflower strips in perennial crops (as well as in annual crops) is covered in detail by Pfiffner and Wyss (see ch. 11).

In terms of the target pest taxon, a relatively large number of studies examined the effectiveness of habitat manipulation methods for the management of aphids (Bondarenko 1975; Gaudchau 1981; White et al. 1995; Costello and Altieri 1995; Singh et al. 2000; Collins et al. 2002). Correspondingly, hoverflies (Diptera: Syrphidae) were one of the most common beneficial taxa studied (Kowalska 1986; Cowgill 1990; White et al. 1995; Hickman and Wratten 1996; Salveter 1998). Two studies that examined the management of the aphid pest *Brevicoryne brassicae* by hoverflies on cabbage drilled the annual plant *Phacelia tanacetifolia* in the vicinity of the cabbage crop (Kowalska 1986; White et al. 1995). In both studies increased predation by syrphids was detected.

Overall, most of the reviewed papers discussed positive results and outcomes; however, it is likely that there have been many studies that yielded inconclusive or negative results and which have not been published. A published study by Aalbersberg et al. (1989), for example, showed that when a border crop of Japanese radish was planted around a wheat field, the border crop enhanced the numbers of natural enemies present within the wheat field but the aphid population was not reduced. Also, the radish border crop hosted another pest, the pentatomid *Bagrada hilaris*, that significantly damaged the wheat crop. Although this study produced an overall negative result in terms of crop production, it is important to publish such results so that future workers can learn from past studies and improve the techniques. Further, meta-analyses of published work may identify valuable trends such as particular crop systems, pest targets or natural enemy taxa/guilds that are associated with relatively high levels of success.

Conclusion

Ecological engineering is a human activity that modifies the environment according to ecological principles. Accordingly, it is a useful conceptual framework for considering the practice of habitat manipulation for arthropod pest management. This form of ecological engineering presents an attractive option for the design of sustainable agroecosystems. Notwithstanding the increase in absolute numbers of habitat manipulation studies published during the 1990s, this took place over a period when research activity as a whole increased. This is shown by a simple CAB Abstracts search for articles containing the words 'arthropod' and 'pest' in any field. The numbers of articles found rose from 320 per year in the 1970s to 1222 per year in the 1990s. When the annual publication rates for habitat manipulation work shown in Figure 1.6 are expressed as a proportion of all work on arthropod pests, the percentage actually falls from approximately 1.6% to around 0.9%. Thus, though the absolute level of research interest in habitat manipulation approaches has increased, the field has remained modest in size and even contracted to a degree in comparison with other forms of pest management. As even a cursory inspection of the literature indicates, this contrasts with a dramatic increase in genetic engineering-related publications over the same period. As argued by Altieri et al. (see ch. 2), greater parity in research investment (intellectual as well as economic) between ecological and genetic engineering approaches is desirable. It is hoped that this book will encourage further interest in ecological engineering for arthropod pest management.

References

Aalbersberg, Y.K., Westhuizen, M.C. van der and Hewitt, P.H. (1989). Japanese radish as a reservoir for the natural enemies of the Russian wheat aphid *Diuraphis noxia* (Hemiptera: Aphididae). *Phytophylactica* 21 (3): 241–245.

Abou-Awad, B.A., El-Sherif, A.A. et al. (1998). Studies on development, longevity, fecundity and predation of *Amblyseius olivi* Nasr & Abou-Awad (Acari: Phytoseiidae) on various kinds of prey and diets. *Zeitschrift fur Pflanzenkrankheiten und Pflanzenschutz* 105 (5): 538–544.

Altieri, M.A. (1991). Increasing biodiversity to improve insect pest management in agro-ecosystems. In *Biodiversity of Microorganisms and Invertebrates: Its Role in Sustainable Agriculture* (D.L. Hawksworth, ed.), pp. 165–182. CAB International, Wallingford.

Baggen, L.R. and Gurr, G.M. (1998). The influence of food on *Copidosoma koehleri*, and the use of flowering plants as a habitat management tool to enhance biological control of potato moth. *Phthorimaea operculella. Biological Control* 11 (1): 9–17.

Barbosa, P. (ed.) (1998). *Conservation Biological Control.* Academic Press, San Diego.

Bondarenko, N.V. (1975). Use of aphidophages for the control of aphids in hothouses. VIII International Plant Protection Congress, Moscow, 1975. Vol. III. Papers at sessions V, VI and VII, 24–29; 14 ref.

Bugg, R.L., Wackers, F.L., Brunsen, K.E., Dutcher, J.D. and Phatak, S.C. (1991). Cool-season cover crops relay intercropped with cantaloupe: influence on a generalist predator, *Geocoris punctipes* (Hemiptera: Lygaeidae). *Journal of Economic Entomology* 84 (2): 408–416.

Collins, K.L., Boatman, N.D., Wilcox, A., Holland, J.M. and Chaney, K. (2002). Influence of beetle banks on cereal aphid predation in winter wheat. *Agriculture, Ecosystems and Environment* 93 (1–3): 337–350.

Collyer, E. and van Geldermalsen, M. (1975). Integrated control of apple pests in New Zealand. 1. Outline of experiment and general results. *New Zealand Journal of Zoology* 2 (1): 101–134.

Costello, M. and Altieri, M.A. (1995). Abundance, growth rate and parasitism of Brevicoryne brassicae and Myzus persicae (Homoptera: Aphididae). *Agriculture, Ecosystems and Environment* 52 (2–3): 187–196.

Cowgill, S. (1990). The ecology of hoverflies on arable land. *Game Conservancy Review* 21: 70–71.

Dangerfield, J.M., McCarthy, T.S. and Ellery, W.N. (1998). The mound-building termite *Macrotermes michaelseni* as an ecosystem engineer. *Journal of Tropical Ecology* 14: 507–520.

Dent, D. (1995). *Integrated Pest Management.* CAB International. Wallingford, UK.

Dong, C.X. and Xu, C.E. (1984). Spiders in cotton fields and their protection and utilization. *China Cotton* 3: 45–47.

Eilenberg, J., Hajek, A. and Lomer, C. (2001). Suggestions for unifying the terminology in biological control. *BioControl* 46: 387–400.

Gaudchau, M. (1981). The influence of flowering plants in intensively cultivated cereal stands on the abundance and efficiency of natural enemies of cereal aphids. *Mitteilungen der Deutschen Gesellschaft fur Allgemeine und Angewandte Entomologie* 3 (1–3): 312–315.

Gurr, G.M. and Wratten, S.D. (1999). 'Integrated biological control': a proposal for enhancing success in biological control. *International Journal of Pest Management* 45 (2): 81–84.

Gurr, G.M., van Emden, H.F. and Wratten, S.D. (1998). Habitat manipulation and natural enemy efficiency: implications for the control of pests. In *Conservation Biological Control* (P. Barbosa, ed.), pp. 155–183. Academic Press, San Diego.

Gurr, G.M., Wratten, S.D. and Barbosa, P. (2000). Success in conservation biological control. In *Biological Control: Measures of Success* (G.M. Gurr and S.D. Wratten, eds), pp. 105–132. Kluwer, Dordrecht.

Halaji, J., Cady, A.B. and Uetz, G.W. (2000). Modular habitat refugia enhance generalist predators and lower plant damage in soybeans. *Environmental Entomology* 29: 383–393.

Halley, S. and Hogue, E.J. (1990). Ground cover influence on apple aphids, Aphis pomi DeGeer (Homoptera: Aphididae), and its predators in a young apple orchard. *Crop Protection* 9 (3): 221–230.

Hengsdijk, H. and van Ittersum, M.K. (2003). Formalising agroecological engineering for future-oriented land use studies. *European Journal of Agronomy* 19: 549–562.

Hickman, J.M. and Wratten, S.D. (1996). Use of Phacelia tanacetifolia strips to enhance biological control of aphids by hoverfly larvae in cereal fields. *Journal of Economic Entomology* 89 (4): 832–840.

Kinkorova, J. and Kocourek, F. (2000). The effect of integrated pest management practices in an apple orchard on Heteroptera community structure and population dynamics. *Journal of Applied Entomology* 124 (9–10): 381–385.

Kowalska, T. (1986). The action of neighbouring plants on populations of entomophagous insects in late cabbage crops. *Colloques de l'INRA* 36: 165–169.

Landis, D.A., Wratten, S.D. and Gurr, G.M. (2000). Habitat management to conserve natural enemies of arthropod pests in agriculture. *Annual Review of Entomology* 45: 175–201.

Liu, G. and Fu, B. (2000). Agro-ecological engineering in China: a way forwards to sustainable agriculture. *Journal of Environmental Sciences* 12: 422–429.

Ma, S. (1985). Ecological engineering: application of ecosystem principles. *Environmental Conservation* 12: 331–335.

Mitsch, W.J. (1991). Ecological engineering: the roots and rationale for a new ecological paradigm. In *Ecological Engineering for Waste Water Treatment. Proceedings of the International Conference at Stensund Folk College, Sweden* (C. Etnier and B. Gutersham, eds), 24–28 March 1991. Bokskogen, Stensund, Sweden.

Mitsch, W.J. and Jørgensen, S.E. (1989). Introduction to ecological engineering. In *Ecological Engineering: An Introduction to Ecotechnology* (W.J. Mitsch and S.E. Jørgensen, eds), pp. 3–19. Wiley, New York.

Mitsch, W.J. and Jørgensen, S.E. (2004). *Ecological Engineering and Ecosystem Restoration*. Wiley, New York. 424 pages.

Morris, M.G., Thomas, J.A., Ward, L.K., Snazell, R.G., Pywell, R.F., Stevenson, M.J. and Webb, N.R. (1994). Recreation of early successional stages for threatened butterflies – an ecological engineering approach. *Journal of Environmental Management* 42: 119–135.

Naranjo, S.E. (2001). Conservation and evaluation of natural enemies in IPM systems for *Bemisia tabaci*. *Crop Protection* 20: 835–852.

Odum, H.T. (1962). Man in the ecosystem. In: Proceedings Lockwood Conference on the Suburban Forest and Ecology. *Bulletin of the Connecticut Agricultural Station 652,* pp. 57–75. Storrs, CT.

Odum, H.T. (1971). *Environment, Power and Society*. John Wiley & Sons, New York. 336 pages.

Parrott, L. (2002). Complexity and the limits of ecological engineering. *Transactions of the American Society of Agriculture Engineers* 45: 1697–1702.

Pickett, C.H. and Bugg, R.L. (eds) (1998). *Enhancing Biological Control: Habitat Management to Promote Natural Enemies of Agricultural Pests*. University of California Press, Berkeley, CA.

Pimentel, D. (1989). Agriculture and ecotechnology. In *Ecological Engineering: An Introduction to Ecotechnology* (W.J. Mitsch and S.E. Jorgensen, eds), pp. 103–125. Wiley, New York.

Risch, S.J., Andow, D. and Altieri, M.A. (1983). Agroecosystem diversity and pest control: data, tentative conclusions, and new research directions. *Environmental Entomology* 12 (3): 625–629.

Root, R.B. (1973). Organisation of a plant–arthropod association in simple and diverse habitats: the fauna of collards (*Brassica oleracea*). *Ecological Monographs* 43: 95–124.

Salveter, R. (1998). The influence of sown herb strips and spontaneous weeds on the larval stages of aphidophagous hoverflies (Diptera: Syrphidae). *Journal of Applied Entomology* 122 (2–3): 103–114.

Singh, H.S., Singh, R. and Singh, R. (2000). Effects of cropping system on the incidence of mustard aphid, Lipaphis erysimi Kalt. and its natural enemies under the agroclimate condition of eastern Uttar Pradesh. *Sashpa* 7 (1): 41–47.

Stephens, M.J., France, C.M., Wratten, S.D. and Frampton, C. (1998). Enhancing biological control of leafrollers (Lepidoptera: Tortricidae) by sowing buckwheat (Fagopyrum esculentum) in an orchard. *Biocontrol Science and Technology* 8 (4): 547–558.

Sutherland, J.P., Sullivan, M.S. and Poppy, G.M. (2001). Distribution and abundance of aphidophagous hoverflies (Diptera: Syrphidae) in wildflower patches and field margin habitats. *Agricultural and Forest Entomology* 3: 57–64.

Sweetman, H.L. (1958). *The Principles of Biological Control*. Wm C. Brown Co., Dubuque, IA.

Thomas, F., Poulin, R., Meeus, T., de Guegan, J.F. and Renaud, F. (1999). Parasites and ecosystem engineering: what roles could they play? *Oikos* 84: 167–171.

Thomas, M.B., Wratten, S.D. and Sotherton, N.W. (1991). Creation of 'island' habitats in farmland to manipulate populations of beneficial arthropods: predator densities and emigration. *Journal of Applied Ecology* 28: 906–917.

van den Bosch, R. and Telford, A.D. (1964). Environmental modification and biological control. In *Biological Control of Insect Pests and Weeds* (P. DeBach, ed.), pp. 459–488. Chapman & Hall, London.

van Emden, H.F. (1989). *Pest Control*. Edward Arnold, London.

Viggiani, G. (2003). Functional biodiversity for the vineyard agroecosystem: aspects of the farm and landscape management in Southern Italy. *Bulletin Oilb/Srop* 26: (4): 197–202.

White, A.J., Wratten, S.D., Berry, N.A. and Weigmann, U. (1995). Habitat manipulation to enhance biological control of Brassica pests by hover flies (Diptera: Syrphidae). *Journal of Economic Entomology* 88 (5): 1171–1176.

Wratten, S. (1992). Farmers weed out the cereal killers. *New Scientist* 135 (1835): 31–35.

Wyss, E. (1995). The effects of weed strips on aphids and aphidophagous predators in an apple orchard. *Entomologia Experimentalis et Applicata* 75 (1): 43–49.

Yan, Y.H. and Duan, J.J. (1988). Effects of cover crops in apple orchards on predator community on the trees. *Acta-Phytophylactica-Sinica* 15 (1): 23–27.

Chapter 2

Genetic engineering and ecological engineering: a clash of paradigms or scope for synergy?

Miguel A. Altieri, Geoff M. Gurr and Steve D. Wratten

Introduction

Genetically engineered (GE) crops, otherwise known as transgenic or genetically modified (GM) crops, are becoming an increasingly dominant feature of agricultural landscapes. Worldwide, the areas planted to transgenic crops have increased dramatically in recent years, from 3 million ha in 1996 to 58.7 million ha in 2002 (see Figure 2.1; James 2002). Globally, the main GE crop species are soybean, occupying 36.5 million ha, and maize (also known as corn) at 12.4 million ha, followed by cotton and canola. Other GM crops available are potato, sugar beet, tobacco and tomato (Hilbeck 2001). In the US, Argentina and Canada, over half of the area planted to major crops such as soybean, corn and canola is occupied by transgenic varieties. Herbicide-tolerant (HT) crops and those expressing insecticidal toxins from the bacterium *Bacillus thuringiensis* (Bt) have consistently been the dominant traits in GE crops, though a range of quality traits has been the subject of much research. These are likely to be used commercially in the near future (Hilbeck 2001).

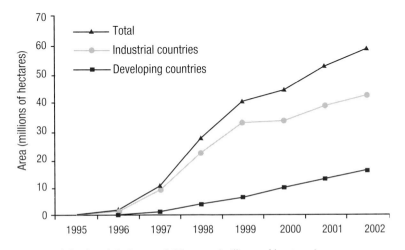

Figure 2.1: Growth in the global area of GE crops (millions of hectares).

Reproduced with permission from James, C. Preview: global status of commercialized transgenic crops: 2002 *International Service for the Acquisition of Agri-Biotech Application* Briefs No. 27; Published by ISAAA 2002.

Transnational corporations, the main proponents of biotechnology, argue that carefully planned introduction of these crops should reduce crop losses due to weeds, insect pests and pathogens. They hold that the use of such crops will also benefit the environment by significantly reducing the use of agrochemicals (Krimsky and Wrubel 1996). It has been suggested that 'if adequately tested', GE crops may promote a sustainable environment (Braun and Ammann 2003). This view is, however, not universal and environmental organisations such as Greenpeace oppose GE crops for a variety of reasons (see below). Scientists are intensely involved in investigating the possible adverse effects of GE crops and the literature in this field is large and growing explosively. Some, such as Herren (2003) and Krebs et al. (1999), question whether we have learnt sufficiently from the past, particularly from the naive optimism with which pesticides were embraced in the mid 20th century. At a recent conference, Tappeser (2003) presented statistics showing the very small fraction, 3% or less, of biotechnology budgets spent on biosafety or biodiversity studies. Wolfenbarger and Phifer (2000) concluded that key experiments on environmental risks and benefits of GM crops are lacking.

Since the turn of the millennium there have been many studies in this area. In October 2003 there was a release of findings from farm-scale evaluations of the effects of GE crops on various aspects of biodiversity (Firbank 2003). Many authors have explored the risks and benefits of GE crops, such as environmental impact (Hails 2003; Jank and Gaugitsch 2001; Dale et al. 2002;), ecosystem services (Lovei 2001), farm biodiversity (Firbank and Forcella 2000; Firbank 2003; Watkinson et al. 2000), changes to plant community structure resulting from gene flow (Pascher and Gollmann 1999) and ethical considerations (Robinson 1999). An extensive literature has also developed on the utility and challenges of Bt crops on target Lepidoptera (e.g. Edge et al. 2001; Cannon 2000; Shelton et al. 2002) in crops such as cotton and corn. This chapter will provide an overview of these issues. However, the possible risks to human health, consumer opposition to the produce from GE crops and the resultant complexities of segregating GE and non-GE produce, international trade implications and market-place product labelling are beyond the scope of this volume. The main objective of this chapter is to consider the opportunities for, and risks to, habitat manipulation of GE crops – effectively the interface of genetic engineering and ecological engineering, two approaches with many points of contrast (Table 2.1). This interface has been the subject of relatively little attention but is informed by the considerable volume of recent publications about the possible effects of GE crops on natural enemies and on farm biodiversity, both key factors in habitat manipulation.

Table 2.1: Comparison of ecological engineering with genetic engineering in agriculture.

Characteristic	Ecological engineering	Genetic engineering
Units engineered	Species	Organisms
Tools for engineering	Ecosystems	Genes
Principles	Ecology	Genetics/ molecular biology
Biotic diversity	Maintained/enhanced	Potentially threatened
Maintenance and development costs	Moderate	High
Public acceptability	High	Low
Level of current use in agriculture	Limited uptake in developed countries, though reflected in many traditional agricultural systems	Widespread in some 'developed' countries

Adapted from and reproduced with permission from Mitsch, W.J. and Jørgensen, S.E. (2004). *Ecological Engineering and Ecosystem Restoration.* Wiley, New York.

Opportunities and threats of GE crops

Opportunities

Experience since the mid-1990s, after which GE crops were widely grown and intensively investigated, suggests some significant advantages over conventional crops. This view is, however, actively contested and, as explored in following sections, many commentators have valid grounds for concern. The use of GE soybean, corn, canola and cotton has been estimated to have reduced pesticide used by 22.3 million kg of formulated product (Phipps and Park 2002). Such reductions are, given the widespread acceptance of the environmental impact of pesticides, likely to have had a beneficial effect on biodiversity. In particular, reductions in pesticide use will reduce the pesticide-induced mortality of natural enemies – an aspect of conservation biological control (Barbosa 1998; Gurr and Wratten 2000) with consequent benefits to pest management. This effect is one of several possible synergistic effects of GE crop use on ecological engineering approaches for pest management (Table 2.2).

Table 2.2: Summary of possible synergistic and antagonistic effects of insect-resistant and herbicide-tolerant genetically engineered crops on habitat manipulation for pest management.

Insect-resistant crops	
Possible synergistic effect	**Possible antagonistic effect**
Reduced pesticide-induced mortality of natural enemies. Resistance management refugia may provide resources for natural enemies. Depressed development rate of pests provides more time for predation/parasitism to occur.	Effects of gene products on natural enemies: directly via polyphagous pests feeding on plant material (e.g. pollen) or indirectly via their prey/hosts. Effects of gene products on soil fauna: food web consequences for natural enemies. Pest resistance to Bt from widespread use of GE crops precludes use of Bt sprays and forces use of insecticides that suppress natural enemy activity.
Herbicide-tolerant crops	
Reduced need for prophylactic herbicide use, greater scope for threshold-driven herbicide use, possible enhancement of weed flora in time and space, leading to: Enhanced resource concentration effect on herbivores. Resources for natural enemies. Reduced need for tillage: less disruption of beneficial soil arthropods. Scope to allow easily managed weed strips connecting margins with field interiors, beetle banks etc. facilitating natural enemy dispersal.	Risk of enhanced 'clean crop' paradigm, impoverished weed flora, leading to: Diminished resource concentration effect on herbivores. Paucity of resources for natural enemies.
Generic issues	
More efficient production on crop lands, greater scope to retain or re-establish areas of non-crop vegetation.	Entrenched monocultures, reduced floral diversity and impoverished natural enemy communities.

Other work has suggested another potential benefit of HT crops; a band-spraying system allowed some weedy areas to remain within GE beet planting, with enhanced arthropod biomass as a result (Dewar et al. 2002). Those authors considered that such HT crops could be a 'powerful tool' for sustainability, a notion broadly echoed by Dale et al. (2002). Theoretically, the ease with which weeds can be managed within HT crops would allow farmers to create precise patterns of weed strips connecting field margins with field interiors and features such as beetle banks (Thomas et al. 1991). Such a network of habitat corridors would allow natural enemies to move easily from their overwintering habitats and disperse readily within crops, enhancing the speed with which a numerical aggregative response to pest foci may take place. However, without legislation, it remains to be seen whether profit-driven agriculturalists will adopt such practices. Also, 'enhanced arthropod biomass' is many steps away from pest population reductions brought about by natural enemies – and even further away from pest populations being reduced below economic thresholds by such practices.

In the farm-scale evaluations of HT crops recently completed in the UK, auditing of herbicide use in GE and conventional crops showed that GE crops generally received only one herbicide active ingredient per crop, later and fewer herbicide sprays and less active ingredient than did the comparable conventional crops. This indicated scope to manage weeds in HT crops with only one well-timed herbicide application (Thomsen et al. 2001). Further, inputs could be tailored to weed pressure and to environmental benefits such as conserving a particular animal species, such as skylark (*Alauda arvensis*) which requires low-growing vegetation (Thomsen et al. 2001). Provided that GE crops lead to spatial or temporal increases in weed populations, their presence within crops is likely to influence arthropod pest densities by resource concentration effects and, whether present within crops or in less severely affected field margins or more widely in the landscape, could benefit natural enemies by providing resources (see ch. 3 in this volume).

The results of the British farm-scale evaluations showed that biomass of weeds was greater in HT corn than in conventionally managed crops, but this effect was reversed in canola and beet (Hawes 2003). Generally, the densities of herbivores, pollinators and natural enemies changed in the same direction as the changes in weed biomass in each crop species. For butterflies in beet and canola, and for Heteroptera and bees in beet, HT crops had lower populations. Effects on soil surface invertebrates such as spiders and carabid beetles were approximately evenly balanced between increases and decreases in the GE crops compared with conventional crops (Roy 2003). Generally, densities were increased in GE corn though decreased in GE canola and beet. Trophic interactions may also involve detritivores (Figure 2.2) – collembolan densities were significantly higher in GE crops – trends that were considered to apply generally across agriculture in the UK (Brooks et al. 2003). The importance of detritivores in pest management is that many are important components of the diets of generalist predators, so their presence could theoretically help maintain within-field communities of natural enemies, even during periods of prey scarcity.

Additional benefits claimed for GE crops include improved levels of control of target pests, increased yield and profits, and reduced farming risk (Edge et al. 2001). One specific benefit from the use of habitat manipulation techniques that could arise from the use of GE Bt crops is the reduced development rates of those herbivores that do not receive a lethal dose of Bt toxin (Johnson 1997). This would allow a longer period of time for predators or parasites to respond to pest patches via aggregative or numerical responses (Solomon 1976).

The major review by Shelton et al. (2002) concluded that, based on available information, the expectation that Bt crops would have lower or equivalent risks than current or alternative technologies (e.g. conventional crops protected by pesticides) was supported. In the future, the development of multi-lines that are uniform for most agronomic characteristics yet differ in genetic resistance traits offers scope to establish within-field crops that will be less prone to 'resistance breakdown' (Conner and Christey 1994). Current GE insect-resistant crops, though

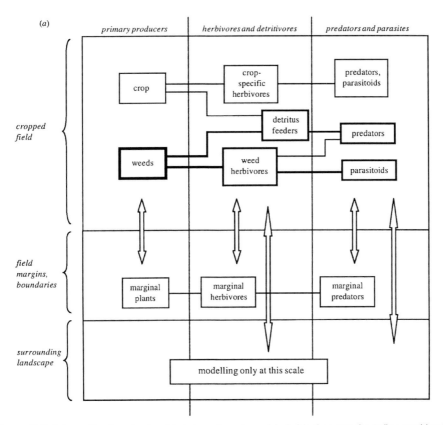

Figure 2.2: Schematic of major trophic interactions in arable fields that may be influenced by GE crops.

Reproduced with permission from Squire et al. (2003). On the rationale of the farm scale evaluations of genetically modified herbicide–tolerant crops. *Philosophical Transactions of the Royal Society of London* B 358: 1779–1799.

they may have multiple ('pyramided') resistance genes, lack multi-lines with differing resistance genes. But these could be used in conjunction with habitat manipulation methods. As noted by Khan and Pickett (ch. 10, this volume), scope exists for ecological engineering approaches, such as the 'push–pull' strategy they describe for stemborer management, to support the durability and effectiveness of Bt crops such as corn. Further, the use of refuge crops – areas of non-GE crop grown as a means of producing adult pests that will mate with the resistant survivors developing in adjacent GE crops – is a form of ecological engineering that has been the subject of much research (e.g. Cannon 2000; Shelton et al. 2002 and references therein). These refuges could also provide resources for natural enemies in ways allied to those explored by Mensah and Sequeira (ch. 12, this volume). Other aspects of how GE crops may support or interfere with ecological engineering (Table 2.2) are yet to be fully explored.

Ultimately, if GE crops can make crop production more efficient, the increasing social requirements for food and fibre will be met with a reduced need for expansion of croplands into natural and semi-natural areas. This will increase scope to conserve, or reintroduce into farm landscapes, areas of non-crop vegetation. As is clear from other chapters, such vegetation can have desirable consequences for pest management (Schmidt et al., ch. 4 this volume), value in wildlife conservation (Kinross et al., ch. 13 this volume), as well as catchment stability, water purification, recreation and aesthetics. The latter processes reinforce the dependence of ecological engineering (Mitsch and Jorgensen 1989) on a sound knowledge of how to enhance 'ecosystem services' (Costanza et al. 1997).

Threats: evidence for and against

To the extent that transgenic crops further entrench the current monocultural system, they discourage farmers from using other pest management methods (Altieri 1996), including the ecological engineering approaches described by contributors to this book. GE crops encourage agricultural intensification and as long as the use of these crops closely follows the high-input, pesticide paradigm, such biotechnological products will reinforce the 'pesticide treadmill' (*sensu* Tait 1987) in agroecosystems.

There are several significant environmental risks associated with the rapid deployment and widespread commercialisation of such crops in large monocultures (Rissler and Mellon 1996; Snow and Moran 1997; Kendall et al. 1997; Altieri 2000). These risks include:

- the spread of transgenes to related weeds or conspecifics via crop–weed hybridisation;
- reduction of the fitness of non-target organisms (especially weeds or local varieties) through the acquisition of transgenic traits via hybridisation;
- the evolution of resistance of insect pests, such as Lepidoptera, to Bt toxins;
- accumulation of the Bt toxins, which remain active in the soil after the crop is ploughed under and bind tightly to clays and humic acids;
- disruption of natural control of insect pests through intertrophic-level effects of the Bt toxin on natural enemies;
- unanticipated effects on non-target herbivorous insects through deposition of transgenic pollen on foliage of surrounding wild vegetation;
- vector-mediated horizontal gene transfer (i.e. to unrelated taxa) and recombination to create new pathogenic organisms;
- the possible escalation of herbicide use in HT crops with consequent environmental impacts including reduced weed populations;
- reduced weed populations leading to declines in bird populations that feed on or shelter in weeds or feed on the arthropods supported by weeds;
- reduced weed populations leading to higher pest damage because of resource concentration (Root 1973; Nicholls and Altieri, ch. 3 this volume) effects or impoverished natural enemy communities.

Much has been written in recent years on the magnitude of such risks and some studies – that concerning monarch butterflies, for example (Losey et al. 1999) – have been criticised by later authors (Conner et al. 2003) on grounds such as limited scale, artificiality and the nature of assumptions made. The aim of this chapter is not to enter the debate concerning the value and limitations of specific studies, but to consider at a larger scale the implications for agroecosystem biodiversity of GE crops, which appear certain to remain a major feature of world agriculture. In accepting this reality, this chapter briefly considers the evidence for and against some of the above threats, but focuses chiefly on the interplay between arthropod pest management strategies based on genetically engineered crops and ecologically engineered crop systems (Table 2.2).

Biotechnology, agrobiodiversity and pest vulnerability

Direct benefits of biodiversity in agriculture lie in the range of ecosystem services provided by the different biodiversity components. These include nutrient cycling, pest regulation and pollination (Gurr et al. 2003). In relation to pest management, the widespread use of crop monocultures and attendant genetic homogeneity are often associated with elevated pest densities (Nicholls and Altieri, ch. 3 this volume). Should the use of GE crops reinforce this simplification of farming systems, a range of negative consequences could affect ecosystem services and agroecosystem function.

There is no doubt that human civilisation in general, and agriculture more specifically, are major causes of the loss of biodiversity (Conner et al. 2003). Agriculture typically represents an extreme form of simplification of terrestrial biodiversity because monocultures, in addition to being genetically uniform and species-poor systems, advance at the expense of non-crop and natural vegetation, key landscape components that provide important ecosystem services (Altieri 1999). Since the onset of agricultural modernisation, farmers and researchers have faced an ecological dilemma arising from the homogenisation of agricultural systems: the increased vulnerability of crops to arthropod pests and diseases which can be devastating when infesting genetically uniform, large-scale monocultures. Worldwide, 91% of the 1.5 billion ha of cropland are under annual crops, mostly monocultures of wheat, rice, maize, cotton and soybeans (Smil 2000). Monocultures may have temporary economic advantages for farmers, but in the long term they do not represent an ecologically sound optimum. Rather, the drastic narrowing of cultivated plant diversity has put the world's food production in greater peril (Robinson 1996). Monocultures also limit the extent to which farmlands – which cover large areas of the world, for example 70% of the UK (Hails 2003) – can contribute to conservation of wildlife (see Kinross et al., ch. 13 this volume).

Experience has repeatedly shown that uniformity in agricultural areas sown to a smaller number of varieties, as in the case of GE crops, is a source of increased risk for farmers, as the genetically homogeneous fields tend to be more vulnerable to disease and pest attack (Robinson 1996). Examples of disease epidemics associated with homogeneous crops abound in the literature, including the $1 billion loss of maize in the US in 1970 and the 18 million citrus trees destroyed by pathogens in Florida in 1984 (Thrupp 1998).

Increasingly, evidence suggests that changes in landscape diversity due to monocultures have led to more insect outbreaks, because of the removal of natural vegetation and decreased habitat diversity (Altieri 1994; Schmidt et al., ch. 4 this volume; Nicholls and Altieri, ch. 3 this volume). A main characteristic of the transgenic agricultural landscape is the large size and homogeneity of crop monocultures that fragment the natural landscape. This can directly affect abundance and diversity of natural enemies, as the larger the area under monoculture the lower the viability of a given population of beneficial fauna. At the field level, decreased plant diversity in agro-ecosystems allows greater chance for invasive species to colonise, leading to enhanced herbivorous insect abundance. Many experiments have shown fairly consistent results: specialised herbivore species usually exhibit higher abundance in monoculture than in diversified crop systems (Andow 1991; Nicholls and Altieri, ch. 3 this volume).

Field margins are widely accepted as important reservoirs of natural enemies of crop pests, as these habitats provide sources of alternative prey/hosts or pollen nectar, and provide shelter. Many studies have demonstrated increased abundance of natural enemies and more effective biological control where crops are bordered by wild vegetation from which natural enemies colonise. Parasitism of the armyworm, *Pseudaletia unipuncta*, was significantly higher in maize fields embedded in a complex landscape than in maize fields surrounded by simpler habitats. In a two-year study, researchers found higher parasitism of *Ostrinia nubilalis* larvae by the parasitoid *Eriborus terebrans* in edges of maize fields adjacent to wooded areas, than in field interiors (Landis et al. 2000). Similarly, in Germany, parasitism of rape pollen beetle was about 50% at the edge of the fields, while at the centre of the fields parasitism dropped significantly to 20% (Thies and Tscharntke 1999). Such landscape effects in biological control are explored further by Schmidt et al. (ch. 4 this volume).

The extent to which negative consequences result from GE crops will be determined by the ways in which they are deployed in farming systems. As argued above, GE crops could *potentially* increase the proportion of non-crop vegetation in farm landscapes and increase agricultural diversity by creating pest- or disease-resistant multi-lines of a given crop or by allowing more rational weed management strategies that provide a net increase in weed presence (Table 2.2). This paradoxical situation suggests that GE technology per se is less important than is the nature of the engineered traits, the ways in which farmers exploit them and the nature of regulatory controls applied by governments. Monocultures of any type of crop, whether GE or conventional, may constitute the most widespread impediment to sustainable pest management. Accordingly, the potential for GE crops to encourage monoculture crop systems represents a possible clash between ecological and genetic engineering.

Risks of herbicide-resistant crops

A concern with transgenes from HT crops is that through gene flow they may give significant biological advantages to deleterious organisms, transforming wild or weed plants into new or worse weeds. Hybridisation of HT crops with populations of free-living relatives would make these plants increasingly difficult to control, especially if they are already recognised as agricultural weeds and if they acquire resistance to widely used herbicides. The GE crop itself may also assume weed status, for example in following crops in a rotational cropping system. In Canada, volunteer canola resistant to three herbicides (glyphosate, imidazolinone and glufosinate) has been detected, a case of 'stacked' or resistance to multiple herbicides (Hall et al. 2000). Reliance on HT crops also perpetuates the weed resistance problems and species shifts that are common to conventional herbicide-based approaches. Herbicide resistance becomes more of a problem as the number of herbicide modes of action to which weeds are exposed becomes fewer and fewer, a trend that HT crops may reinforce due to market forces. Given industry pressures to increase herbicide sales, areas treated with broad-spectrum herbicides could expand, exacerbating the resistance problem.

Many weeds are important components of agroecosystems because they positively affect the biology and dynamics of beneficial insects. Weeds offer many important requisites for natural enemies such as alternative prey/hosts, pollen or nectar as well as microhabitats that are unavailable in weed-free monocultures (Landis et al. 2000; Altieri and Whitcomb 1979; Jervis et al., ch. 5 this volume). Many insect pests are not continuously present in annual crops, and their predators and parasitoids must survive elsewhere during their absence. Weeds can provide such resources, thus aiding the survival of viable natural enemy populations. Much of the research reviewed in contributions to this book has shown that outbreaks of certain types of crop pests are less likely to occur in weed-diversified crop systems than in weed-free fields, mainly due to increased mortality imposed by natural enemies. Crop fields with a dense weed cover and high diversity usually have more predacious arthropods than do weed-free fields. The successful establishment of parasitoids usually depends on the presence of weeds that provide nectar for the adult female wasps. Relevant examples of cropping systems in which the presence of specific weeds has enhanced the biological control of particular pests have been reviewed by Altieri (1994).

Accordingly, perhaps the greatest problem associated with the use of HT crops is that the related broad-spectrum herbicides offer scope to completely remove weeds from fields, thus reducing plant diversity in agroecosystems. This contrasts with herbicidal weed management approaches in conventional crops where selective herbicide use may leave some weed taxa present. Many studies have produced evidence that the manipulation of a specific weed species or a particular weed control practice can affect the ecology of insect pests and associated natural enemies (Altieri and Letourneau 1982).

The situation is not clear-cut, however. The ability to use an effective, broad-spectrum herbicide such as glyphosate to control within-crop weeds may reduce herbicide use and increase weed populations. Such effects would apply if HT crops and their associated herbicide replaced the use of pre-emergence herbicides. Currently these pre-emergence herbicides are widely used in a prophylactic fashion as soil-applied, residual products. They cannot be applied after crop emergence; farmers cannot wait until weeds have emerged within the growing crop then apply weed treatment only if an appropriate threshold density is exceeded. Even in HT crops that do eventually require spraying, the presence of weeds early in the season, until the time that the need for herbicidal application is clear, could have beneficial effects on fauna. Presence of weeds early in the season would cause minimal competition with the developing crop, protect it to some degree by resource concentration effects and provide a moderated microclimate and possibly other resources to natural enemies of pests. Such early season enhancement of natural enemies may be especially important in preventing pest population build-up.

Of course, the fact that an HT crop/herbicide package *can* allow more rational weed management with benefits for arthropod pest management does not mean that some farmers will not use the technology inappropriately, for example by seeking to achieve total weed control in all crops throughout the season. Though such an approach may not be rational in terms of thresholds and weed management theory, other considerations such as ease of management, risk averseness and peer pressure for 'clean' crops may influence individual decision making.

British farm-scale evaluations (Haughton et al. 2003; Roy et al. 2003) showed that reduction of weed biomass and the flowering and seeding of plants under HT crop management within and in the margins of beet and spring oilseed rape involved changes in resource availability, with knock-on effects on higher trophic levels. There was reduced abundance of relatively sedentary and host-specific herbivores including Heteroptera, as well as butterflies and bees. Counts of predacious carabid beetles that feed on weed seeds were also smaller in HT crop fields. In beet and oilseed rape, weed densities were lower in the HT crops than in conventional crops, while the biomass in GE beet and oilseed rape was one-sixth and about one-third, respectively, of that in conventional plots. Researchers also recorded lower biomass for many species of weeds among the two HT crops. This led them to conclude that, compounded over time, these differences would result in large decreases in population densities of arable weeds. The abundance of invertebrates – which are food for mammals, birds and other invertebrates, and important for controlling pests or recycling nutrients within the soil – was also found to be generally lower in HT beet and canola. Specifically, a reduction in bees was associated with fewer flowering weeds in the GE beet. If such a trend applied more generally and to a severe extent it could have negative consequences for natural enemies of pests, such as aphidophagous syrphids and parasitoids that, like bees, require weed flowers for nectar and pollen.

It is noteworthy that though the British farm-scale evaluations were ambitious in scale and rigorous in design, like all scientific investigations, they were naturally contained. This limits the extent to which findings can be generalised (Firbank 2003). For example, organic systems where biodiversity levels may be considerably higher were not included. Further, though densities of natural enemies were measured, process rates such as predation and parasitism of pests were not investigated.

Ecological risks of Bt crops

Pest resistance
Based on the fact that more than 500 species of pests have already evolved resistance to conventional insecticides, pests can also evolve resistance to Bt toxins present in transgenic crops (e.g. Sayyed et al. 2003). No one questions whether Bt resistance can develop; the question is how best to manage it by use of refuge crops and integrated approaches. Susceptibility to Bt toxins can therefore be viewed as a natural resource that could be quickly depleted by inappropriate use of Bt crops, especially after such a rapid roll-out of Bt crops (7.6 million ha worldwide in 2002; Mellon and Rissler 1998).

There is a parallel between current Bt crops and primitive (c. 1950s–70s) calendar spraying, in which insecticides were applied regularly during crop growth irrespective of pest presence or density. Bt and other insect-resistant crops express toxins more or less uniformly over the plant and continuously over their lives, continually exposing the pest population to a selection pressure. In contrast, the use of Bt sprays (though these face challenges such as their UV stability) are generally applied in response to monitoring of pest densities. They may be alternated with other pest management strategies, for example other pesticides or inundative biological control products, to minimise the development of resistance in the pest. To move beyond the 'pesticide paradigm' mirrored in GE Bt crops, it may be possible to manipulate the natural 'induced defence' mechanisms of plants so that they turn on toxin-producing genes only when subject to pest damage. Alternatively, technologies could be designed to increase tolerance to pests rather than resistance to pests. Tolerance does not rely on toxicity to kill pests, is not eroded by resistance in pest populations and does not negatively affect non-target organisms (Welsh et al. 2002). Such approaches might also avoid some possible non-target risks of insect-resistant GE crops, for example to soil fauna, discussed below.

Despite the pressures for US farmers to use insect-resistant GE varieties, the benefits of using transgenic corn are not assured because population densities of the key pest, European corn borer, are not predictable. Due to this and other factors, the use of transgenic corn has not significantly reduced insecticide use in most of the corn-growing areas of the Midwest US. During the past few years, since the late 1990s, the percentage of field corn treated with insecticides in the US has remained at approximately 30%, despite a significant increase in the hectares of Bt corn planted (Obrycki et al. 2001).

Those who face the greatest risk from the development of insect resistance to Bt are neighbouring organic farmers who grow corn and soybeans without agrochemicals. Once resistance appears in insect populations, organic farmers will not be able to use Bt in its microbial insecticide form to control the lepidopteran pests that move in from neighbouring transgenic fields. In addition, genetic pollution of organic crops resulting from gene flow (pollen) from transgenic crops can jeopardise the certification of organic crops, forcing organic farmers to lose premium markets.

Bt crops and beneficial insects

Bt proteins are becoming ubiquitous, bioactive substances in agroecosystems present for many months. Most, if not all, non-target herbivores colonising Bt crops in the field, although not lethally affected, ingest plant tissue containing Bt protein which they can pass on to their natural enemies. Polyphagous natural enemies that move between crops are likely to encounter Bt containing non-target herbivorous prey in more than one crop over the course of a season.

According to Groot and Dicke (2002), natural enemies may come in contact more often with Bt toxins via non-target herbivores, because the toxins do not bind to receptors on the midgut membrane of non-target herbivores. This is a major ecological concern, given studies that documented that the Cry1 Ab Bt toxin adversely affected the predacious lacewing *Chrysoperla carnea* reared on Bt corn-fed prey larvae (Hilbeck et al. 1998; Hilbeck 2001) (though see Romeis et al. 2004). In another study, *C. carnea* fed three different herbivore species exposed to Bt-maize showed a significant increase in mortality and delayed development when fed *Spodoptera littoralis* reared on Bt-maize (Dutton et al. 2002). In that study, however, reduced prey quality is likely to have contributed to the negative effects on *C. carnea*. That sub-lethal effect shows that natural enemies may be indirectly affected by Bt toxins exposed in GE crops via feeding on sub-optimal food or because of host death and scarcity (Groot and Dicke 2002). Moreover, the toxins produced by Bt plants may be passed on to predators and parasites in plant material (pollen and sometimes, such as in the case of *Geocoris* spp, leaf tissue). Nobody has analysed the conse-

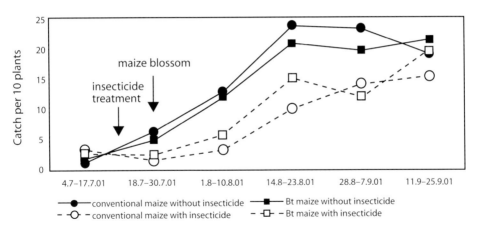

Figure 2.3: Seasonal development of spider abundance in Bt and conventional corn with and without insecticides (mean of four sites).

Reproduced with permission from Meissle, M. and Lang, A. *Proceedings of the Conference on Biodiversity Implications of Genetically Modified Plants*. 7–12 September 2003, Switzerland.

quences of such transfers for the myriad of natural enemies that depend on pollen for reproduction and longevity. Further, though nectar does not contain insecticidal gene products, parasitoids inadvertently ingest pollen when taking nectar and this directly exposes them to toxins in the pollen. Finally, because of the development of a new generation of Bt crops with much higher expression levels, the effects on natural enemies reported so far are likely to be an underestimate.

Among the natural enemies that live exclusively on insects which the Bt crops are designed to kill (chiefly Lepidoptera), egg and larval parasitoids would be most affected because they are totally dependent on live hosts for development and survival, whereas some predators could theoretically thrive on dead or dying prey (Schuler et al. 1999). Though the Bt toxin expression is the insect resistance trait most widely used in GE crops, expression of the snowdrop lectin GNA has also been engineered into potato. Birch et al. (1999) showed that this toxin had a deleterious effect on the fecundity, egg viability and longevity of the two-spot ladybird (*Adalia bipunctata*). Subsequent studies suggest that these effects on the predator result from the reduced weight of individual aphids reared on GNA-expressing plants, rather than a direct effect of the toxin on the third trophic level (Connor et al. 2003).

In considering the magnitude of any negative effects of insect-resistant GE crops on natural enemies, it is important to consider that, in the majority of cases, the alternative to their use is an insecticidal spray program. The area of farmland where habitat manipulation approaches and organic systems are in use is limited. Accordingly, the most appropriate reference point for the effects of insect-resistant GE crop use is insecticidal control. In a recently reported study, a 2×2 treatment experiment compared the effects of +/-Bt toxin and +/-insecticide (a total of four treatments; see Figure 2.3). This showed that the adverse effect on spiders of insecticide use (whether applied to GE or conventional corn) was greater than the effect of GE crop (whether sprayed or unsprayed) (Meissle and Lang 2003). This finding agreed with the broad conclusion of Cannon (2000) regarding use of Bt corn as well as results of Liu et al. (2003) that spider fauna was increased in Bt cotton compared with insecticide-treated conventional cotton.

Despite the above findings, adverse effects of Bt toxins have not been apparent in several recent studies of thrips (Obrist 2003), coccinellids (Szekeres et al. 2003), parasitoids (Steinbrecher and Vidal 2003), predatory mites (Zemek et al. 2003) and various other arthropods including stapylinids and syrphids (Manachini 2000, Figure 2.4).

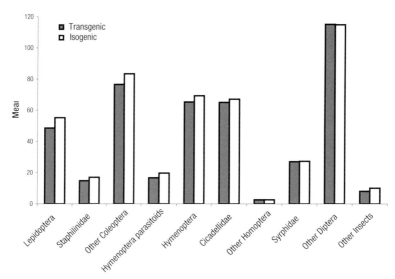

Figure 2.4: Malaise trap catches of aerial insects in transgenic (GE Bt corn) and isogenic (non-GE corn).

Reproduced with permission from Manachini, B. (2000). Ground beetle assemblages (Coleoptera, Carabidae) and plant dwelling non-target arthropods in isogenic and transgenic corn crops. *Bollettino di Zoologia agraria e di Bachicoltura II* 32 (3): 193.

A more fundamental problem associated with widespread use of insect-resistant GE crops is that, by keeping pest populations at extremely low levels, GE crops could potentially starve natural enemies. Many predators and parasitic wasps that feed on pests need a small amount of prey to survive in the agroecosystem. This suggests a way in which GE technology could be supported by ecological engineering approaches, which offer scope to increase the availability of alternative foods to allow at least generalist predators and parasitoids to persist.

These findings are problematic for small farmers in developing countries and for diversified organic farmers whose insect pest control depends on the rich complex of predators and parasites associated with their mixed cropping systems (Altieri 1994). Research results showing that natural enemies can be affected directly through inter-trophic level effects of the toxin present in Bt crops raises concerns about the potential disruption of natural pest control, as polyphagous predators that move within and between crop cultivars will encounter Bt-containing, non-target prey throughout the crop season. Disrupted biological control mechanisms will likely result in increased crop losses due to pests, or to increased use of pesticides, with consequent health and environmental hazards.

Effects on the soil ecosystem

The possibilities for soil biota to be exposed to transgenic products are high. The little research conducted in this area has already demonstrated persistence of insecticidal products (Bt and proteinase inhibitors) in soil after exposure to decomposing microbes (Donegan and Seidler 1999). The insecticidal toxin produced by *Bacillus thuringiensis* subsp. *kurstaki* remains active in the soil, where it binds rapidly and tightly to clays and humic acids. The bound toxin retains its insecticidal properties and is protected against microbial degradation by being bound to soil particles, persisting in various soils for 234 days (Palm et al. 1996). Palm et al. (1996) found that 25–30% of the Cry1A proteins produced by Bt cotton leaves remained bound in the soil even after 140 days. In another study, researchers confirmed the presence of the toxin in exudates from Bt corn and verified that it was active in an insecticidal bioassay using larvae of the tobacco hornworm (Saxena et al. 1999). In a recent study, after 200 days of exposure, adult earthworms,

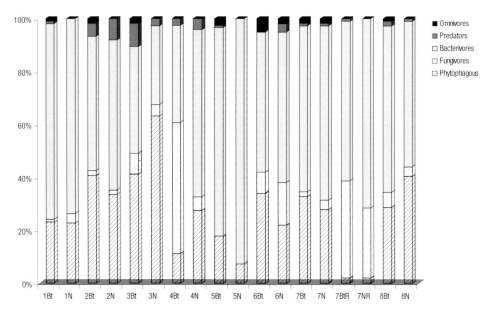

Figure 2.5: Trophic composition of nematode communities from GE corn (Bt) and conventional corn (N) on a range of sites (1–8) showing a difference in trophic groups only for site 4 where fungal feeders were more numerous and bacterial feeders were less numerous in the GE corn.

Reproduced with permission from Manachini, B. and Lozzia, G.C. (2002). First investigations into the effects of Bt corn crop on Nematofauna. *Bollettino di Zoologia agraria e di Bachicoltura.II* 34 (1): 85–96.

Lumbricus terrestris, experienced a significant weight loss when fed Bt corn litter compared with earthworms fed on non-Bt corn litter (Zwahlen et al. 2003). These earthworms may serve as intermediaries through which Bt toxins may be delivered to their predators. Given the persistence and the possible presence of exudates, the microbial and invertebrate community may suffer prolonged exposure to such toxins, and studies should therefore evaluate the effects of transgenic plants on both microbial and invertebrate communities and the ecological processes they mediate (Altieri 2000).

If transgenic crops substantially alter soil biota and affect processes such as soil organic matter decomposition and mineralisation, this would be of serious concern to organic farmers and most poor farmers in the developing world who cannot purchase or do not want to use expensive chemical fertilisers, relying instead on local residues, organic matter and especially soil organisms for soil fertility (i.e. key invertebrate, fungal or bacterial species) which can be affected by the soil-bound toxin. Soil fertility could be dramatically reduced if crop leachates inhibit the activity of the soil biota and slow the natural rates of decomposition and nutrient release. Due to accumulation of toxins during degradation of plant biomass, the doses of Bt toxin to which these soil organisms are exposed may increase with time, so impacts on soil biology could be worse and longer-term. Again, very little information is available on the potential effects of such toxins on soil-inhabiting predacious fauna (beetles, spiders etc.) and the pest consequences associated with potential reductions of beneficial soil fauna.

Studies in tropical Asian irrigated rice agroecosystems by Settle et al. (1996) showed that, by increasing organic matter in test plots, researchers could boost populations of detritivores and plankton-feeders, and in turn significantly boost the abundance of generalist predators. Surprisingly, organic matter management proved to be a key mechanism in the support of high levels of natural biological control. Bt toxins can potentially disrupt such mechanisms, thus indirectly promoting pest outbreaks.

Nematodes are another important component of soil ecosystems and the effects of Bt toxins from GE plants on these have been little studied. In the case of a study of nematode fauna (Manachini and Lozzia 2002), there was no significant effect of Bt corn cultivation (Figure 2.5), though a change in trophic groups was noted for one region and the need for longer-term studies was pointed out.

HT crops can also affect soil biota indirectly through effects of glyphosate, the application of which may be encouraged by some HT crops. This herbicide appears to act as antibiotic in the soil, inhibiting mycorrizae, antagonists and nitrogen-fixing bacteria. Root development, nodulation and nitrogen fixation is impaired in some HT soybean varieties, which exhibit lower yields. Effects are worse under drought stress or in infertile soils (Benbrook 2001). Elimination of antagonists could render GE soybean more susceptible to soil-borne pathogens.

Conclusions and recommendations

The literature suggests that widespread use of transgenic crops poses substantial potential risks. The environmental effects are not limited to pest resistance and creation of new weeds or virus strains (Kendall et al. 1997). As argued above, transgenic crops produce toxins that can move through the food chain and end up in the soil where they bind to colloids and retain their toxicity, affecting invertebrates and possibly nutrient cycling (Altieri 2000). It is virtually impossible to quantify or predict the long-term impacts on agrobiodiversity, and the processes they mediate, resulting from widespread use of GE crops.

There is a clear need for greater resources to be allocated to assessing the severity, magnitude and scope of risks associated with the use of transgenic crops. Much of the evaluation of risks must move beyond comparing GE fields and conventionally managed systems to include alternative cropping systems featuring crop diversity and low external-input approaches. These systems express higher levels of biological diversity and thus allow scientists to capture the full range of GE crops' impacts on biodiversity and agroecosystem processes.

Moreover, the increased landscape homogenisation that could result from GE crops will exacerbate the ecological problems already associated with monoculture agriculture (Altieri 2000). Unquestioned expansion of this technology into developing countries may not be wise nor desirable, particularly into tropical areas where centres of biodiversity could be threatened (Kathen 1996). There is strength in the agricultural diversity of many of these countries, and it should not be jeopardised by extensive monoculture, especially when consequences of doing so results in serious social and environmental problems (Altieri 1996). Options such as the push–pull strategy for protection of maize from stemborer pests (Khan and Pickett, ch. 10 this volume) constitute a practical alternative for growers as well as an exciting precedent for researchers to develop similarly elegant ecological engineering approaches. Though there is scope for GE crops to work synergistically with ecological engineering for pest management, much of this is theoretical and needs to be tested.

The repeated use of transgenic crops in an area may result in cumulative effects such as those resulting from the build-up of toxins in soils. For this reason, risk assessment studies not only have to be ecological in nature to capture effects on ecosystem processes, but also of sufficient duration so that probable accumulative effects can be detected. Manachini and Lozzia (2002) have stressed the need for longer-term risk assessment. A decade of carefully monitored field ecology is necessary to assess the full potential of environmental risks from GE crops. Decreases in pesticide use alone are not acceptable as proxies for environmental benefits. The application of multiple diagnostic methods to assess multitrophic effects and impacts on agroecosystem function will provide the most sensitive and comprehensive assessment of the potential ecological impact of GE crops.

Until these studies are completed, a moratorium on transgenic crops based on the precautionary principle should be imposed in the US – where most GE crops are grown (James 2002) – and other regions, as is currently the case for GE canola across Australia. This principle advises that instead of using as a criterion the 'absence of evidence' of serious environmental damage, the proper decision criterion should be the 'evidence of absence', in other words avoiding a 'type II' statistical error: the error of assuming that no significant environmental risk is present when in fact risk exists.

Although biotechnology may be a powerful and intellectually stimulating tool, GE crops are developed largely for profit motives and, as argued in this chapter, carry significant yet hard-to-quantify risks. Equivalent levels of research and development investment have not been made in ecological engineering approaches, at least partly because the solutions generated by habitat manipulation approaches are management-based rather than product-based. This presents few opportunities for patenting and revenue generation from intellectual property, so private investment is unlikely to become significant. This signals a need for government and university researchers to invest resources in such research because development of ecological engineering approaches may be less risky than alternative genetic engineering-based options.

The UN Food and Agriculture Organisation considers that GE crops have merit for use together with other technologies but, as Herren (2003) stated, GE crops are often viewed as unilateral solutions to agricultural problems rather than as a tool for use where and when appropriate. Given that GE crops are likely to be in widespread use in many countries, it is important that work be done to quantify the extent of negative consequences for biodiversity. This will indicate the need for steps to ameliorate any effects, for example, by additional area being allocated to non-crop vegetation where biodiversity may be preserved (Sutherland and Watkinson 2001). With the high level of research interest in this topic, the next few years will yield much knowledge on the extent to which the interplay between genetic engineering and ecological engineering approaches for pest management offer scope for tangible, real-world synergy. The alternative scenario – as reflected in the rather polarised public and scientific discourse – is that these two forms of engineering are irreconcilable, with GE crops constituting a threat to ecological engineering.

References

Altieri, M.A. (1994). *Biodiversity and Pest Management in Agroecosystems*. Haworth Press, New York.

Altieri, M.A. (1996). *Agroecology: The Science of Sustainable Agriculture*. Westview Press, Boulder.

Altieri, M.A. (1999). The ecological role of biodiversity in agroecosystems. *Agriculture, Ecosystems and Environment*. 74: 19–31.

Altieri, M.A. (2000). The ecological impacts of transgenic crops on agroecosystem health. *Ecosystem Health* 6: 13–23.

Altieri, M.A. and Letourneau, D.K. (1982). Vegetation management and biological control in agroecosystems. *Crop Protection* 1: 405–430.

Altieri, M.A. and Whitcomb, W.H. (1979). The potential use of weeds in the manipulation of beneficial insects. *Hortscience* 14 (1): 12–18.

Andow, D.A. (1991). Vegetational diversity and arthropod population response. *Annual Review of Entomology* 36: 561–586.

Barbosa, P. (1998). *Conservation Biological Control*. Academic Press, San Diego. 396 pages.

Benbrook, C. (2001). Troubled times amid commercial success for roundup ready soybeans. Ag Bio Tech InfoNet Technical Paper No. 4. Online at Ag BioTech InfoNet Website http://www.biotech-info.net

Birch, A.N.E., Geoghegan, I.E., Majerus, M.E.N., McNicol, J.W., Hackett, C.A., Gatehouse, A.M.R. and Gatehouse, J. (1999). A tri-trophic interaction involving pest aphids, predatory 2-spot ladybirds and transgenic potatoes expressing snowdrop lectin for aphid resistance. *Molecular Breeding* 5: 75–83.

Braun, R. and Ammann, C. (2003). Introduction: Biodiversity – the impact of biotechnology. In *Methods for Risk Assessment of Transgenic Plants. IV Biodiversity and Biotechnology* (K. Ammann, Y. Jacot and R. Braun, eds), pp. vii–xv. Birkhauser Verlag Basel, Switzerland.

Brooks, D.R., Bohan, D.A., Champion, G.T., Haughton, A.J., Hawes, C., Heard, M.S., Clark, S.J., Dewar, A.M., Firbank, L.G., Perry, J.N., Rothery, P., Scott, R.J., Woiwod, I.P., Birchall, C., Skellern, M.P., Walker, J.H., Baker, P., Bell, D., Browne, E.L., Dewar, A.J.G., Fairfax, C.M., Garner, B.H., Haylock, L.A., Horne, S.L., Hulmes, S.E., Mason, N.S., Norton, L.P., Nuttall, P., Randall, Z., Rossall, M.J., Sands, R.J.N., Singer, E.J. and Walker, M.J. (2003). Invertebrate responses to the management of genetically modified herbicide-tolerant and conventional spring crops. I. Soil-surface-active invertebrates. *Philosophical Transactions of the Royal Society of London* B 358: 1847–1862.

Cannon, R.J.C. (2000). *Bt* transgenic crops: risks and benefits. *Integrated Pest Management Reviews* 5: 151–173.

Conner, A.J., Glare, T.R. and Nap, J.P. (2003). The release of genetically modified crops into the environment. *Plant Journal* 33: 19–46.

Conner, J.J. and Christey, M.C. (1994). Plant-breeding and seed marketing options for the introduction of transgenic insect resistant crops. *Biocontrol Science and Technology* 4: 463–474.

Costanza, R., D'Arge, R., de Groot, R., Farber, S., Grasso, M., Hannon, B., Limburg, K., Naeem, S., O'Neill, R.V., Paruelo, J., Raskin, R. G., Sutton, P. and van den Belt, M. (1997). The value of the world's ecosystem services and natural capital. *Nature* 387 (6630): 253–260.

Dale, P.J., Clarke, B. and Fontes, E.M.G. (2002). Potential for the environmental impact of transgenic crops. *Nature Biotechnology* 20: 567–574.

Dewar, A.M., May, M.J., Woiwod, I.P., Haylock, L.A., Champion, G.T., Garner, B.H., Sands, R.J.N., Qui, A. and Pidgeon, J.D. (2002). A novel approach to the use of genetically modified herbicide tolerant crops for environmental benefit. *Proceedings of the Royal Society of London* B 270: 335–340.

Donegan, K.K. and Seidler, R.J. (1999). Effects of transgenic plants on soil and plant microorganisms. *Recent Research Developments in Microbiology* 3 (2): 415–424.

Donegan, K.K., Palm, C.J., Fieland, V.J., Porteous, L.A., Ganis, L.M., Scheller, D.L. and Seidler, R.J. (1995). Changes in levels, species, and DNA fingerprints of soil micro organisms associated with cotton expressing the *Bacillus thuringiensis* var. *kurstaki* endotoxin. *Applied Soil Ecology* 2: 111–124.

Dutton, A., Klein, H., Romeis, J. and Bigler, F. (2002). Uptake of Bt toxin by herbivores feeding on transgenic maize and consequences for the predator *Chrysoperla carnea*. *Ecological Entomology* 27: 441–447.

Edge, J.M., Benedict, J.H., Carroll, J.P. and Reding, H.K. (2001). Bollgard cotton: an assessment of global economic, environmental and social benefits. *Journal of Cotton Science* 5: 121–136.

Firbank, L.G. (2003). Introduction. *Philosophical Transactions of the Royal Society of London* B 358: 1777–1778.

Firbank, L.G. and Forcella, F. (2000). Genetically modified crops and farmland biodiversity. *Science* 289: 1481–1482.

Groot, A.T. and Dicke, M. (2002). Insect-resistance transgenic plants in a multi-trophic context. *The Plant Journal* 31: 387–406.

Gurr, G. and Wratten, S. (2000). *Biological Control: Measures of Success*. Kluwer Academic, Dordrecht. 429 pages.

Gurr, G.M., Wratten, S.D. and Luna, J. (2003). Multi-function agricultural biodiversity: pest management and other benefits. *Basic and Applied Ecology* 4: 107–116.

Hails, R.S. (2003). Transgenic crops and their environmental impact. *Antenna* 27: 313–319.

Hall, L.M., Huffman, J. and Topinka, K. (2000). Pollen flow between herbicide tolerant canola in the cause of multiple resistant canola volunteers. *Weed Science Society Abstracts* vol. 40. Annual Meeting of the Weed Science Society of America, Toronto, Canada.

Haughton, A.J., Champion, G.T., Hawes, C., Heard, M.S., Brooks, D.R., Bohan, D.A., Clark, A.M., Dewar, L.G., Firbank, L.G., Osborne, J.L., Perry, J.N., Rothery, P., Roy, D.B., Scott, R.J., Woiwod, I.P., Birchall, C., Skellern, M.P., Walker, J.H., Baker, P., Browne, E.L. Dewar, A.J.G., Garner, B.H., Haylock, L.A., Horne, S.L., Mason, N.S., Sands, R.J.N. and Walker, M.J. (2003). Invertebrate responses to the management of genetically modified herbicide-tolerant and conventional spring crops. II. Within-field epigeal and aerial arthropods. *Philosophical Transactions of the Royal Society of London* B 358: 1863–1877.

Hawes, C., Haughton, A.J., Osborne, J.L., Roy, D.B., Clark, S.J., Perry, J.N., Rothery, P., Bohan, D.A., Brooks, D.J., Champion, G.T., Dewar, A.M., Heard, M.S., Woiwod, I.P., Daniels, R.E., Yound, M.W., Parish, A.M., Scott, R.J., Firbank, L.G. and Squire, G.R. (2003). Responses of plants and invertebrate trophic groups to contrasting herbicide regimes in the farm scale evaluations of genetically modified herbicide-tolerant crops. *Philosophical Transactions of the Royal Society of London* B 358: 1899–1913.

Herren, H.R. (2003). Genetically engineered crops and sustainable agriculture. In *Methods for Risk Assessment of Transgenic Plants, IV Biodiversity and Biotechnology* (K. Ammann, Y. Jacot and R. Braun, eds). Birkhauser Verlag Basel, Switzerland.

Hilbeck, A. (2001). Implications of transgenic, insecticidal plants for insect and plant biodiversity. *Perspectives in Plant Ecology, Evolution and Systematics* 4 (1): 43–61.

Hilbeck, A., Moar, W.J., Puszatai-Carey, M., Filippini, A. and Bigler, F. (1998). Toxicity of *Bacillus thuringiensis* Cryl Ab toxin to the predator *Chrysoperla carnea* (Neuroptera: Chrysopidae). *Environmental Entomology* 27: 1255–1263.

James, C. (2002). Preview: global status of commercialized transgenic crops: 2002. *International Service for the Acquisition of Agri-Biotech Application* Briefs No. 27. ISAAA, Ithaca.

Jank, B. and Gaugitsch, H. (2001). Assessing the environmental impacts of transgenic plants. *Trends in Biotechnology* 19 (9): 371–372.

Johnson, M.T. (1997). Interaction of resistant plants and wasps parasitoids of tobacco budworm (Lepidoptera: Noctuidae). *Environmental Entomology* 26: 207–214.

Kathen, A. de (1996). The impact of transgenic crop releases on biodiversity in developing countries. *Biotechnology and Development Monitor* 28: 10–14.

Kendall, H.W., Beachy, R., Eismer, T., Gould, F., Herdt, R., Ravon, P.H., Schell, J. and Swaminathan, M.S. (1997). Bioengineering of crops. *Report of the World Bank Panel on Transgenic Crops*, p. 30. World Bank, Washington, D.C.

Krebs, J.R., Wilson, J.D., Bradbury, R.B.and Siriwardena, G.M. (1999). The second silent spring? *Nature* 400: 611–612.

Krimsky, S. and Wrubel, R.P. (1996). *Agricultural Biotechnology and the Environment: Science, Policy and Social Issues*. University of Illinois Press, Urbana.

Landis, D., Wratten, S.D. and Gurr, G.M. (2000) Habitat management for natural enemies. *Annual Review of Entomology* 45: 175–201.

Liu, W.X., Wan, F.H., Guo, J.Y. and Lovei, G.L. (2003). Spider diversity and seasonal dynamics in transgenic Bt vs. conventionally managed cotton fields in China. *Proceedings of the Conference on Biodiversity Implications of Genetically Modified Plants*, abstract book, p. 28. 7–12 September, Monte Verita, Ascona, Switzerland.

Losey, J.J.E., Rayor, L.S. and Cater, M.E. (1999). Transgenic pollen harms monarch larvae. *Nature* 399: 241.

Lovei, G.L. (2001). Ecological risks and benefits of transgenic plants. *New Zealand Plant Protection* 54: 93–100.

Manachini, B. (2000). Ground beetle assemblages (Coleoptera, Carabidae) and plant dwelling non-target arthropods in isogenic and transgenic corn crops. *Bollettino di Zoologia agraria e di Bachicoltura II* 32 (3): 181–198.

Manachini, B. and Lozzia, G.C. (2002). First investigations into the effects of Bt corn crop on Nematofauna. *Bollettino di Zoologia agraria e di Bachicoltura II* 34 (1): 85–96.

Meissle, M. and Lang, A. (2003). Sampling methods to evaluate the effects of Bt maize and insecticide application on spiders. *Proceedings of the Conference on Biodiversity Implications of Genetically Modified Plants*, abstract book, pp. 34–35. 7–12 September, Monte Verita, Ascona, Switzerland.

Mellon, M. and Rissler, J. (1998). *Now or Never: Serious Plans to Save a Natural Pest Control*. Union of Concerned Scientists, Washington, D.C.

Mitsch, W.J. and Jørgensen, S.E. (1989). Introduction to ecological engineering. In *Ecological Engineering: An Introduction to Ecotechnology* (W.J. Mitsch and S.E. Jørgensen, eds), pp. 3–19. Wiley, New York.

Mitsch, W.J. and Jørgensen, S.E. (2004). *Ecological Engineering and Ecosystem Restoration*. Wiley, New York. 424 pages.

Obrist, L., Klein, H., Dutton, A. and Bigler, F. (2003). Are natural enemies at risk when feeding on thrips on Bt maize? *Proceedings of the Conference on Biodiversity Implications of Genetically Modified Plants*, abstract book, p. 9. 7–12 September, Monte Verita, Ascona, Switzerland.

Obrycki, J.J., Losey, J.E., Taylor, O.R. and Jesse, L.C.H (2001). Transgenic insecticidal corn: beyond insecticidal toxicity to ecological complexity. *Bioscience* 51: 353–361.

Palm, C.J., Schaller, D.L., Donegan, K.K. and Seidler, R.J. (1996). Persistence in soil of transgenic plant produced *Bacillus thuringiensis* var. *kurstaki* endotoxin. *Canadian Journal of Microbiology* 42: 1258–1262.

Pascher, K. and Gollmann, G. (1999). Ecological risk assessment of transgenic plant releases: an Austrian perspective. *Biodiversity and Conservation* 8: 1139–1158.

Phipps, R.H. and Park, J.R. (2002). Environmental benefits of genetically modified crops: global and European perspectives on their ability to reduce pesticide use. *Journal of Animal Feed Science* 11: 1–18.

Rissler, J. and Mellon, M. (1996). *The Ecological Risks of Engineered Crops*. MIT Press, Cambridge.

Robinson, J. (1999). Ethics and transgenic crops: a review. *EJB Electronic Journal of Biotechnology* 2 (2): 71–81.

Robinson, R.A. (1996). *Return to Resistance: Breeding Crops to Reduce Pesticide Resistance*. AgAccess, Davis.

Romeis, J., Dutton, A. and Bigler, F. (2004). *Bacillus thuringiensis* toxin (Cry1 Ab) has no direct effect on larvae of the green lacewing *Chrysoperla carnea* (Stephens) (Neuroptera: Chrysopidae). *Journal of Insect Physiology* 50: 175–183.

Root, R.B. (1973). Organization of a plant–arthropod association in simple and diverse habitats: the fauna of collards (*Brassica oleraceae*). *Ecological Monographs* 43: 94–125.

Roy, D.B. et. al. (2003). Invertebrates and vegetation of field margins adjacent to crops subject to contrasting herbicide regimes in the farm scale evaluations of genetically modified herbicide-tolerant crops. *Philosophical Transactions of the Royal Society of London* B 358: 1879–1898.

Saxena, D., Flores, S. and Stotzky, G. (1999) Insecticidal toxin in root exudates from Bt corn. *Nature* 401: 480.

Sayyed, A.H., Schuler, T.H. and Wright, D.J. (2003) Inheritance of resistance to Bt canola in a field-derived population of *Plutella xylostella*. *Pest Management Science* 59: 1197–1202.

Schuler, T.H., Potting, R.P.J., Dunholm, I. and Poppy, G.M. (1999). Parasitoid behavior and Bt plants. *Nature* 400: 825.

Settle, W.H., Ariawan, H., Astuti, E.T., Cahyana, W., Hakim, A.L., Hindayana, D., Lestari, A.S. and Pajarningsih, S. (1996). Managing tropical rice pests through conservation of generalist natural enemies and alternative prey. *Ecology* 77: 1975–1988.

Shelton, A.M., Zhao, J.-Z. and Roush, R.T. (2002). Economic, ecological, food safety, and social consequences of the deployment of Bt transgenic plants. *Annual Review of Entomology* 47: 845–881.

Smil, V. (2000). *Feeding the World: A Challenge for the Twenty-first Century*. MIT Press, Cambridge.

Snow, A.A. and Moran, P. (1997). Commercialization of transgenic plants: potential ecological risks. *BioScience* 47: 86–96.

Solomon, M.E. (1976). *Animal Populations*. Edward Arnold, London.

Squire, G.R., Brooks, D.R., Bohan, D.A., Champion, G.T., Daniels, R.E., Haughton, A.J., Hawes, C., Heard, M.S., Hill, M.O., May, M.J., Osborne, J.L., Perry, J.N., Rothery, P., Roy, D.B., Woiwod, I.P. and Firbank, L.G. (2003). On the rationale and interpretation of the farm scale evaluations of genetically modified herbicide-tolerant crops. *Philosophical Transactions of the Royal Society of London* B 358: 1779–1799.

Steinbrecher, I. and Vidal, S. (2003). Potential non-target effects of Bt transgenic plants on higher trophic levels: case studies using parasitoids. *Proceedings of the Conference on Biodiversity Implications of Genetically Modified Plants*, abstract book, pp. 44–45. 7–12 September, Monte Verita, Ascona, Switzerland.

Sutherland, W.J. and Watkinson, A.R. (2001). Policy making within ecological uncertainty: lessons from badgers and GM crops. *Trends in Ecology and Evolution* 16: 261–263.

Szekeres, D., Szentkiralyi, F., Kiss, J. and Kadar, F. (2003). Comparison of characteristics of coccinellid assemblages studied in experimental Bt- and isogenic maize fields in Hungary. *Proceedings of the Conference on Biodiversity Implications of Genetically Modified Plants*, abstract book, p. 46. 7–12 September, Monte Verita, Ascona, Switzerland.

Tait, E.J. (1987). Planning an integrated pest management system. In *Integrated Pest Management* (A.J. Burn, T.H. Coaker and P.C. Jepson, eds). Academic Press, London.

Tappeser, B. (2003). Biosafety research programmes and the biodiversity issue. *Proceedings of the Conference on Biodiversity Implications of Genetically Modified Plants*, abstract book, p. 49. 7–12 September, Monte Verita, Ascona, Switzerland.

Thies, C. and Tscharntke, T. (1999). Landscape structure and biological control in agroecosystems. *Science* 285: 893–895.

Thomas, M.B., Wratten, S.D. and Sotherton, N.W. (1991). Creation of island habitats in farmland to manipulate populations of beneficial arthropods: predator densities and emigration. *Journal of Applied Ecology* 28: 906–917.

Thomsen, S., Wratten, S.D. and Frampton, C.M. (2001). Skylark (*Alauda arvensis*) winter densities and habitat associations in Canterbury, New Zealand. In *The Ecology and Conservation of Skylarks* Alauda arvensis (P.F. Donald and J.A. Vickery, eds), pp. 139–148. Royal Society for the Protection of Birds.

Thrupp, L.A. (1998). *Cultivating Diversity: Agrobiodiversity and Food Security*. Institute for World Resource Research, Woodridge, USA.

Watkinson, A.R., Freckleton, R.P., Robinson, R.A. and Sutherland, W.J. (2000). Predictions of biodiversity response to genetically modified herbicide-tolerant crops. *Science* 289: 1554–1556.

Welsh, R., Hubbell, B., Ervin, D.E. and Jahn, M. (2002). GM crops and the pesticide paradigm. *Nature Biotechnology* 20: 548–549.

Wolfenbarger, L.L. and Phifer, P.R. (2000). The ecological risks and benefits of genetically engineered plants. *Science* 290: 2088–2093.

Zemek, R., Rovenska, G.Z., Schmidt, J.E.U. and Hilbeck, A. (2003). Prey-mediated effects of transgenic plants expressing Bt toxins on predatory mites (Acari: Phytoseiidae). *Proceedings of the Conference on Biodiversity Implications of Genetically Modified Plants*, abstract book, pp. 55–56. 7–12 September, Monte Verita, Ascona, Switzerland.

Zwahlen, C., Hilbeck, A., Howald, R. and Nentwig, W. (2003). Effects of transgenic Bt corn litter on the earthworm *Lumbricus terrestris*. *Molecular Ecology* 12: 1077–1082.

Chapter 3

Agroecological bases of ecological engineering for pest management

Clara I. Nicholls and Miguel A. Altieri

Introduction

The integrated pest management concept (IPM) arose in the 1960s in response to concerns about impacts of pesticides on the environment. By providing an alternative to the strategy of unilateral intervention with chemicals, it was hoped that IPM would change the practice of crop protection to one that entailed a deeper understanding of insect and crop ecology, resulting in a strategy which relied on the use of several complementary tactics. It was envisioned that ecological theory should provide a basis for predicting how specific changes in production practices and inputs might affect pest problems. It was also thought that ecology could help to design agricultural systems less vulnerable to pest outbreaks. In such systems, pesticides would be used as occasional supplements to natural regulatory mechanisms. In fact many authors wrote papers and reviews depicting the ecological basis of pest management (Southwood and Way 1970; Price and Waldbauer 1975; Pimentel and Goodman 1978; Levins and Wilson 1979). But despite all this early work that provided much of the needed ecological foundations, most IPM programs instead became schemes of 'intelligent pesticide management' and failed to put ecologically based theory into practice.

Lewis et al. (1997) argue that the main reason why IPM science has been slow to provide an understanding that will assist farmers move beyond the current production methods is that IPM strategies have long been dominated by quests for 'silver bullet' products to control pest outbreaks. The emphasis has been on tactics to suppress pests and reduce crop damage, rather than on why agroecosystems are vulnerable and how to make them more pest-resilient. Agroecosystem redesign through ecological engineering involves a shift from linear, one-to-one relationships between target pests and a particular management tactic, to webs of relationships between insect pests, associated natural enemies and crop diversification schemes. Emphasis is on preventing pest problems by enhancing the 'immunity' of the agroecosystem, and on integrating pest management activities with other farming practices that maintain soil productivity and crop health, while ensuring food security and economic viability. Although it is important to understand autoecological factors that explain why pests quickly adapt and succeed in agroecosystems, it is more crucial to pinpoint what makes agroecosystems susceptible to pests. By designing agroecosystems that both work against the pests' performance and are less vulnerable to pest invasion, farmers can substantially reduce pest numbers.

This chapter argues that long-term solutions to pest problems can be only achieved by restructuring and managing agricultural systems in ways that maximise the array of built-in preventive strengths, with therapeutic tactics serving strictly as backups of natural regulator processes. Lewis et al. (1997) suggested three main methods of bringing pest populations within acceptable bounds. Of these, ecological engineering is the most promising in harnessing the inherent strengths that emerge when agroecosystems are designed following agroecological principles.

Agroecology and pest management

One way of further advancing the ecosystem management approach in IPM is through the understanding that crop health and sustainable yields derive from the proper balance of crops, soils, nutrients, sunlight, moisture and coexisting organisms. The agroecosystem is productive and healthy when this balance of rich growing conditions prevails, and when crop plants remain resilient to tolerate stress and adversity. Occasional disturbances can be overcome by vigorous agroecosystems, which are adaptable and diverse enough to recover once the stress has passed (Altieri and Rosset 1996). If the cause of disease, pest, soil degradation etc. is imbalance, then the goal of ecological engineering is to recover the balance. This is known in ecology as resilience – the maintenance of the system's functions to compensate for external stress factors – and requires a thorough understanding of the nature of the agroecosystems and the principles by which they function. Agroecology provides basic ecological principles on how to study, design and manage agroecosystems that are productive, enduring and natural resource-conserving (Altieri 1995). Agroecology goes beyond a one-dimensional view of agroecosystems – their genetics, agronomy and edaphology – to embrace an understanding of ecological and social levels of coevolution, structure and function. Instead of focusing a particular component of the agroecosystem, agroecology emphasises the interrelatedness of all agroecosystem components and the complex dynamics of ecological processes such as nutrient cycling and pest regulation (Gliessman 1999).

From a management perspective, the agroecological objective is to provide a balanced environment, sustainable yields, biologically mediated soil fertility and natural pest regulation through the design of diversified agroecosystems and the use of low-input technologies (Altieri 1994). The strategy is based on ecological principles of optimal recycling of nutrients and organic matter turnover, close energy flows, water and soil conservation, and balanced pest–natural enemy populations. The strategy exploits the complementation that results from the various combinations of crops, trees and animals in spatial and temporal arrangements (Altieri and Nicholls 1999). These combinations determine the establishment of a functional biodiversity which, when correctly assembled, delivers key ecological services which subsidise agroecosystem processes that underlie agroecosystem health.

In other words, the ecological concepts favour natural processes and biological interactions that optimise synergies, so that diversified farms can sponsor their own soil fertility, crop protection and productivity through the activation of soil biology, recycling of nutrients and enhancement of beneficial arthropods and antagonists. Based on these principles, agroecologists involved in pest management have developed a framework to achieve crop health through agroecosystem diversification and soil quality enhancement, key pillars of agroecosystem health. The main goal is to enhance the immunity of the agroecosystem (i.e. natural pest-control mechanisms) and regulatory processes (i.e. nutrient cycling and population regulation) through management practices and agroecological designs that enhance plant species and genetic diversity, and organic matter accumulation and biological activity of the soil (Altieri 1999).

Agroecosystems can be manipulated to improve production and produce more sustainably, with fewer negative environmental and social impacts and fewer external inputs (Altieri 1995). The design of such systems is based on the application of the following ecological principles (Reinjtes et al. 1992):

- enhancing recycling of biomass and optimising nutrient availability and balancing nutrient flow;
- securing favourable soil conditions for plant growth, particularly by managing organic matter and enhancing soil biotic activity;
- minimising losses due to flows of solar radiation, air and water by way of microclimate management, water harvesting and soil management through increased soil cover;
- species and genetic diversification of the agroecosystem in time and space;
- enhancing beneficial biological interactions and synergisms among agrobiodiversity components, resulting in the promotion of key ecological processes and services.

These principles can be applied by way of various techniques and strategies. Each of these will have different effects on productivity, stability and resiliency within the farm system, depending on the local opportunities, resource constraints and, in most cases, the market. The ultimate goal of agroecological design is to integrate components so that overall biological efficiency is improved, biodiversity is preserved, and agroecosystem productivity and its self-sustaining capacity are maintained.

Understanding pest vulnerability in agroecosystems

Over the past half-century, crop diversity has declined precipitously in conventional high-input farming systems in the US and other industrialised countries, as well as in the agroexport regions of the developing world. Such reduction in crop diversity has resulted in the simplification of the landscape. The expansion of monocultures has decreased abundance and activity of natural enemies due to the removal of critical food resources and overwintering sites. Many scientists are concerned that, with accelerating rates of habitat removal, the contribution to pest suppression by biological control agents using these habitats is declining and consequently agroecosystems are becoming increasingly vulnerable to pest invasion and outbreaks.

A key task for agroecologists is to understand why modern agroecosystems are so vulnerable to insect pests, in order to reverse such vulnerability by increasing vegetational diversity in agricultural landscapes. Human manipulation and alteration of ecosystems for the purpose of establishing agricultural production makes agroecosystems structurally and functionally very different from natural ecosystems. Cultivation brings about many changes including horticultural simplicity, phenological uniformity, fertilisation-mediated nutritional changes in plant foliage, changes in plant characteristics through breeding and so on.

In general, monocultures do not constitute good environments for natural enemies (Andow 1991). Such simple crop systems lack many of the resources such as refuge sites, pollen, nectar and alternative prey and hosts, that natural enemies need to feed, reproduce and thrive. Normal cultural activities such as tillage, weeding, spraying and harvesting can have serious effects on farm insects. To the pests, the monocrop is a dense and pure concentration of its basic food resource – many insect herbivores boom in such fertilised, weeded and watered fields. According to the plant vigour hypothesis (Price 1991), such an effect would particularly apply to herbivore guilds most closely associated with plant growth processes, such as the endophytic gallers and shoot borers. For the natural enemies, such overly simplified cropping systems are less hospitable because natural enemies require more than prey and hosts to complete their life-cycles. Many parasitoid adults, for instance, require pollen and nectar to sustain themselves while they search for hosts (Jervis et al., ch. 5 this volume).

Given the major differences between mechanised agroecosystems and natural ecosystems, especially the prevalence of monocultures and the high levels of disturbance, modern systems lack a suitable ecological infrastructure to resist pest invasions and outbreaks (Altieri 1994; Landis et al. 2000). As explained below, many factors underlie the vulnerability of monocultures to pest invasions.

Decreased landscape diversity

The spread of modern agriculture has resulted in tremendous changes in landscape diversity. There has been a consistent trend toward simplification that entails (1) the enlargement of fields, (2) the aggregation of fields, (3) an increase in the density of crop plants, (4) an increase in the uniformity of crop population age structure and physical quality, and (5) a decrease in inter- and intraspecific diversity within the planted field.

Although these trends appear to exist worldwide, they are more apparent, and certainly best documented, in industrialised countries. Increasingly, evidence suggests that these changes in landscape diversity have led to more insect outbreaks due to the expansion of monocultures at the expense of natural vegetation, through decreasing habitat diversity. One of the main characteristics of the modern agricultural landscape is the large size and homogeneity of crop monocultures, which fragment the natural landscape. This can directly affect the abundance and diversity of natural enemies, as the larger the area under monoculture the lower the viability of a given population of beneficial fauna. The issue of colonisation of crop 'islands' by insects is also relevant. In the case of annual crops, insects must colonise from the borders each season; the larger the field, the greater the distance that must be covered. Several studies suggest (not surprisingly) that natural enemies tend to colonise after their hosts/prey and that the lag time between the arrival of pest and natural enemy increases with distance from border (source pool). For instance, Price (1976) found that the first occurrence of a herbivore and that of a predatory mite in a soybean field were separated by one week on the edge versus a three-week lag in the centre. To the extent that this is a general phenomenon, increased field size should lead to more frequent insect outbreaks. Landscape-level effects on pest management are explained in further detail by Schmidt et al. (ch. 4 this volume).

Decreased on-farm plant diversity

Many ecologists have conducted experiments testing the theory that decreased plant diversity in agroecosystems allows greater chance for invasive species to colonise, leading to enhanced herbivorous insect abundance. Many of these experiments have shown that mixing certain plant species with the primary host of a specialised herbivore gives a fairly consistent result: specialised species usually exhibit higher abundance in monoculture than in diversified crop systems (Andow 1983).

Several reviews have documented the effects of habitat heterogeneity on insects (Altieri and Letourneau 1984; Risch et al. 1983). There are two main hypotheses about why herbivore populations tend to be greater in monocultures and how, in agroecosystems, insect populations can be stabilised by constructing vegetational architectures that support natural enemies and/or inhibit pest attack. These hypotheses are the enemies hypothesis (Pimentel 1961) and the resource concentration hypothesis (Root 1973). A recent study in Portugal illustrated the effects of decreased field-plant diversity on increased pest incidence. As new policy and market forces prompt the conversion of traditional complex agroforest vineyard systems to monocultures, Altieri and Nicholls (2002) found higher prevalence of grape herbivores and *Botrytis* bunch rot. Although monocultures may be productive, gains occurred at the expense of biodiversity and agricultural sustainability, reflected in higher pest vulnerability.

Pesticide-induced insect outbreaks
Many examples are reported in the literature of insect pest outbreaks and/or resurgence following insecticide applications (Pimentel and Perkins 1980). Pesticides either fail to control the
target pests or create new pest problems. Development of resistance in insect pest populations is
the main way in which pesticide use can lead to pest control failure. More than 500 species of
arthropods have become resistant to a series of insecticides and acaricides (Van Driesche and
Bellows 1996).

Another way in which pesticide use can foster outbreaks of pests is through the elimination of
the target pest's natural enemies. Predators and parasites often experience higher mortality than
herbivores, following a given spray (Morse et al. 1987). This is partly due to the greater mobility
of many natural enemies, which exposes them to more insecticide per unit time following a spray.

In addition, natural enemies appear to evolve resistance to insecticides much more slowly
than do herbivores. This results from a smaller probability that some individuals in populations
of natural enemies will have genes for insecticide resistance. This in turn is due to the much
smaller size of the natural-enemy population relative to the pest population and the different
evolutionary history of natural enemies and herbivores.

Pesticides also create new pest problems when natural enemies of ordinarily non-economic
species are destroyed by chemicals. These secondary pests then reach higher density than normal
and begin to cause economic damage (Pimentel and Lehman 1993).

Fertiliser-induced pest outbreaks
Luna (1988) suggested that crops' physiological susceptibility to insects may be affected by the
form of fertiliser used (organic vs chemical). In most studies evaluating aphid and mite response
to nitrogen fertilisation, increases in nitrogen rates dramatically increased aphid and mite
numbers. According to van Emden (1966), increases in fecundity and developmental rates of the
green peach aphid, *Myzus persicae*, were highly correlated to increased levels of soluble nitrogen
in leaf tissue. In reviewing 50 years of research relating to crop nutrition and insect attack,
Scriber (1984) found 135 studies showing increased damage and/or growth of leaf-chewing
insects or mites in N-fertilised crops, and fewer than 50 studies in which herbivore damage was
reduced by normal fertilisation regimes. These results suggest a hypothesis with implications for
fertiliser use patterns in agriculture, namely that high nitrogen inputs can precipitate high levels
of herbivore damage in crops. As a corollary, crop plants would be expected to be less prone to
insect pests and diseases if organic soil amendments are used, as these generally result in lower
nitrogen concentrations in the plant tissue.

Studies (see Altieri and Nicholls 2003) documenting lower abundance of several insect herbivores in organic farming systems have partly attributed such reduction to low nitrogen content
in the organically farmed crops. In comparative studies, conventional crops (treated with chemical fertiliser) tend to develop a larger infestation of insects (especially Homoptera) than organic
counterparts.

Interestingly, it has been found that certain pesticides can also alter the nutritional biochemistry of crop plants by changing the concentrations of nitrogen, phosphorus and potassium, by
influencing the production of sugars, free amino acids and proteins, and by influencing the
ageing process which affects surface hardness, drying and wax deposition (Oka and Pimentel
1976; Rodriguez et al. 1957).

Weather-induced insect pest outbreaks
It has been argued that weather can be the most important factor triggering insect outbreaks
(Milne 1957). For example, Miyashita (1963), in reviewing the dynamics of seven of the most

serious insect pests in Japanese crops, concluded that weather was the principal cause of the outbreaks in each case. There are several ways in which weather can trigger insect outbreaks. Perhaps the most straightforward mechanism is direct stimulation of the insect and/or host plant physiology. The development and widespread use of degree-day models to predict outbreaks of particular pests and appropriate control strategies are an indication of the importance of the linkage between temperature and growth and the development of herbivorous insects and their host plants. Gutierrez et al. (1974) have shown that weather plays a key role in the development of cowpea-aphid populations in south-east Australia. There, a series of climatic events favours complex changes in aphid physiological development, migration and dispersal in such a way as to cause localised outbreaks. Clearly, climate change and global warming have the potential to have major – though hard to predict – effects on pest outbreaks.

Changes induced by plant breeding

Domestication and breeding can induce changes in plant quality and other crop characteristics that may render crops more susceptible to pests. Chen and Welter (2002) found populations of the moth *Homoeosoma electellum* (Lepidoptera: Pyralidae) to be consistently more abundant on sunflower cultivars grown in agriculture than on wild sunflower species in native habitats. Agricultural sunflowers were much larger than wild sunflowers and also exhibited uniformity in flowering that influenced both herbivory and also parasitism by Hymenopteran parasitoids. Wild sunflowers were less susceptible to the herbivore.

Genetically engineered crops and insect pest outbreaks

In the last six years (since the late 1990s) transgenic crops have expanded in area, reaching today about 58 million ha worldwide. Such areas are dominated by monocultures of few crop varieties, mainly herbicide-resistant soybeans and Bt corn, with a clear tendency towards decreased agricultural habitat diversity (Marvier 2001). Several agroecologists argue that such massive and rapid deployment of genetically engineered (GE) crops will exacerbate the problems of conventional modern agriculture (Rissler and Mellon 1996; Altieri 2000). At issue is the genetic homogeneity of agroecosystems with GE crops that, in turn, can make such systems increasingly vulnerable to pest and disease problems (NAS 1972).

GE crops may affect natural enemies in several ways: the enemy species may feed directly on corn tissues (e.g. pollen), or on hosts that have fed on Bt corn, or host populations may be reduced. By keeping Lepidoptera pest populations at extremely low levels, Bt crops could potentially starve natural enemies, as predators and parasitic wasps that feed on pests need a small amount of prey to survive in the agroecosystem. Among the natural enemies that live exclusively on insects that the GE crops are designed to kill (Lepidoptera), egg and larval parasitoids would be most affected because they are totally dependent on live hosts for development and survival, whereas some predators could theoretically thrive on dead or dying prey (Schuler et al. 1999).

Natural enemies could also be affected directly through intertrophic-level effects of the toxin. The potential for Bt toxins to move through arthropod food chains poses serious implications for natural biological control in agricultural fields. Recent evidence shows that the Bt toxin can affect beneficial insect predators that feed on insect pests present in Bt crops (Hilbeck et al. 1998). Studies in Switzerland showed that mean total mortality of predaceous lacewing larvae (Chrysopidae) raised on Bt fed prey was 62%, compared to 37% when raised on Bt-free prey. These Bt prey-fed Chrysopidae also exhibited prolonged development time throughout their immature life stage (Hilbeck et al. 1998). Intertrophic-level effects of the Bt toxin raise serious concerns about the potential disruption of natural pest control.

In the case of herbicide-tolerant crops, the biomass of weeds in agroecosystems is usually reduced with knock-on effects on higher trophic levels via reductions in resource availability

(Hawes et al. 2003). Elimination of weeds, within or around fields, that provide nectar or alternative prey/hosts for natural enemies can significantly affect the abundance and diversity of predators and parasitoids in crop fields (Altieri 1994). The GE crop–biological control nexus, including possible beneficial interactions, is explored in more detail by Altieri et al. (ch. 2 this volume).

Habitat manipulation: restoring soil health and plant diversity

The instability of agroecosystems, manifesting as the worsening of most insect pest problems (and therefore increased dependence on external inputs), is increasingly linked to the expansion of crop monocultures (Altieri 1994). Plant communities that are modified to meet the special needs of humans become subject to heavy pest damage and, generally, the more intensely such communities are modified, the more abundant and serious the pests. The inherent self-regulating characteristics of natural communities are lost when humans modify such communities by promoting monocultures. Some agroecologists maintain that this breakdown can be repaired by adding or enhancing plant biodiversity at the field and landscape level (Gliessman 1999; Altieri 1999) – forms of ecological engineering.

Emergent ecological properties develop in diversified agroecosystems, allowing biodiversity to thrive and establish complex food webs and interactions. But biodiversification must be accompanied by improvement of soil quality, as the link between healthy soils and healthy plants is fundamental to ecologically based pest management. The lower pest levels reported in some organic-farming systems may, in part, arise from insect resistance mediated by biochemical or mineral-nutrient dynamics typical of crops under such management practices (e.g. Rao et al. 2001) as well as via enhanced natural enemy densities (e.g. Miliczky et al. 2000). Such studies provide evidence to support the view that the long-term joint management of plant diversity and soil organic matter can lead to better protection against insect pests.

Harmonising soil and plant health in agroecosystems

Although the integrity of the agroecosystem relies on synergies of plant diversity and the continuing function of the soil microbial community, and its relationship with organic matter (Altieri and Nicholls 1999), the evolution of IPM and integrated soil fertility management (ISFM) have proceeded separately. This has prevented many scientists realising that many pest management methods used by farmers can also be considered soil fertility management strategies, and vice versa. There are positive interactions between soils and pests that, once identified, can provide guidelines for optimising total agroecosystem function. New research increasingly suggests that the ability of a crop plant to resist or tolerate insect pests and diseases is tied to optimal physical, chemical and mainly biological properties of soils. Soils with high organic matter and active soil biological activity generally exhibit good soil fertility as well as complex food webs and beneficial organisms that prevent infection (Magdoff and van Es 2000).

Much of what we know about the relationship between crop nutrition and pest incidence comes from studies comparing the effects of organic agricultural practices and modern conventional methods on specific pest populations. Soil fertility practices can affect the physiological susceptibility of crop plants to insect pests by affecting the resistance of individual plants to attack or by altering plant acceptability to certain herbivores. Some studies have also documented how the shift from organic soil management to chemical fertilisers has increased the potential of certain insects and diseases to cause economic losses.

Effects of nitrogen fertilisation on insect pests

The indirect effects of fertilisation practices, acting through changes in the nutrient composition of the crop, have been reported to influence plant resistance to many insect pests. Among the nutritional factors that influence the level of arthropod damage in a crop, total nitrogen (N) has been considered critical for both plants and their consumers (Mattson 1980; Scriber 1984; Slansky and Rodriguez 1987). Several other authors have also indicated that aphid and mite populations increase from nitrogen fertilisation. Herbivorous insect populations associated with *Brassica* crop plants are particularly known to increase in response to higher soil nitrogen levels (Luna 1988, Altieri and Nicholls 2003).

In a two-year study, Brodbeck et al. (2001) found that populations of the thrips *Frankliniella occidentalis* were significantly higher on tomatoes that received higher rates of nitrogen fertilisation. Seasonal trends in *F. occidentalis* on tomato were found to be correlated to the number of flowers per host plant and changed with the nitrogen status of flowers. Plants subjected to higher fertilisation rates produced flowers that had higher nitrogen content as well as variations in several amino-acid profiles that coincided with peak thrip population density. Abundance of *F. occidentalis* (particularly adult females) was most highly correlated to flower concentrations of phenylalanine during population peaks. Other insect populations found to increase following nitrogen fertilisation included fall armyworm in maize, corn earworm on cotton, pear psylla on pear, Comstock mealybug (*Pseudococcus comstocki*) on apple, and European cornborer (*Ostrinia nubilalis*) on field corn (Luna 1988).

Because plants are a source of nutrients to herbivorous insects, an increase in the nutrient content of the plant may be argued to increase its acceptability as a food source to pest populations. Variations in herbivore response may be explained by differences in the feeding behaviour of the herbivores themselves (Pimentel and Warneke 1989). For example, with increasing nitrogen concentrations in creosotebush (*Larrea tridentata*) plants, populations of sucking insects were found to increase, but the number of chewing insects declined. With higher nitrogen fertilisation, the amount of nutrients in the plant increases, as well as the amount of secondary compounds that may selectively affect herbivore feeding patterns. Thus, protein digestion inhibitors that accumulate in plant cell vacuoles are not consumed by sucking herbivores, but will harm chewing herbivores (Mattson 1980).

Letourneau (1988) questions if the nitrogen-damage hypothesis, based on Scriber's review, can be extrapolated to a general warning about fertiliser inputs associated with insect pest attack in agroecosystems. Of 100 studies of insects and mites on plants treated experimentally with high and low N fertiliser levels, Letourneau found that two-thirds showed an increase in growth, survival, reproductive rate, population densities or plant damage levels in response to increased N fertiliser. The remaining third of the arthropods studied showed either a decrease in damage with N fertiliser or no significant change. The author also noted that experimental design can affect the types of responses observed, suggesting that more reliable data emerged in experiments conducted in field plots, using damage level, population levels and reproductive rate in individual insect species as best predictors of insect response to increased nitrogen.

Dynamics of insect herbivores in organically managed systems

Studies documenting lower abundance of several insect herbivores in low-input systems have partly attributed such reductions to the lower nitrogen content in organically farmed crops (Lampkin 1990). In Japan, density of immigrants of the planthopper species *Sogatella furcifera* was significantly lower, and the settling rate of female adults and survival rate of immature stages of ensuing generations were generally lower, in organic than in conventional rice fields. Consequently, the density of planthopper nymphs and adults in later generations decreased in organically farmed fields (Kajimura 1995). In India, the introduction of high-yielding rice varieties by the Green Revolution was accompanied by increased and frequent inputs of fertilis-ers. These changes unexpectedly influenced mosquito breeding and thereby affected the incidence of mosquito-borne disease. Researchers found that the application of urea in rice fields significantly increased the population densities of mosquito larvae and pupae (anophe-lines as well as culicines) in a dose-related manner. In contrast, fields treated with organic fertilisers such as farmyard manure exhibited significantly lower population densities of mosquito immatures (Greenland 1997).

In England, conventional winter wheat fields exhibited a larger infestation of the aphid *Metopolophium dirhodum* than their organic counterparts. The conventionally fertilised wheat crop also had higher levels of free protein amino acids in its leaves during June, which were attributed to a nitrogen top-dressing applied early in April. However, the difference in aphid infestations between crops was attributed to the aphids' response to the relative proportions of certain non-protein to protein amino acids present in the leaves at the time of aphid settling on crops (Kowalski and Visser 1979). The authors concluded that chemically fertilised winter wheat was more palatable than its organically grown counterpart; hence the higher level of infestation.

In greenhouse experiments, when given a choice of maize (also known as corn) grown on organic or chemically fertilised soils collected from nearby farms, European cornborer (*Ostrinia nubilalis*) females laid significantly more eggs in the chemically fertilised plants (Phelan et al. 1995). There was significant variation in egg-laying among chemical fertiliser treatments within the conventionally managed soil, but in plants under organic soil management, egg-laying was uniformly low. Pooled results across all three farms showed that variance in egg-laying was approximately 18 times higher among plants in conventionally managed soil than among plants grown under an organic regimen. The authors suggested that this difference is evidence for a form of biological buffering found more commonly in organically managed soils.

Altieri et al. (1998) conducted a series of comparative experiments on various growing seasons between 1989 and 1996 in which broccoli was subjected to varying fertilisation regimes (conventional versus organic). The goal was to test the effects of different nitrogen sources on the abundance of the key insect pests, cabbage aphid (*Brevicoryne brassicae*) and flea beetle (*Phyllotreta cruciferae*). Conventionally fertilised monoculture consistently developed a larger infestation of flea beetles and (sometimes) cabbage aphids, than did the organically fertilised broccoli systems. The reduction in aphid and flea beetle infestations in the organically fertilised plots was attributed to lower levels of free nitrogen in plant foliage. This further supports the view that insect pest preference can be moderated by altering the type and amount of fertiliser used.

By contrast, a study comparing the population responses of *Brassica* pests to organic versus synthetic fertilisers, measured higher *Phyllotreta* flea beetles populations on sludge-amended collard (*Brassica oleracea*) plots early in the season than on mineral-fertiliser-amended and unfertilised plots (Culliney and Pimentel 1986). However, later in the season in these same plots, insect population levels were lowest in organic plots for beetles, aphids and lepidopteran pests. This suggests that the effects of fertiliser type vary with plant growth stage and that organic fertilisers do not necessarily diminish pest populations – at times they may increase them. For example, in a survey of California tomato producers, Letourneau et al. (1996) found that despite the pronounced differences in plant quality (N content of leaflets and shoots) both within and among tomato fields, there was no indication that greater concentrations of tissue N in tomato plants were associated with higher levels of insect damage.

Links between below- and above-ground food webs
Agroecology encourages practices that enhance the greatest abundance and diversity of above- and below-ground organisms. While these aerial and epigeal components have usually been considered in isolation, they are dependent upon each other. Producers provide the organic

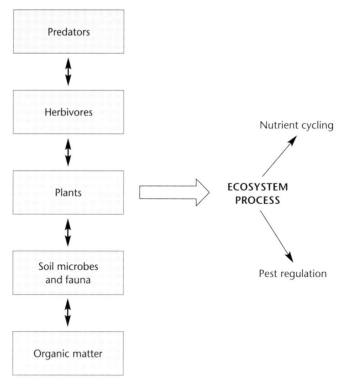

Figure 3.1: Summary of multitrophic interactions and their effects on agricultural productivity via ecosystem processes.

carbon sources that drive the decomposer activity, which is in turn responsible for mineralising nutrients required for maintaining growth of the producers. On the other hand, mutualists, herbivores, pathogens, predators and parasites affect producer–decomposer interactions by directing changes in the flow of energy and resources or by imposing selective forces.

Research has demonstrated often-complex and unexpected feedbacks between the elements of below- and above-ground trophic systems, with implications for the structure and functioning of the entire food web. For example, spiders that prey on important detritivores and fungivores can depress rates of litter decomposition and potentially reduce plant growth. If decomposition and grazing food webs are linked by common top predators, it is possible that increased inputs of detritus could elevate the biomass of primary producers by reducing herbivory through predation. Such links could also have implications in pest regulation. Studies in tropical Asian irrigated-rice agroecosystems by Settle et al. (1996) showed that by increasing organic matter in test plots, researchers could boost populations of detritivores and plankton-feeders, and in turn significantly boost the abundance of generalist predators. Surprisingly, organic matter management proved to be a key mechanism in the support of high levels of natural biological control. Such mechanisms have previously been ignored by scientists as important elements in rice pest management.

Figure 3.1 suggests likely feedbacks between plants, ecosystem processes and biota at other trophic levels. In agroecosystems, plant species' richness and functional diversity have been shown to affect both herbivore abundance and diversity. But increased diversity of plants may also influence organisms and processes in the soil and vice versa, though many aspects of this issue remain unexplored. More research is required to confirm feedbacks, as obviously the better researchers and farmers understand the intricate relationships among soils, microbes, crops, pests and natural enemies, the more skilfully they can incorporate the many elements of bio-diversity in ecological engineering to optimise key processes essential to sustain agroecosystem productivity and health.

Conversion

In reality, the implementation of an ecologically based pest management strategy usually occurs while an agroecosystem is undergoing conversion from a high-input conventional management system to a low external-input system. This conversion can be conceptualised as a transitional process with three marked phases (McRae et al. 1990):

1 increased efficiency of input use as emphasised by traditional integrated pest management;
2 input substitution or substitution of environmentally benign inputs for agrochemical inputs as practised by many organic farmers;
3 system redesign: diversification with an optimal crop/animal assemblage, which encourages synergism so that the agroecosystem may sponsor its own soil fertility, natural pest regulation and crop productivity.

Many of the practices currently being promoted as components of IPM fall in categories 1 and 2. Both of these stages offer clear benefits in terms of lower environmental impacts, as they decrease agrochemical input and often provide economic advantages compared to conventional systems. Incremental changes are likely to be more acceptable to farmers than drastic modifications that may be viewed as highly risky or that complicate management. But can the adoption of practices that increase the efficiency of input use or that substitute biologically based inputs for agrochemicals, but that leave the monoculture structure intact, really lead to the productive redesign of agricultural systems?

In general, the fine-tuning of input use through IPM does little to move farmers toward an alternative to high-input systems. In most cases, IPM translates to 'intelligent pesticide management' as it results in selective use of pesticides according to a predetermined economic threshold, which pests often surpass in monoculture situations. On the other hand, input substitution follows the same paradigm of conventional farming; overcoming the limiting factor but this time with biological or organic inputs. Many of these alternative inputs have become commodified, therefore farmers continue to be dependent on input suppliers, many of a corporate nature (Altieri and Rosset 1996). Clearly, as it stands today, input substitution has lost much of its ecological potential. System redesign, however, arises from the transformation of agroecosystem function and structure by promoting management focused on ensuring fundamental agroecosystem processes. Promotion of biodiversity within agricultural systems is the cornerstone strategy of system redesign, as research has demonstrated that higher diversity (genetic, taxonomic, structural, resource) within the cropping system leads to higher diversity in associated biota, usually leading to more effective pest control and tighter nutrient cycling.

As more information about specific relationships between biodiversity, ecosystem processes and productivity in a variety of agricultural systems is accumulated, design guidelines can be developed further and used to improve agroecosystem sustainability and resource conservation.

Syndromes of production

One of the frustrations of research in sustainable agriculture has been the inability of low-input practices to outperform conventional practices in side-by-side experimental comparisons, despite the success of many organic and low-input production systems in practice (Vandermeer 1997). A potential explanation for this paradox was offered by Andow and Hidaka (1989) in their description of 'syndromes of production'. These researchers compared the traditional shizeñ system of rice (*Oryza sativa*) production with the contemporary Japanese high-input system. Although rice yields were comparable in the two systems, management practices differed in almost every respect: irrigation practice, transplanting technique, plant density, fertility source and quantity, and management of insects, diseases and weeds. Andow and Hidaka (1989) argued that systems like shizeñ function in a qualitatively different way from conventional systems. The array of cultural technologies and pest management practices results in functional differences that cannot be accounted for by any single practice.

A production syndrome is a set of management practices that are mutually adaptive and lead to high performance. However, subsets of the collection of practices may be substantially less adaptive; that is, the interaction among practices leads to improved system performance that cannot be explained by the additive effects of individual practices. In other words, each production system represents a distinct group of management techniques and, by implication, ecological relations. This re-emphasises the fact that agroecological designs (i.e. pest-suppressive crop combinations) are site-specific and what may be applicable elsewhere are not the techniques, but rather the ecological principles that underlie sustainability. It is no use transferring technologies from one site to another, if the set of ecological interactions associated with such techniques cannot be replicated.

Diversified agroecosystems and pest management

Diversified cropping systems, such as those based on intercropping and agroforestry or cover cropping of orchards, have recently been the target of much research. This interest is largely based on the emerging evidence that these systems are more stable and resource-conserving (Vandermeer 1995). Many of these attributes are connected to the higher levels of functional biodiversity associated with complex farming systems. As diversity increases, so do opportunities for coexistence and beneficial interference between species that can enhance agroecosystem sustainability (van Emden and Williams 1974). Diverse systems encourage complex food webs which entail more potential connections and interactions among members, and many alternative paths of energy and material. For this and other reasons a more complex community exhibits more stable production and less fluctuations in numbers of undesirable organisms. Studies further suggest that the more diverse the agroecosystems and the longer this diversity remains undisturbed, the more internal links develop to promote greater insect stability. It is clear, however, that the stability of the insect community depends not only on its trophic diversity, but also on the actual density-dependence nature of the trophic levels (Southwood and Way 1970). In other words, stability will depend on the precision of the response of any particular trophic link to an increase in the population at a lower level.

Recent studies conducted in grassland systems suggest, however, that there are no simple links between species diversity and ecosystem stability. What is apparent is that functional characteristics of component species are as important as the total number of species in determining processes and services in ecosystems (Tilman et al. 1996). This finding has practical implications for agroecosystem management. If it is easier to mimic specific ecosystem processes than to duplicate all the complexity of nature, then the focus should be placed on a specific biodiversity component that plays a specific role, such as a plant that fixes nitrogen, provides cover for soil protection or harbours resources for natural enemies. In the case of farmers without major economic and resource limits, who can allow some risk of crop failure, a crop rotation or simple polyculture may be all it takes to achieve a desired level of stability. But in the case of resource-poor farmers, who cannot tolerate crop failure, highly diverse cropping systems would probably be the best choice. The obvious reason is that the benefit of complex agroecosystems is low-risk; if a species falls to disease, pest attack or weather, another species is available to fill the void and maintain full use of resources. Thus there are potential ecological benefits to having several species in an agroecosystem: compensatory growth, full use of resources and nutrients, and pest protection (Ewel 1999).

Plant diversity and insect pest incidence
An increasing body of literature documents the effects of plant diversity on the regulation of insect herbivore populations by favouring the abundance and efficacy of associated natural enemies (Landis et al. 2000). Research has shown that mixing certain plant species usually leads to density reductions of specialised herbivores. In a review of 150 published investigations, Risch et al. (1983) found evidence to support the notion that specialised insect herbivores were less numerous in diverse systems (53% of 198 cases). In another comprehensive review of 209 published studies that dealt with the effects of vegetation diversity in agroecosystems on herbivores arthropod species, Andow (1991) found that 52% of the 287 total herbivore species examined were less abundant in polycultures than in monocultures, while only 15.3% (44 species) exhibited higher densities in polycultures. In a more recent review of 287 cases, Helenius (1998) found that the reduction of monophagous (specialist) pests was greater in perennial systems, and that the reduction of polyphagous pest numbers was less in perennial than in annual systems. Helenius (1998) concluded that monophagous insects are more susceptible to crop diversity than polyphagous insects. He warned of the increased risk of pest attack if the dominant herbivore fauna in a given agroecosystem is polyphagous. In examining numerous studies testing the responses of pest and beneficial arthropods to plant diversification in cruciferous crops, Hooks and Johnson (2003) concluded that biological parameters of herbivores affected by crop diversification were mainly related to the behaviour of the insect studied. Mechanisms accounting for herbivore responses to plant mixtures include reduced colonisation, reduced adult tenure time in the crop, and oviposition interference. They suggest that lower herbivore populations in mixed *Brassica* plantings are due to lower plant size and quality of these crops in diverse systems than in monocultures. Hooks and Johnson (2003) urged changes on *Brassica* agronomy so that mixed cropping can offer crop protection benefits without yield reduction.

The ecological theory relating to the benefits of mixed versus simple cropping systems revolves around two possible explanations of how insect pest populations attain higher levels in monoculture systems than in diverse ones. The two hypotheses proposed by Root (1973) are:

1 the enemies hypothesis, which argues that pest numbers are reduced in more diverse systems because the activity of natural enemies is enhanced by environmental opportunities prevalent in complex systems;
2 the resource concentration hypothesis, which argues that the presence of a more diverse flora has direct negative effects on the ability of insect pests to find and utilise the host plant and to remain in the crop habitat.

The resource concentration hypothesis predicts lower pest abundance in diverse communities because a specialist feeder is less likely to find its host plant due to the presence of confusing masking chemical stimuli, physical barriers to movement or other environmental effects such as shading. It will tend to remain in the intercrop for a shorter time simply because the probability of landing on a non-host plant is increased. It may also have a lower survivorship and/or fecundity (Bach 1980). The extent to which these factors operate will depend on the number of host plant species present and the relative preference of the pest for each, the absolute density and spatial arrangement of each host species and the interference effects from more host plants.

The enemies hypothesis attributes lower pest abundance in intercropped or more diverse systems to a higher density of predators and parasitoids (Bach 1980). The greater density of natural enemies is caused by an improvement in conditions for their survival and reproduction, such as a greater temporal and spatial distribution of nectar and pollen sources, which can increase parasitoid reproductive potential and abundance of alternative host/prey when the pest species are scarce or at inappropriate stages (Risch 1981; Jervis et al., ch. 5 this volume). These factors can in theory combine to provide more favourable conditions for natural enemies and thereby enhance their numbers and effectiveness as control agents.

The relative importance of these hypotheses has been investigated in two ways: reviews of the literature relating to crop diversity and pest abundance, and by experimentation. Risch et al. (1983) concluded that the resource concentration hypothesis was the most likely explanation for reductions in pest abundance in diverse systems. However, 19 studies that tested the natural enemy hypothesis were reviewed by Russell (1989), who found that mortality rates from predators and parasitoids in diverse systems were higher in nine, lower in two, unchanged in three and variable in five. Russell (1989) concluded that the natural enemy hypothesis is an operational mechanism, but he considered the two hypotheses complementary. In studies of crop/weed systems, Baliddawa (1985) found that 56% of pest reductions in weed-diversified cropping systems were caused by natural enemies.

One of the major problems has been predicting which cropping systems will reduce pest abundance, since not all combinations of crops will produce the desired effect. Blind adherence to the principle that a more diversified system will reduce pest infestation is clearly inadequate and often totally wrong (Gurr et al. 1998). To some researchers this indicates the need for caution and a greater understanding of the mechanisms involved to explain how, where and when such exceptions are likely to occur. It will only be through more detailed ecological studies that such an understanding can be gained and an appropriate predictive theory developed. This means that greater emphasis has to be placed on ecological experiments than on purely descriptive comparative studies.

Recent practical case studies

Despite some of the above-mentioned knowledge gaps, many studies have transcended the research phase and have found applicability to regulate specific pests.

- Researchers working with farmers in 10 townships in Yumman, China, covering an area of 5350 ha, encouraged farmers to switch from rice monocultures to planting various mixtures of local rice with hybrids. Enhanced genetic diversity reduced blast incidence by 94% and increased total yields by 89%. By the end of two years, it was concluded that fungicides were no longer required (Wolfe 2000; Zhu et al. 2000).

- In Africa, scientists at the International Center of Insect Physiology and Ecology (ICIPE) developed a habitat manipulation system (push–pull system) that uses plants in the borders of maize fields, which act as trap crops (Napier grass and Sudan grass) attracting stemborer colonisation. The system involves the maize (the 'pull') and two plants intercropped with maize (molasses grass and silverleaf) that repel the stemborers (the 'push') (Khan et al. 1998; Khan and Pickett, ch. 10 this volume). Border grasses also enhance the parasitisation of stemborers by the wasp *Cotesia sesamiae*, and are important fodder plants. The leguminous silverleaf (*Desmodium uncinatum*) suppresses the parasitic weed Striga by a factor of 40 when compared with maize monocrop. *Desmodium*'s N-fixing ability increases soil fertility, and it is excellent forage. As a bonus, selling *Desmodium* seed has provided a new income-generating opportunity for women in the project areas. The push–pull system has been tested on over 450 farms in two districts of Kenya and has now been released for uptake by the national extension systems in East Africa. Participating farmers in the breadbasket of Trans Nzoia are reporting a 15–20% increase in maize yield. In the semi-arid Suba district, plagued by both stemborers and Striga, a substantial increase in milk yield occurred in four years, with farmers now able to support higher numbers of dairy cows on the fodder produced. When farmers plant maize together with the push–pull plants, there is a return of $US2.30 for every dollar invested, compared to only $US1.40 obtained by planting maize as a monocrop.

• Several researchers have introduced flowering plants as strips within crops as a way to enhance the availability of pollen and nectar, necessary for optimal reproduction, fecundity and longevity of many natural enemies of pests. *Phacelia tanacetifolia* strips have been used in wheat, sugar beets and cabbage, leading to greater abundance of aphidophagous predators (especially syrphid flies) and reduced aphid populations. In England, in an attempt to provide suitable overwintering habitat within fields for aphid predators, researchers created 'beetle banks' sown with perennial grasses such as *Dactylis glomerata* and *Holcus lanatus*. When these banks run parallel with the crop rows, great enhancement of predators (up to 1500 beetles per square metre) can be achieved in only two years (Landis et al. 2000).

• In perennial cropping systems, the presence of flowering undergrowth enhances the biological control of a series of insect pests. The beneficial insectary role of *Phacelia* in apple orchards was well demonstrated by Russian and Canadian researchers more than 30 years ago (Altieri 1994). Organic farmers maintained floral diversity throughout the growing season in California vineyards, in the form of summer cover crops of buckwheat (*Fagopyrum esculentum*) and sunflower (*Helianthus annuus*). This had a substantial impact on the abundance of western grape leafhopper, *Erythroneura elegantula* (Homoptera: Cicadellidae), western flower thrips, *Frankliniella occidentalis* (Thysanoptera: Thripidae) and associated natural enemies. During two consecutive years, vineyard systems with flowering cover crops were characterised by lower densities of leafhoppers and thrips, and larger populations and more species of general predators, including spiders. Although *Anagrus epos* (Hymenoptera: Mymaridae), the most important parasitoid, achieved high numbers and inflicted noticeable mortality of grape leafhopper eggs, no differences in egg parasitism rates were observed between cover-cropped and monoculture systems. Mowing the cover crops forced *Anagrus* and predators to move to adjacent vines, resulting in the lowering of leafhopper densities in such vines. Results indicated that habitat diversification using summer cover crops that bloom most of the growing season, supports large numbers of predators and parasitoids, thereby favouring enhanced biological control of leafhoppers and thrips in vineyards (Nicholls et al. 2001).

• In Washington state, USA, researchers reported that organic apple orchards that retained some level of plant diversity in the form of weeds mowed as needed, gave similar apple yields to conventional and integrated orchards. Their data showed that the low external-input organic system ranked first in environmental and economic sustainability as it offered higher profitability, greater energy efficiency and lower negative environmental impact (Reganold et al. 2001).

• In Central America, Staver et al. (2001) designed pest-suppressive multistrata shade-grown coffee systems, selecting tree species and associations, density and spatial arrangement as well as shade management regimes, with the main goal of creating optimum shade conditions for pest suppression. For example, in low-elevation coffee zones, 35–65% shade promotes leaf retention in the dry seasons and reduces brown eye spot disease (*Cercospora coffeicola*), weeds and *Planococcus citri* (Homoptera: Pseudococcidae). At the same time, it enhances the effectiveness of microbial and parasitic organisms without contributing to increased rust disease (*Hemileia vastatrix*) levels or reducing yields.

- Several entomologists have concluded that the abundance and diversity of predators and parasites within a field are closely related to the nature of the vegetation in the field margins. There is wide acceptance of the importance of field margins as reservoirs of the natural enemies of crop pests. Many studies have demonstrated increased abundance of natural enemies and more effective biological control where crops are bordered by wild vegetation from which natural enemies colonise. Parasitism of the armyworm, *Pseudaletia unipuncta*, was significantly higher in maize fields embedded in a complex landscape than in maize fields surrounded by simpler habitats. In a two-year study researchers found higher parasitism of *Ostrinia nubilalis* larvae by the parasitoid *Eriborus terebrans* in edges of maize fields adjacent to wooded areas, than in field interiors (Marino and Landis 1996). Similarly, in Germany, parasitism of rape pollen beetle was about 50% at the edge of the fields, while at the centre of the fields parasitism dropped significantly to 20% (Thies and Tscharntke 1999).

- One way to introduce the beneficial biodiversity from surrounding landscapes into large-scale monocultures is by establishing vegetationally diverse corridors that allow the movement and distribution of useful arthropod biodiversity into the centre of monocultures. Nicholls et al. (2001) established a vegetational corridor which connected to a riparian forest and cut across a vineyard monoculture. The corridor allowed natural enemies emerging from the riparian forest to disperse over large areas of otherwise monoculture vineyard systems. The corridor provided a constant supply of alternative food for predators, effectively decoupling predators from a strict dependence on grape herbivores and avoiding a delayed colonisation of the vineyard. This complex of predators continuously circulated into the vineyard interstices, establishing a set of trophic interactions leading to a natural enemy enrichment, which led to lower numbers of leafhoppers and thrips on vines located up to 30–40 m from the corridor.

All these examples constitute forms of habitat diversification that provide resources and environmental conditions suitable for natural enemies. The challenge is to identify the type of biodiversity that is desirable to maintain and/or enhance in order to carry out ecological services of pest control, then to determine the best practices that will encourage such desired biodiversity components.

Designing pest-stable agroecosystems

The key challenge for 21st-century pest managers is to translate ecological principles into practical alternative systems to suit the specific needs of farming communities in different agroecological regions of the world. This chapter emphasises that a major strategy in designing a more sustainable agriculture is restoring agricultural diversity in time and space by following key agroecological guidelines:

- increase species diversity in time and space through multiple cropping and agroforestry designs;
- increase genetic diversity through variety mixtures, multi-lines and use of local germplasm and varieties exhibiting horizontal resistance;
- include and improve fallow through legume-based rotations, use of green manures, cover crops and/or livestock integration;
- enhance landscape diversity with biological corridors and vegetationally diverse crop-field boundaries or create a mosaic of agroecosystems and maintain areas of natural or secondary vegetation as part of the agroecosystem matrix.

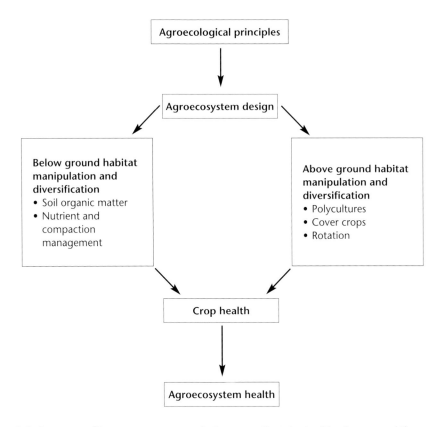

Figure 3.2: Summary of how agroecosystem design may affect the health of crops and the agroecosystem.

As mentioned above, diversification schemes should be complemented by soil organic management as both strategies form the pillars of agroecosystem health (Figure 3.2).

There are different options to diversify cropping systems, depending on whether the monoculture systems to be modified are based on annual or perennial crops. Diversification can also take place outside the farm, for example in crop-field boundaries with windbreaks, shelter-belts and living fences, which can improve habitat for wildlife and beneficial insects, provide resources of wood, organic matter, resources for pollinating bees, and also modify wind speed and microclimate (Altieri and Letourneau 1982). Plant diversification can be considered a form of conservation biological control with the goal of creating a suitable ecological infrastructure within the agricultural landscape to provide resources such as pollen and nectar for adult natural enemies, alternative prey or hosts, and shelter from adverse conditions. These resources must be integrated into the landscape in a way that is spatially and temporally favourable to natural enemies, and practical to implement.

Landis et al. (2000) recommended the following guidelines when implementing habitat manipulation strategies:

- selection of the most appropriate plant species;
- the spatial and temporal arrangement of such plants within and/or around the fields;
- the spatial scale over which the habitat enhancement operates, with implications at the field or landscape level;

- the predator/parasitoid behavioural mechanisms which are influenced by the habitat manipulation;
- potential conflicts when adding new plants to the agroecosystem (i.e. in California, *Rubus* blackberries around vineyards increases populations of grape leafhopper parasitoids but can also enhance abundance of the sharpshooter which serves as a vector for Pierce's disease);
- develop ways in which added plants do not upset other agronomic management practices, and select plants that preferentially have multiple effects such as improving pest regulation but at the same time improve soil fertility, weed suppression, etc.

Hooks and Johnson (2003) further identified a need for more categorical research on the use of crop diversification, recommending that more attention should be devoted to:

- defining ways to suppress pests through diversity without significant yield reductions;
- determining how mixed cropping systems affect the population dynamics and searching behaviours of natural enemies;
- discovering methods to make mixed plantings more economically feasible and compatible with conventional farm operations;
- determining how mixed cropping systems can be effectively combined with other pest control tactics.

Crop diversification may not suffice as a stand-alone pest management tactic. However, if compatible with other pest management tactics (e.g. biological control, host plant resistance) some of the shortcomings associated with habitat manipulation may be overcome. This will depend on the type of crop, the nature of surrounding habitats, the diversity of beneficial biota and the prevalent pest. In diversified farms that have undergone agroecological conversion for three or more years, diversity usually provides all the needed protection. The objective is to apply management practices that will enhance or regenerate the kind of biodiversity that not only maximises the sustainability of agroecosystem by providing ecological services such as biological control, but also nutrient cycling, water and soil conservation.

If one or more alternative diversification schemes are used, the possibilities of complementary interactions between agroecosystem components are enhanced, resulting in one or more of the following effects:

- continuous vegetation cover for soil protection;
- constant production of food, ensuring a varied diet and several marketing items;
- closing nutrient cycles and effective use of local resources;
- soil and water conservation through mulching and wind protection;
- enhanced biological pest control by providing diversification resources to beneficial biota;
- increased multiple-use capacity of the landscape;
- sustained crop production without relying on environmentally degrading chemical inputs.

In summary, key ecological principles for the design of diversified and sustainable agroecosystems include the following.

- *Increasing species diversity,* as this promotes fuller use of resources (nutrient, radiation, water etc.), pest protection and compensatory growth. Many researchers have highlighted the importance of various spatial and temporal plant combinations to facilitate complementary resource use or to provide intercrop advantage, for example legumes facilitating the growth of cereals by supplying extra nitrogen. Compensatory growth is also a desirable trait because if one species succumbs to pests, weather or harvest; another species fills the void, maintaining full use of available resources.

- *Enhance longevity* through the addition of perennials that contain a thick canopy, thus providing continual cover that can also protect the soil from erosion. Constant leaf-fall builds organic matter and allows uninterrupted nutrient circulation. Dense and deep root systems of long-lived woody plants are an effective mechanism for nutrient capture, offsetting the negative losses through leaching. Perennial vegetation also provides more habitat permanence, contributing to more stable pest–enemy complexes.
- *Impose a fallow* to restore soil fertility through biologically mediated mechanisms, and to reduce agricultural pest populations as life cycles are interrupted with forest regrowth or legume-based rotations.
- *Enhance additions of organic matter* by including high biomass-producing plants, as organic matter forms the foundation of complex food webs which may indirectly influence the abundance and diversity of natural enemies.
- *Increase landscape diversity* by having a mosaic of agroecosystems representative of various stages of succession. Risk of complete failure is spread among, as well as within, the various cropping systems. Improved pest control is also linked to spatial heterogeneity at the landscape level.

References

Altieri, M.A. (1994). *Biodiversity and Pest Management in Agroecosystems*. Haworth Press, New York.

Altieri, M.A. (1995). *Agroecology: The Science of Sustainable Agriculture*. Westview Press, Boulder.

Altieri, M.A. (1999). The ecological role of biodiversity in agroecosystems. *Agriculture, Ecosystems and Environment* 74: 19–31.

Altieri, M.A. (2000). The ecological impacts of transgenic crops on agroecosystem health. *Ecosystem Health* 6: 13–23.

Altieri, M.A. and Letourneau, D.K. (1982). Vegetation management and biological control in agroecosystems. *Crop Protection* 1: 405–430.

Altieri, M.A. and Letourneau, D.K. (1984). Vegetation diversity and outbreaks of insect pests. CRC, *Critical Reviews in Plant Science*s 2: 131–169.

Altieri, M.A. and Nicholls, C.I. (1999). Biodiversity, ecosystem function and insect pest management in agricultural systems. In *Biodiversity in Agroecosystems* (W.W. Collins and C.O. Qualset, eds), pp. 69–84. CRC Press, Boca Raton.

Altieri, M.A. and Nicholls, C.I. (2002). The simplification of traditional vineyard base agroforests in northwestern Portugal: some ecological implications. *Agroforestry Systems* 56 (3): 185–191.

Altieri, M.A. and Nicholls, C.I. (2003). Soil fertility management and insect pests: harmonizing soil and plant health in agroecosystems. *Soil and Tillage Research* 72: 203–211.

Altieri, M.A. and Rosset, P. (1996). Agroecology and the conversion of large-scale conventional systems to sustainable management. *International Journal of Environmental Studies* 50: 165–185.

Altieri, M.A., Rosset, P. and Thrupp, L.A. (1998). The potential of agroecology to combat hunger in the developing world. 2020 Brief. IFPRI, Washington, D.C.

Andow, D.A. (1983). The extent of monoculture and its effects on insect pest populations with particular reference to wheat and cotton. *Agriculture, Ecosystems and Environment* 9: 25–35.

Andow, D.A. (1991). Vegetational diversity and arthropod population response. *Annual Review of Entomology* 36: 561–586.

Andow, D.A. and Hidaka, K. (1989). Experimental natural history of sustainable agriculture: syndromes of production. *Agriculture, Ecosystems and Environment* 27: 447–462.

Bach, C.E. (1980). Effects of plant diversity and time of colonisation on an herbivore–plant interaction. *Oecologia* 44: 319–326.

Baliddawa, C.W. (1985). Plant species diversity and crop pest control: an analytical review. *Insect Sci. Appl.* 6: 479–487.

Brodbeck, B., Stavisky, J., Funderburk, J., Andersen, P. and Olson, S. (2001). Flower nitrogen status and populations of *Frankliniella occidentalis* feeding on *Lycopersicon esculentum*. *Entomologia Experimentalis et Applicata* 99 (2): 165–172.

Chen, Y.H. and Welter, S.C. (2002). Abundance of native moth *Homeosoma electellum* (Lepidoptera: Pyralidae) and activity of indigenous parasitoids in native and agricultural sunflower habitats. *Environmental Entomology* 31: 626–636.

Corbett, A. and Rosenheim, J.A. (1996). Impact of natural enemy overwintering refuge and its interaction with the surrounding landscape. *Ecological Entomology* 21: 155–164.

Culliney, T.W. and Pimentel, D. (1986). Ecological effects of organic agricultural practices on insect populations. *Agriculture, Ecosystems and Environment* 15: 253–266.

Ewel, J.J. (1999). Natural systems as models for the design of sustainable systems of land use. *Agroforestry Systems* 45: 1–21.

Gliessman, S.R. (1999). *Agroecology: Ecological Processes in Agriculture.* Ann Arbor Press, Mich.

Greenland, D.J. (1997). *The Sustainability of Rice Farming.* CAB International Wallingford, UK.

Gurr, G.M., Wratten, S.D., Irvin, N.A., Hossain, Z., Baggen, L.R. et al. (1998). Habitat manipulation in Australasia: recent biological control progress and prospects for adoption. In *Pest Management— Future Challenges: Proc. 6th Aust. App. Entomol. Res. Conv. 29 Sept.—2 Oct. Vol. 2* (M.P. Zaluki, R.A.I. Drew and G.G. White, eds), pp. 225–235. University of Queensland, Brisbane.

Gutierrez, A.P., Havenstein, D.E., Nix, H.A. and Moore, P.A. (1974). The ecology of Aphis craccivora Koch and subterranean clover stunt virus in Southeast Australia. II. A model of cowpea aphid populations in temperate pastures. *Journal of Applied Ecology* 11: 1–20.

Hawes, C. et al. (2003). Responses of plants and invertebrate trophic groups to contrasting herbicide regimens in the farm scale evaluations of genetically modified herbicide-tolerant crops. *Philosophical Transactions of the Royal Society, London* 358: 1899–1913.

Helenius, J. (1998). Enhancement of predation through within-field diversification. In *Enhancing Biological Control* (E. Pickett and R.L. Bugg, eds), pp. 121–160. University of California Press, Berkeley.

Hilbeck, A., Baumgartner, M., Fried, P.M. and Bigler, F. (1998). Effects of transgenic *Bacillus thuringiensis* corn fed prey on mortality and development time of immature *Chrysoperla carnea* (Neuroptera: Chrysopidae). *Environmental Entomology* 27: 460–487.

Hooks, C.R.R. and Johnson, M.W. (2003). Impact of agricultural diversification on the insect community of cruciferous crops. *Crop Protection* 22: 223–238.

Kahn, Z.R., Ampong-Nyarko, K., Hassanali, A. and Kimani, S. (1998). Intercropping increases parasitism of pests. *Nature* 388: 631–632.

Kajimura, T. (1995). Effect of organic rice farming on planthoppers: reproduction of white backed planthopper, *Sogatella furcifera* (Homoptera: Delphacidae). *Res. Popul. Ecol.* 37: 219–224.

Kowalski, R. and Visser, P.E. (1979). Nitrogen in a crop–pest interaction: cereal aphids. In *Nitrogen as an Ecological Parameter* (J.A. Lee, ed.), pp. 67–74. Blackwell Scientific, Oxford.

Lampkin, N. (1990). *Organic Farming.* Farming Press Books, Ipswich, UK.

Landis, D.A., Wratten, S.D. and Gurr, G.M. (2000). Habitat management to conserve natural enemies of arthropod pests in agriculture. *Annual Review of Entomology* 45: 175–201.

Letourneau, D.K. (1988). Soil management for pest control: a critical appraisal of the concepts. In: Global perspectives on agroecology and sustainable agricultural systems. *Proceedings of the Sixth International Scientific Conference of IFOAM*, pp. 581–587. Santa Cruz, CA.

Letourneau, D.K., Drinkwater, L.E. and Shennon, C. (1996). Effects of soil management on crop nitrogen and insect damage in organic versus conventional tomato fields. *Agriculture, Ecosystems and Environment* 57: 174–187.

Levins, R. and Wilson, M. (1979). Ecological theory and pest management. *Annual Review of Entomology* 25: 7–29.

Lewis, W.J., van Lenteren, J.C., Phatak, S.C. and Tumlinson, J.H. (1997). A total system approach to sustainable pest management. *Proceedings of the National Academy of Sciences USA* 94: 12243–12248.

Luna, J.M. (1988). Influence of soil fertility practices on agricultural pests. In: Global perspectives on agroecology and sustainable agricultural systems. *Proceedings of the Sixth International Scientific Conference of IFOAM,* pp.589–600. Santa Cruz, CA.

Magdoff, F. and van Es, H. (2000). *Building Soils for Better Crops.* SARE, Washington, D.C.

Marino, P.C. and Landis, D.A. (1996). Effect of landscape structure on parasitoid diversity and parasitism in agroecosystems. *Ecological Applications* 6: 276–284.

Marvier, M. (2001). Ecology of transgenic crops. *American Scientist* 89: 160–167.

Mattson, W.J. Jr (1980). Herbivory in relation to plant nitrogen content. *Annual Review of Ecology and Systematics* 11: 119–161.

McRae, R.J., Hill, S.B., Mehuys, F.R. and Henning, J. (1990). Farm scale agronomic and economic conversion from conventional to sustainable agriculture. *Adv. Agron.* 43: 155–198.

Miliczky, E.R., Calkins, C.O. and Horton, D.R. (2000). Spider abundance and diversity in apple orchards under three insect pest management programmes in Washington State, U.S.A. *Agricultural and Forest Entomology* 2: 203–215.

Milne, A. (1957). The natural control of insect populations. *Canadian Entomologist Entomology* 89: 193–213.

Miyashita, K. (1963). Outbreaks and population fluctuations of insects, with special reference to agricultural insect pests in Japan. *Bull. National Sust. Agric. Sci. Ser.* C15: 99–170.

Morse, J.G., Bellows, T.S. and Gaston, L.K. (1987). Residual toxicity of acaricides to three beneficial species on California citrus. *Journal of Experimental Entomology* 80: 953–960.

National Academy of Sciences (1972). *Genetic Vulnerability of Major Crops.* NAS, Washington, D.C.

Nicholls, C.I, Parrella, M.P. and Altieri, M.A. (2000). Reducing the abundance of leafhoppers and thrips in a northern California organic vineyard through maintenance of full season floral diversity with summer cover crops. *Agricultural and Forest Entomology* 2: 107–113.

Nicholls, C.I, Parrella, M.P. and Altieri, M.A. (2001). The effects of a vegetational corridor on the abundance and dispersal of insect biodiversity within a northern California organic vineyard. *Landscape Ecology* 16: 133–146.

Oka, I.N. and Pimentel, D. (1976). Herbicide (2,4-D) increases insect and pathogen pests on corn. *Science* 143: 239–240.

Phelan, P.L., Mason, J.F. and Stinner, B.R. (1995). Soil fertility management and host preference by European corn borer, *Ostrinia nubilalis,* on *Zea mays*: a comparison of organic and conventional chemical farming. *Agriculture, Ecosystems and Environment* 56: 1–8.

Pimentel, D. (1961). Species diversity and insect population outbreaks. *Annals of the Entomological Society of America* 54: 76–86.

Pimentel, D. and Goodman, N. (1978). Ecological basis for the management of insect populations. *Oikos* 30: 422–437.

Pimentel, D. and Lehman, H. (1993). *The Pesticide Question.* Chapman & Hall, New York.

Pimentel, D. and Perkins, J.H. (1980). Pest control: cultural and environmental aspects. *AAAS Selected Symposium 43.* Westview Press, Boulder.

Pimentel, D. and Warneke, A. (1989). Ecological effects of manure, sewage sludge and and other organic wastes on arthropod populations. *Agricultural Zoology Reviews* 3: 1–30.

Price, P.W. (1976). Colonization of crops by arthropods: non-equilibrium communities in soybean fields. *Environmental Entomology* 5: 605–612.

Price, P.W. (1991). The plant vigor hypothesis and herbivore attack. *Oikos* 62: 244–251.

Price, P.W. and Waldbauer, G.P. (1975). Ecological aspects of pest management. In *Introduction to Insect Pest Management* (R.L. Metcalf and W. Luckmann, eds). John Wiley & Sons, New York.

Rao, K.R., Rao, P.A. and Rao, K.T. (2001). Influence of organic manures and fertilizers on the incidence of groundnut leafminer, *Aproaerema modicella* Dev. *Annals of Plant Protection Sciences* 9: 12–15.

Reganold, J.P., Glover, J.D., Andrews, P.K. and Hinman, H.R. (2001). Sustainability of three apple production systems. *Nature* 410: 926–930.

Reinjtes, C., Haverkort, B. and Waters-Bayer, A. (1992). *Farming for the Future.* Macmillan, London.

Risch, S.J. (1981). Insect herbivore abundance in tropical monocultures and polycultures: an experimental test of two hypotheses. *Ecology* 62: 1325–1340.

Risch, S.J., Andow, D. and Altieri, M.A. (1983). Agroecosystem diversity and pest control: data, tentative conclusions, and new research directions. *Environmental Entomology* 12: 625–629.

Rissler, J. and Mellon, M. (1996). *The Ecological Risks of Engineered Crops*. MIT Press, Cambridge.

Rodriguez, J.G., Chen, H.H. and Smith, W.T. (1957). Effects of sol insecticides on beans, soybeans, and cotton and resulting effects on mite nutrition. *Journal of Economic Entomology* 50: 587–593.

Root, R.B. (1973). Organization of a plant–arthropod association in simple and diverse habitats: the fauna of collards (*Brassica oleraceae*). *Ecological Monographs* 43: 94–125.

Russell, E.P. (1989). Enemies hypothesis: a review of the effect of vegetational diversity on predatory insects and parasitoids. *Environmental Entomology* 18: 590–599.

Schuler, T.H., Potting, R.P.J., Dunholm, I. and Poppy, G.M. (1999). Parasitoid behavior and Bt plants. *Nature* 400: 525.

Scriber, J.M. (1984). Nitrogen nutrition of plants and insect invasion. In *Nitrogen in Crop Production* (R.D. Hauck, ed.). American Society of Agronomy, Madison, WI.

Settle, W., Ariawan, H., Tri Astuti, E., Cahyana, W., Hakim, A.L., Hindayana, D., Lestari, A.S. and Pajarningsih (1996). Managing tropical rice pests through conservation of generalist natural enemies and alternative prey. *Ecology* 77 (7): 1975–1988.

Slansky, F. and Rodriguez, J.G. (1987). *Nutritional Ecology of Insects, Mites, Spiders and Related Invertebrates*. Wiley, New York.

Southwood, T.R.E. and Way, M.J. (1970). Ecological background to pest management. In *Concepts of Pest Management*. (R.L. Rabb and F.E. Guthrie, eds). North Carolina State University, Raleigh, NC.

Staver, C., Guharay, F., Monterroso, D. and Muschler, R.G. (2001). Designing pest-suppressive multistrata perennial crop systems: shade grown coffee in Central America. *Agroforestry Systems* 53: 151–170.

Thies, C. and Tscharntke, T. (1999). Landscape structure and biological control in agroecosystems. *Science* 285: 893–895.

Tilman, D., Wedin, D. and Knops, J. (1996). Productivity and sustainability influenced by biodiversity in grassland ecosystems. *Nature* 379: 718–720.

Vandermeer, J. (1995). The ecological basis of alternative agriculture. *Annual Review of Ecological Systems* 26: 210–224.

Vandermeer, J. (1997). Syndromes of production: an emergent property of simple agroecosystem dynamics. *Journal of Environmental Management* 51: 59–72.

Vandermeer, J. and Perfecto, I. (1995). *Breakfast of Biodiversity*. Food First Books, Oakland, CA.

Van Driesche, R.G. and Bellows, T.S. Jr (1996). *Biological Control*. Chapman & Hall. New York.

van Emden, H.F. (1966). Studies on the relations of insect and host plant. III. A comparison of the reproduction of *Brevicoryne brassicae* and *Myzus persicae* (Hemiptera: Aphididae) on brussels sprout plants supplied with different rates of nitrogen and potassium. *Entomological Experiments Applied* 9: 444–460.

van Emden, H.F. and Williams, G.F. (1974). Insect stability and diversity in agroecosystems. *Annual Review of Entomology* 19: 455–475.

Victor, T.J. and Reuben, R. (2000). Effects of organic and inorganic fertlisiers on mosquito populations in rice fields on southern India. *Medical and Veterinary Entomology* 14: 361–368.

Wolfe, M. (2000). Crop strength through diversity. *Nature* 406: 681–682.

Zhu, Y., Fen, H., Wang, Y., Li, Y., Chen, J., Hu, L. and Mundt, C.C. (2000). Genetic diversity and disease control in rice. *Nature* 406: 718–772.

Chapter 4

Landscape context of arthropod biological control

Martin H. Schmidt, Carsten Thies and Teja Tscharntke

Introduction

Historically, biological control has primarily relied on local facilitation of natural enemy populations, but recent studies have found strong effects of landscape structure on the dynamic interactions between pests and their natural enemies. Case studies in arable crops show how
predators and parasitoids of major pests have been enhanced by the presence of a high percentage of non-crop habitats in the landscape. In this chapter we explore the major challenge of
understanding the coaction of landscape effects on pests and their multiple enemies, and of
identifying applicable measures for landscape manipulation that enhance natural pest control.

The importance of the spatial context for interactions between plants, herbivores and their
enemies has been increasingly recognised during the past decades (Ricklefs 1987; Kareiva
1990a, b). Because of the high mobility of many organisms, the spatial scale of many studies has
widened to include processes active over whole landscapes. Theory is successfully expanding to
consider population dynamics and plant-pest-enemy interactions at a landscape scale (Kareiva
and Wennegren 1995; Holt and Hochberg 2001; Holt 2002; Tscharntke and Brandl 2004). In
contrast, conducting manipulative experiments with independent landscape replicates is difficult, though scaling-up from the local to the landscape level may provide new insights (e.g.
Kruess and Tscharntke 1994; With and Christ 1995; Roland and Taylor 1997; Cappuccino et al.
1998; Kruess and Tscharntke 2000; van Nouhuys and Hanski 2002; With et al. 2002). Here, we
focus on the current development of large-scale studies demonstrating landscape effects on
processes relevant for biological control.

The applicability of classical spatial concepts such as island biogeography or metapopulation
theory to annual crops is limited (Tscharntke and Brandl 2004). Impoverishment of biodiversity
is typical for simplified agricultural landscapes dominated by few types of annual crops (Stoate
et al. 2001; Benton et al. 2003). In such landscapes, crop fields are not isolated from each other,
but may be distant from less disturbed habitats that are potential sources of immigrants. Many
organisms depend, at some life stage, on resources that annual crops lack and such species may
be rare or absent in crop-dominated landscapes if their dispersal power is smaller than the
distance to these resources (Tscharntke and Brandl 2004). For example, dispersal limitation of
many parasitoids leads to reduced build-up of populations in crop fields. When natural enemies
of herbivores become locally impoverished, their host or prey may be released from natural
control and increasingly damage crops. Essential resources for biological control agents that are

usually not available in annually ploughed fields are pollen and nectar sources for adults of parasitoid Hymenoptera and Diptera, or perennial plant cover and an undisturbed soil surface for overwintering and larval development of a wide array of insects and spiders (Landis et al. 2000; Tscharntke 2000; Gurr et al. 2003). The few existing studies with replicated landscapes mostly found that landscape context had strong effects on local community structure and inter-actions (Roland and Taylor 1997; Thies and Tscharntke 1999; Östman et al. 2001; Elliott et al. 2002; Steffan-Dewenter et al. 2002; but see Menalled et al. 1999). Considering the landscape context adds exciting new aspects to biological control, some of which will be described in this chapter.

Two case studies: cereals and oilseed rape

Study area and organisms

The case studies presented here deal with pest–natural enemy interactions in oilseed rape and winter wheat, which are two major crop species in many temperate areas. These case studies are of particular relevance to developing an understanding of the landscape context of biological control because of the annual nature of both crops, leading to high levels of disturbance and botanical impoverishment. In the study region around the city of Göttingen, Germany, struc-turally simple landscapes dominated by annual crops (covering >95%) are geographically inter-spersed with structurally complex landscapes comprising a mixture of crop fields, grasslands, fallows, forests and other non-crop elements (>50% non-crop area). Overall, arable land covered 43% of the area, 36% of which was planted to winter wheat (*Triticum aestivum*) and 14% to winter oilseed rape (*Brassica napus*). Forests covered 33%, and grasslands 7.5% of the area (Niedersächsisches Landesamt für Statistik, pers. comm.). Interactions between major insect pests and their natural enemies were studied in relation to landscape context in 15 (rape) and 18 (wheat) non-overlapping landscape sectors, respectively, covering a gradient of structural complexity. Thereby, the percentage of (annually ploughed) arable land was a good indicator for structural complexity, showing close correlations with other parameters of landscape composi-tion and configuration (Table 4.1). Low percentages of arable land were associated with high values of the perimeter-to-area ratio, meaning that fields are small and large amounts of field edges could enhance immigration of species whose occurrence in fields is limited by dispersal. At the same time, percentage of arable land was negatively correlated to the diversity of land-use types and to percentage grassland. Depending on species, grassland or other non-crop habitats provide potential overwintering habitats (Landis et al. 2000; Sunderland and Samu 2000).

In oilseed rape, the pollen beetle *Meligethes aeneus* (Coleoptera: Nitidulidae) is one of the major pest species. Pollen feeding of the adults inhibits pod and seed development, leading to economically important crop damage. Larvae of the pollen beetles, which develop in the rape

Table 4.1: Correlation coefficients between the percentage of arable land, percentage grassland, perimeter-to-area ratio (PAR) and the Shannon-Wiener diversity of land-use types (H_s) in the 18 studied landscape sectors of 1 km radius.

	PAR	H_s	Grassland (%)
Arable land (%)	-0.82	-0.95	-0.73
Grassland (%)	0.85	0.73	
H_s	0.75		

All correlations are significant at p<0.001.

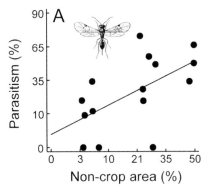
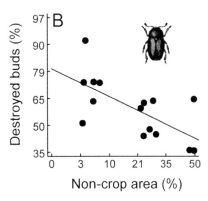

Figure 4.1: (a) Percentage parasitism of rape pollen beetles by wasps in relation to the percentage of non-crop habitats in a circular sector of 1.5 km diameter around the experimental plot. Y = 7.75 + 0.87 X, n = 15, r = 0.57, p = 0.02. (b) Percentage of buds destroyed by rape pollen beetles in relation to the percentage of non-crop habitats in a circular sector of 1.5 km diameter around the experimental plot. Y = 70.8 – 0.51 X, n = 15, r = 0.66, p = 0.007.

Reprinted with permission from Thies and Tscharntke (1999), *Science* 285: 893–895. © 1999 AAAS.

flowers, are mainly attacked by parasitoid wasps (Hymenoptera: Ichneumonidae). Pollen beetle herbivory was measured as percentage of destroyed buds, and its mortality due to parasitism was recorded by dissecting last instar larvae for the presence of parasitoid eggs (see Thies et al. 2003).

Cereal aphids (Homoptera: Aphididae) are dominant herbivores in winter wheat, causing economic damage in some years, partly due to transmission of viruses (BYDV, Bruehl 1961). Three species are common in the study region, of which *Sitobion avenae* is usually the most abundant, followed by *Metopolophium dirhodum* and *Rhopalosiphum padi*. Natural enemies of cereal aphids were quantified. Parasitoid wasps (Hymenoptera: Aphidiidae) and spiders (Araneae: mostly Linyphiidae) were found to be important antagonists of cereal aphids in the study region (Schmidt et al. 2003).

Enhanced biological control in complex landscapes

Plant damage by the rape pollen beetle and its parasitism by wasps were studied in winter rape fields within the 15 landscapes varying from structurally simple to complex. Additionally, potted summer rape plants were raised under standardised conditions and exposed in grassy field margin strips in each landscape. This was done to circumvent confounding influences such as local soil type, which may vary between landscapes. The full-flowering period of rape was in May (fields) and June (potted plants) in this northern hemisphere study. Parasitism of pollen beetles increased with the percentage of non-crop area in the surrounding landscape on both the potted rape plants (Figure 4.1a) and in the rape fields (Thies and Tscharntke 1999). Correspondingly, the percentage of buds destroyed by the beetle declined with increasing landscape complexity (Figure 4.1b). In landscapes with more than 20% non-crop area, parasitism exceeded the threshold value of 32–36%, below which a success in classical biological control has never been found (Hawkins and Cornell 1994.)

Functional scales of landscape processes

A basic question in the study of landscape effects is the scale at which they are manifested. Dispersal and foraging distances are hard to measure for many agricultural pests and antagonists, because they are too small, too abundant or too mobile for mark–recapture experiments. A possible approach for specialists is to expose food plants or hosts at known distances to the next population, and measure their colonisation (e.g. Kruess and Tscharntke 1994, 2000). However, rape pollen beetles, cereal aphids and their natural enemies also live on wild crucifers and grasses, respectively, and hence in many non-crop habitats (Dean 1974; Horstmann 1981; Charpentier 1985). Their widespread occurrence complicates determining the distance to the nearest population. However, the response of parameters such as parasitism to non-crop habitats in the surrounding landscape renders an alternative approach possible. Landscape metrics can be calculated for various diameters around the study sites (Figure 4.2a), most straightforwardly with a geographic information system (GIS). Correlations for the organisms' response to landscape features can be calculated across a range of spatial scales. Results of this approach show that the strength of such relations, measured by the coefficient of determination r^2, peaks at distinct scales (Figure 4.2b). This is the functional spatial scale at which a species responds to the landscape context, and likely reflects its foraging range. The approach can be used to test the hypothesis that the spatial scale experienced by organisms widens with increasing trophic level (Holt 1996) or body size (Ritchie and Olff 1999). For the rape pollen beetle system, a larger spatial domain for the parasitoids than for herbivores was not found. Functional scales of both herbivory and their parasitism by wasps lay at 1.5 km diameter (Figure 4.2b; Thies et al. 2003). In contrast, parasitoids of cereal aphids are influenced by landscape structure at smaller scales (0.5–1 km diameter), probably related to their smaller body size (CT, unpublished data), whereas strong dispersers such as sheetweb weaving spiders (Araneae: Linyphiidae) respond to landscape structure at scales larger than 2 km (MHS, unpublished data). Densities of pollinators revealed a differentiation of functional scales from small (0.5 km for solitary wild bees) to large (6 km for honey bees) (Steffan-Dewenter et al. 2002.)

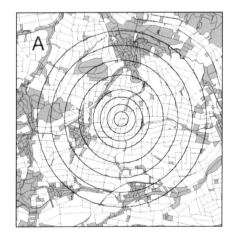 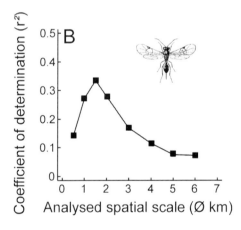

Figure 4.2: (a) Calculating landscape metrics for nested sectors of varied size around 1 out of 15 study fields (white = arable crops; grey = non-crop habitats, diameters ranging from 0.5–6 km). (b) Comparing coefficients of determination from correlations between parasitism and percentage of non-crop within diameters of 0.5–6 km around the study plots (single correlations analogous to Figure 4.1a). The functional scale for parasitism can be inferred at 1.5 km.

Modified and reprinted with permission from Thies et al. (2003).

Temporal dynamics

Agricultural landscapes are subject to yearly changes in land use resulting from crop rotations and changes in political schemes and subsidies. This may affect pest and natural enemy populations, which further vary with climate, annual phenologies and interspecific interactions (Holt and Barfield 2003). Particularly, crop rotations continuously change the arrangement and composition of crop types to prevent selective depletion of soil nutrients and local build-up of (mostly soil-borne) pest populations. Large-scale consequences of crop rotations and the resulting change in habitat area were studied for the rape pollen beetle and its parasitoids. Herbivory (% destroyed buds) and parasitism were negatively correlated to the change in rape-crop area relative to the preceding year (Thies et al. in review). In landscapes where the percentage of rape crop had decreased, rape fields showed higher levels of both attack by pollen beetles and parasitism. In landscapes where rape-crop area had expanded, herbivory and parasitism were reduced. This indicates crowding and dilution effects resulting from changed availabilities of food resources at the landscape level. A large-scale bottom-up effect of resource availability acts on pollen beetles in addition to the top-down effect induced by landscape structure via parasitism.

   An example of a very dynamic system, probably driven by the availability of overwintering habitat and weather, is the colonisation of arable fields by sheetweb-weaving spiders. Spider densities in wheat fields were low relative to grasslands in early spring, but rose considerably in late spring, especially in grassland-rich landscapes. This temporarily led to more than four times higher spider abundances in wheat fields with a high percentage of grassland in the surrounding landscape than in wheat fields with a grassland-poor surrounding (MHS, unpublished data). A similar variation in spider densities has been shown to affect cereal aphid populations in field experiments (Schmidt et al. 2003), thereby suggesting that aphid population growth can be reduced by high spider densities in grassland-rich landscapes. However, the differences in spider abundances between simple and complex landscapes were only temporary, because populations in simple landscapes caught up within few weeks. This may have been due to reproduction or ongoing long-distance dispersal in concert with limited carrying capacity of the habitat (competition for web sites). Moreover, the timing and intensity of spider immigration into crops varied considerably between years, probably being influenced by the suitable weather for the ballooning dispersal of these species (Vugts and van Wingerenden 1976).

Interactions between local and regional diversification

Non-crop vegetation, such as field margins, sown flower strips, beetle banks or conservation headlands can enhance populations of beneficials in adjacent crops (Landis et al. 2000; Marshall and Moonen 2002; Pfiffner and Wyss, ch. 11 this volume). It contributes to diversification at the landscape scale. However, its effect may depend on landscape structure. This has been found in the rape pollen beetle system (Tscharntke et al. 2002). Parasitism of pollen beetles was higher near the crop field edge, near the parasitoids' overwintering habitats, and decreased towards the centre of the field. However, this was only true in landscapes dominated by annual crops (<20% non-crop area). In landscapes with a high percentage of permanent non-crop area, such edge effects disappeared (Figure 4.3), presumably due to the high overall density of the parasitoids. Thus, decisions about the placement of such diversification measures should consider that the local benefit is probably higher in structurally poor landscapes. Local orientation of management practices appears to be necessary and useful only in simple landscapes. However, for the conservation of biodiversity including the many rare or sensitive species, creation of new habitats may be more effective in structurally rich landscapes where such species are still present. Consequently, evaluations of local diversification measures should take the landscape context into account.

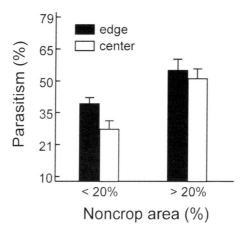

Figure 4.3: Parasitism (± 1 SE) of the rape pollen beetle in relation to the percentage of non-crop area in the agricultural landscape. Parasitism significantly decreased from the edge to the centre of the field in landscapes with <20% of non-crop area (n = 16, F = 6. 0, p = 0.028), whereas parasitism did not decrease from the edge to the centre in landscapes with >20% of non-crop area (n = 14, F = 0.38, p = 0.55).
Modified and reprinted with permission from Tscharntke et al. (2002).

Counteracting processes

The landscape context can influence pests and their multiple enemies at the same time. Under certain conditions, the resulting interactions may impede biological control. For example, cereal aphids overwinter predominantly in perennial habitats (Dean 1974). Despite the fact that these aphids can bridge tens or even hundreds of kilometres as aerial plankton, infestation levels in wheat were higher in structurally complex landscapes in June (CT, unpublished data). Hence, two counteractive processes are caused by structural complexity in the landscape: increased aphid colonisation, but also increased aphid control by parasitoid wasps and spiders. In both the aphids and their enemies, the increased availability of alternative resources may have enhanced early populations, but aphid abundances were no longer correlated to landscape complexity later in the season. Possibly, enhanced control by their natural enemies adjusted the initially higher aphid abundances in complex landscapes to similar levels as in structurally poor landscapes. These preliminary results point at extensive interactions that are likely to be driven by landscape complexity. Counterintuitive outcomes may also occur in the case of strong interference between multiple predators (Sih et al. 1998). Snyder and Ives (2001) showed how ground beetles (Coleoptera: Carabidae), which themselves were able to reduce aphid densities, disrupted aphid control by parasitoid wasps through preying selectively on the parasitised aphids (mummies) in alfalfa. In the long term, parasitoids alone suppressed aphid populations more effectively than did ground beetles alone, or both enemies in combination. Thereby, landscape manipulation in favour of both enemy groups would prove less efficient than selective facilitation of the parasitoid wasps. Contrastingly, in the case of cereal aphid suppression, no interference between parasitoid wasps and spiders was found, and both enemy groups together produced the highest level of aphid control (Schmidt et al. 2003). Direct impacts of landscape diversification on pests and emergent impacts of multiple enemies should be considered in landscape manipulation for pest management.

Conclusions

The recent focus on landscape context of pest–natural enemy interactions suggests that a purely local orientation of biological control is not sufficient. Examples from oilseed rape and winter wheat underline the importance of non-crop habitats for natural enemies of major pest species. Also, the effectiveness of local habitat diversification (e.g. field margin strips) was affected by the structure of the surrounding landscape. Our results indicate that many natural enemies of major pest species depend on other, less disturbed, habitats than the crop itself. This is a substantial argument for the conservation or creation of refuge habitats such as grasslands, fallows or other species-rich habitats in agricultural landscapes. Despite these emerging patterns, our knowledge is still fragmentary. A major challenge arises when pests and their multiple enemies are concurrently affected by landscape structure, leading to counteractive processes. In such cases, specific conditions for pest inhibition and facilitation of beneficials need to be detected. We need additional well-designed, long-term field studies on a landscape scale, accompanied by theoretical work, to better understand patterns and mechanisms of food–web interactions in time and space.

Acknowledgements

Geoff M. Gurr, Robert D. Holt and Douglas A. Landis provided valuable comments on an earlier draft of this chapter. We are grateful to Christoph Bürger, Doreen Gabriel, Indra Roschewitz and Ulrich Thewes for help in the field and processing of landscape data. The work greatly profited from collaboration with Jens Dauber, Tobias Purtauf and Volkmar Wolters within the BIOPLEX project (Biodiversity and spatial complexity in agricultural landscapes under global change), funded by the German Ministry for Research and Education (BMBF). Further support came from the German Science Foundation (Deutsche Forschungsgemeinschaft) and from the German National Academic Foundation (Studienstiftung des deutschen Volkes).

References

Benton, T.G., Vickery, J.A. and Wilson, J.D. (2003). Farmland biodiversity: is habitat heterogeneity the key? *TREE* 18: 182–188.

Bruehl, G.W. (1961). Barley yellow dwarf, a virus disease of cereals and grasses. *Monograph of the American Phytopathological Society* I. 52 pages.

Cappuccino, N., Lavertu, D., Bergeron, Y. and Régnière, J. (1998). Spruce budworm impact, abundance and parasitism rate in a patchy landscape. *Oecologia* 114: 236–242.

Charpentier, R. (1985). Host plant-selection by the pollen beetle *Meligethes aeneus*. *Entomologia Experimentalis et Applicata* 38: 277–285.

Dean, G.J. (1974). The overwintering and abundance of cereal aphids. *Annals of Applied Biology* 76: 1–7.

Elliott, N.C., Kieckhefer, R.W., Michels, G.J. and Giles, K.L. (2002). Predator abundance in alfalfa fields in relation to aphids, within-field vegetation, and landscape matrix. *Environmental Entomology* 31: 253–260.

Gurr, G.M., Wratten, S.D. and Luna, J.M. (2003). Multi-function agricultural biodiversity: pest management and other benefits. *Basic and Applied Ecology* 4: 107–116.

Hawkins, B.A. and Cornell, H.V. (1994). Maximum parasitism rate and successful biological control. *Science* 262: 1886.

Holt, R.D. (1996). Food webs in space: an island biogeographic perspective. In *Food Webs: Integration of Patterns and Dynamics* (G.A. Polis and K.O. Winemiller, eds), pp. 313–323. Chapman & Hall, New York.

Holt, R.D. (2002). Food webs in space: on the interplay of dynamic instability and spatial processes. *Ecological Research* 17: 261–273.

Holt, R.D. and Barfield, M. (2003). Impacts of temporal variation on apparent competition and coexistence in open ecosystems. *Oikos* 101: 49–58.

Holt, R.D. and Hochberg, M.E. (2001). Indirect interactions, community modules and biological control: a theoretical perspective. In *Evaluating Indirect Ecological Effects of Biological Control* (E. Wajnberg, J.K. Scott and P.C. Quimby, eds), pp. 13–37. CAB International, Wallingford, UK.

Horstmann, K. (1981). Revision der Europäischen Tersilochinen II (Hymenoptera: Ichneumonidae). *Spixiana Suppl.* 4: 1–76.

Kareiva, P. (1990a). Population dynamics in spatially complex environments: theory and data. *Philosophical Transactions of the Royal Society of London*, Series B: Biological Sciences 330: 175–190.

Kareiva, P. (1990b). The spatial dimension in pest–enemy interactions. In *Critical Issues in Biological Control* (M. Machauer, L.E. Ehler and J. Roland, eds), pp. 213–226. Intercept Press, Andover, UK.

Kareiva, P. and Wennegren, U. (1995). Connecting landscape patterns to ecosystem and population processes. *Nature* 373: 299–302.

Kruess, A. and Tscharntke, T. (1994). Habitat fragmentation, species loss, and biological control. *Science* 264: 1581–1584.

Kruess, A. and Tscharntke, T. (2000). Species richness and parasitism in a fragmented landscape: experiments and field studies with insects on *Vicia sepium*. *Oecologia* 122: 129–137.

Landis, D.A., Wratten, S.D. and Gurr, G.M. (2000). Habitat management to conserve natural enemies of arthropod pests in agriculture. *Annual Review of Entomology* 45: 175–201.

Marshall, E.J.P. and Moonen, A.C. (2002). Field margins in northern Europe: their functions and interactions with agriculture. *Agriculture, Ecosystems and Environment* 89: 5–21.

Menalled, F.D., Marino, P.C., Gage, S.H. and Landis, D.A. (1999). Does agricultural landscape structure affect parasitism and parasitoid diversity? *Ecological Applications* 9: 634–641.

Östman, Ö., Ekbom, B. and Bengtsson, J. (2001). Farming practice and landscape heterogeneity influence biological control. *Basic and Applied Ecology* 2: 365–371.

Ricklefs, R.E. (1987). Community diversity: relative roles of local and regional processes. *Science* 235: 167–171.

Ritchie, M.E. and Olff, H. (1999). Spatial scaling laws yield a synthetic theory of biodiversity. *Nature* 400: 557–560.

Roland, J. and Taylor, P.D. (1997). Insect parasitoid species respond to forest structure at different spatial scales. *Nature* 386: 710–713.

Schmidt, M.H., Lauer, A., Purtauf, T., Thies, C., Schaefer, M. and Tscharntke, T. (2003). Relative importance of predators and parasitoids for cereal aphid control. *Proceedings of the Royal Society of London*, Series B: Biological Sciences 270: 1905–1909.

Sih, A., Englund, G. and Wooster, D. (1998). Emergent impacts of multiple predators on prey. *Trends in Ecology and Evolution* 13: 350–355.

Snyder, W.E. and Ives, A.R. (2001). Generalist predators disrupt biological control by a specialist parasitoid. *Ecology* 82: 705–716.

Steffan-Dewenter, I., Munzenberg, U., Burger, C., Thies, C. and Tscharntke, T. (2002). Scale-dependent effects of landscape context on three pollinator guilds. *Ecology* 83: 1421–1432.

Stoate, C., Boatman, N.D., Borralho, R.J., Carvalho, C.R., de Snoo, G.R. and Eden, P. (2001). Ecological impacts of arable intensification in Europe. *Journal of Environmental Management* 63: 337–365.

Sunderland, K. and Samu, F. (2000). Effects of agricultural diversification on the abundance, distribution, and pest control potential of spiders: a review. *Entomologia Experimentalis et Applicata* 95: 1–13.

Thies, C. and Tscharntke, T. (1999). Landscape structure and biological control in agroecosystems. *Science* 285: 893–895.

Thies, C., Steffan-Dewenter, I. and Tscharntke, T. (2003). Effects of landscape context on herbivory and parasitism at different spatial scales. *Oikos* 101: 18–25.

Thies, C., Steffan-Dewenter, I. and Tscharntke, T. (in review). Plant pest-natural enemy interactions in spatio-temporally changing agricultural landscapes.

Tscharntke, T. (2000). Parasitoid populations in the agricultural landscape. In *Parasitoid Population Biology* (M.E. Hochberg and A.R. Ives, eds), pp. 235–253. Princeton University Press, Princeton.

Tscharntke, T. and Brandl, R. (2004). Plant–insect interactions in fragmented landscapes. *Annual Review of Entomology* 49: 405–430.

Tscharntke, T., Steffan-Dewenter, I., Kruess, A. and Thies, C. (2002). Contribution of small habitat fragments to conservation of insect communities of grassland–cropland landscapes. *Ecological Applications* 12: 354–363.

van Nouhuys, S. and Hanski, I. (2002). Multitrophic interactions in space: metacommunity dynamics in fragmented landscapes. In *Multitrophic Level Interactions* (T. Tscharntke and B.A. Hawkins, eds), pp. 127–147. Cambridge University Press, Cambridge.

Vugts, H.F. and van Wingerenden, W.K.R.E. (1976). Meteorological aspects of aeronautic behaviour of spiders. *Oikos* 27: 433–444.

With, K.A. and Christ, T.O. (1995). Critical thresholds in species' response to landscape structure. *Ecology* 76: 2446–2459.

With, K.A., Pavuk, D.M., Worchuck, J.L., Oates, R.K. and Fisher, J.L. (2002). Threshold effects of landscape structure on biological control in agroecosystems. *Ecological Applications* 12: 52–65.

Chapter 5

Use of behavioural and life-history studies to understand the effects of habitat manipulation

Mark A. Jervis, Jana C. Lee and George E. Heimpel

Introduction

Simply increasing diversity per se in agroecosystems does not guarantee a desirable outcome. It can prove to be a waste of effort, time and money if it does not lead to significantly improved pest control. Worse still, it can even exacerbate some pest problems (Andow and Risch 1985; Collins and Johnson 1985; Sheehan 1986; Baggen and Gurr 1998; Gurr et al. 1998, 2004). Therefore, biological control workers ought ideally to concentrate on manipulating the specific elements of diversity that are most likely to have a positive influence on the survival and reproduction, and thus the impact, of natural enemies (Landis et al. 2000). Fundamental research into the specific non-pest resource requirements (supplementary foods, prey and hosts, physical refugia) of natural enemies is the key to this. Although research of this kind may be difficult, costly and time-consuming, it should provide information that will at best increase the chances of success of a habitat manipulation program, and will at the very least provide useful insights into natural enemy biology. This chapter considers the approaches available to researchers for improving our understanding of the effects of habitat manipulation on pest–natural enemy interactions. The review is organised as follows:

- the application of selection criteria to insect natural enemies in manipulation programs;
- ways of identifying and comparing the limiting resources of natural enemies by using behavioural, morphological and anatomical studies. Discussion will mainly focus on whether natural enemies demonstrate a need for supplemental food, but we also consider the requirements of natural enemies for alternative hosts/prey and refugia;
- how manipulation of such resources might alter the natural enemy's per capita searching efficiency;
- how habitat manipulation might alter the abundance of natural enemies by attracting parasitoids and predators to the crop and retaining them there, and by improving natural enemy growth, development, survival and fecundity.

Such work requires knowledge, often of a highly detailed nature, of the life history of the natural enemy. As noted by Gurr et al. (2004), empirically derived life-history information draws from, and also contributes to, ecological theory. It can also be used in parameterising both prospective (pre-manipulation) and retrospective (post-manipulation) biological control models (e.g. see Waage 1990; Kidd and Jervis 2004).

Selection criteria

'Selection criteria' is a term normally reserved for 'classical' biological control (Waage 1990; Kidd and Jervis 2004), but here we apply it to conservation biological control where a subset, rather than the full complex, of natural enemy species is chosen for manipulation. Although some programs have involved manipulating natural enemy guilds rather than particular species (Hagen et al. 1971; Evans and England 1993), we advocate investigating the effects of manipulation on the component members of a complex. We recommend a 'directed' rather than a 'shotgun' approach (Gurr et al. 2004), which not only increases the chances of obtaining a positive result, but provides a better understanding of the mechanisms underlying successes or failures. This kind of approach, which is very similar to 'active adaptive management', has recently been advocated for a wide range of problems in pest management (Shea et al. 2002).

The key selection criterion in conservation biological control is that the natural enemy must be amenable to habitat manipulation. For example, it is pointless attempting to manipulate a parasitoid species by providing supplemental food if the adults have no need for such food. The following may seem obvious, but it needs to be asked from the outset of a manipulation program: 'Are supplemental foods and/or alternative hosts/prey and/or shelter important limiting factors for this species?' The answer may already exist in the literature. For example, examination of published host-rearing records given in taxonomic catalogues may reveal a parasitoid species to be monophagous. If so, the parasitoid will not be amenable to manipulation using an alternative host species, although it could be tractable with respect to another resource such as supplemental food. If there is no existing answer to the above question, then criteria useful for ranking and eliminating candidate species from a manipulation program can be developed. This task requires first identifying the limiting, manipulable resources for those particular insects.

If the limiting resource can be identified for a natural enemy known to effect significant control of pests under unmanipulated conditions, then that natural enemy should be given priority over species whose contribution is either weak or unknown. Investigators should, however, remain open-minded about the pest control potential of species whose contribution to pest mortality is considered insignificant or is unknown. Under environmental conditions achieved through manipulation, these natural enemies might also control pests substantially.

Identifying limiting resources

When a range of resources, such as flowering plant species, is identified, the research should ideally culminate in ranking. This can be based on responses of individual insects, gauged by laboratory experiments and on counts of insects in each resource type. Field counts should be related to the abundance of the resource, so as to provide true measures of preference (Cowgill et al. 1993). Some plant species can be eliminated from consideration if they are difficult to propagate or produce undesirable side-effects. For example, among a set of candidate flowering plants, some species are invasive weeds or provide food for the pest (Baggen and Gurr 1998; Baggen et al. 1999; Romeis and Wäckers 2000), or enhance enemies of biocontrol agents (Stephens et al. 1998). On the other hand, a plant may itself be a crop (e.g. maize provides pollen to coccinellids) in which case there is a good case for using it (e.g. as an intercrop, as discussed by Mensah and Sequeira, ch. 12 this volume.)

Before embarking on practical research, it is useful to first consider current and past farming practices in relation to the known life-histories of the predators and parasitoids. Clues may be gleaned as to which limiting resource has undergone changes to the extent that natural enemy effectiveness has become constrained. For example, Hodek (1971) made the connection between burning of old grass on headlands, dykes and other habitats, and the knowledge that entomophagous coccinellids use such habitats as overwintering refugia. Farmers may relate experiences that indicate a connection between natural enemy abundance and farming practice, such as retention/disposal of crop residues. Taxonomic publications and collections (through data labels on specimens) can give data on the natural enemy's characteristic habitat and foods. Discussions with taxonomists can provide insights into natural enemy life-histories and resource requirements.

Supplemental food

Feeding behaviour

Floral nectar and pollen. Searching the literature for records of the foods that a natural enemy consumes is a worthwhile undertaking. Nevertheless, with flower visitors, it is often the case that while the plant species involved is specified by an author, no information is given on the materials consumed, and it may even be unclear as to whether feeding actually occurred. Allen (1929)

Figure 5.1: (a) Diverse flower strip in the margin of a British cereal crop capable of providing floral resources to hoverflies (Photograph: G.M. Gurr). (b) Hoverflies (Photograph: M. Bowie). (c) Parasitoids (Photograph: L.R. Baggen). (d) Predatory coccinellids (Photograph: Z. Hossain).

and Jervis et al. (1993) investigated more precisely the basic categories of materials (nectar, pollen, honeydew) fed upon by parasitoid faunas (dipteran and hymenopteran respectively). Published reports of feeding on extrafloral nectar tend to be reliable, but most records of honeydew feeding require verification, given the possibility that some parasitoids use honeydew solely as a kairomone in host location behaviour. Records in old literature must always be treated with caution because of the possibility, with both insect and plant species, of misidentifications and of nomenclatural changes.

Floral anatomy texts (e.g. Hickey and King 1981) and botanical 'floras' (e.g. Clapham et al. 1989) can indicate which plant species lack nectaries – floral and/or extrafloral – and which might therefore be eliminated from further consideration in a manipulation program. Flowers are relatively conspicuous resources, and are therefore easy, compared with other food sources, for investigators to locate, identify and monitor for the presence of entomophagous insect visitors (Figure 5.1). In the case of exposed floral nectaries, it is usually obvious whether the insects, even minute ones, are feeding, but in the case of concealed floral nectaries it can be hard to determine both whether feeding is taking place and, if it is occurring, what materials are being fed upon (e.g. see Wäckers et al. 1996).

Flowers may be presented to the insects in the laboratory or greenhouse, and observations on behaviour carried out (Györfi 1945; Leius 1960; Syme 1975; Shahjahan 1974; Idris and Grafius 1997; Patt et al. 1997; Drumtra and Stephen 1999). However, the results of such tests should be treated with caution, as under field conditions the insects may not visit the plant species presented in the laboratory. Particular caution needs to be applied to the results of laboratory tests that involve presenting nectars independently of flower sources, as has been done for ants (Feinsinger and Swarm 1978; Haber et al. 1981).

Studies involving artificial nectars and flower 'dummies'/models can help identify the range of flowering plants that may be more suitable for manipulation of entomophagous insects. Patt et al. (1997) showed, using artificial flowers, that the rapidity with which wasps locate artificial nectar droplets varies with the degree of exposure of the food as well as with parasitoid species. This finding accords with what one would expect intuitively regarding rapidity of location of floral nectar.

In the absence of field surveys or experimental data indicating the identity of flowers exploited by natural enemies, it is advisable to use plant species with exposed floral nectaries, notably umbellifers (Apiaceae) or spurges (Euphorbiaceae), provided they are not also exploited by the pest (Jervis et al. 1993; Baggen and Gurr 1998).

Pollen feeding is very common among predators (Majerus 1994 on Coccinellidae; Canard 2001 on Chrysopidae, Gilbert 1986 on Syrphidae), but is rare among parasitoids (Gilbert and Jervis 1998; Jervis 1998). Pollen feeding is often much easier to observe than is nectar and honeydew feeding. Usually, pollen grains are collected from the anthers using the mouthparts (Grinfel'd 1975; Gilbert 1981; Jervis 1998), although some bombyliids collect pollen using their foretarsi (Deyrup 1988).

Extrafloral nectar. Natural enemy feeding at extrafloral nectaries is relatively easily observed, provided the precise location of the nectaries has previously been established (see Limburg and Rosenheim 2001).

Honeydew. Homopteran honeydew is widely exploited as food by insect predators and parasitoids (Majerus 1994 on Coccinellidae; Canard 2001 on Chrysopidae; Gilbert and Jervis 1998 on parasitoid flies; Jervis 1998 on parasitoid wasps; Evans 1993 on natural enemies of aphids). Honeydew solidifies very rapidly after deposition (especially if the ambient humidity is low), becoming a crystalline sugar film, which Diptera such as tachinids and syrphids easily exploit using their fleshy labella (Allen 1929; Downes and Dahlem 1987). Feeding on wet honeydew is likely to be practised by many parasitoid wasp species, particularly in arid habitats, but feeding on dried honeydew has been observed in only a few wasps (Bartlett 1962; Jervis 1998). Obtaining direct evidence of feeding by parasitoids on honeydew deposits can be problematic because: (a) the honeydew films, although often highly abundant, are colourless and so are difficult for human observers to detect; (b) observers are less likely to observe honeydew-feeding in the field if a given-sized natural enemy population is dispersed among a large number of honeydew patches, compared with a smaller number of flower patches; and (c) some parasitoid wasps apply their mouthparts to honeydew films for the purpose of detecting host-finding kairomones (Budenberg 1990), and the possibility exists that honeydew feeding does not occur in some of these species.

One question that has not been resolved involves the need to provide parasitoids of honeydew-producing hosts with supplemental nectar. Many of these species feed upon their host's honeydew, and may therefore not experience sugar-limitation in the field. However, honeydew is typically an inferior food source compared with artificial sugar-rich foods (Leius 1961; Elliott et al. 1987; Idoine and Ferro 1988; Hagley and Barber 1992) and floral nectar. It may contain melezitose and raffinose, sugars which are nutritionally less suitable for parasitoids (Wäckers 1999, 2000, 2001), and it may even have toxic properties (Avidov et al. 1970) (but note that some floral nectars also can be toxic, Hausmann et al. 2003). Also, host density may not be sufficiently high for all parasitoids to obtain ample honeydew meals. In one study of the aphid parasitoid *Aphelinus albipodus*, over a broad range of host densities, an estimated 20% of field-caught parasitoids had fed upon host honeydew (Heimpel et al. 2004). Finally, honeydew production itself may vary with environmental conditions (Klingauf 1987; Henneberry et al. 2002).

Interactions with other foragers. When making direct observations of feeding behaviour in the field, it is important to explore possible negative or positive effects on other foragers arising from interference or resource exploitation. These foragers include higher-order predators (Kevan 1973; Jervis 1990; Heimpel et al. 1997b), hyperparasitoids and bees. Bees have the potential to seriously reduce floral nectar levels for parasitoids (Lee and Heimpel 2003), but they may also increase the accessibility of nectaries by 'tripping' flowers (Gurr et al. 2004).

Indirect evidence of feeding on supplementary materials

Pollen. In the absence of observational evidence, indirect evidence of feeding can be sought. The body surface, including mouthparts, of natural enemies collected in the field may be examined for the presence of pollen grains (Holloway 1976; Stelleman and Meeuse 1976; Gilbert 1981). The sculpturing on the surface of the exine of pollen grains often differs markedly among flowering plant species, and grains can be attributed to plant species either by using identification works (Sawyer 1981, 1988; Faegri and Iversen 1989; Moore et al. 1991; Erdtman 1969; Reitsma 1966) or by comparing the grains with a reference collection, preferably one comprising grains collected by the investigator from the flora in and around the target crop.

The presence of pollen grains on the body surface, even on the mouthparts, does not, however, provide conclusive proof of pollen consumption (Wäckers et al. 1996). Insects may have become contaminated with grains while searching solely for nectar.

Gut dissection, which can be coupled with staining, reveals the presence of pollen grains. This technique has been used with hoverflies (van der Goot and Grabant 1970; Holloway 1976; Haslett and Entwistle 1980; Leereveld 1982; Haslett 1989; Wratten et al. 1995, Hickman et al. 2001), parasitoid wasps (Györfi 1945; Leius 1963; Hocking 1967), coccinellid beetles (Hemptinne and Desprets 1986; Hoogendoorn and Heimpel 2004) and green lacewings (Sheldon and MacLeod 1971). It can also be applied to insects that have been deep-frozen, to dried, pinned specimens and to specimens preserved in ethanol (Holloway 1976; Leereveld 1982; Haslett 1989, Hickman et al. 2001). The exines of pollen grains are resistant to decay or chemical treatment, and in many cases are refractory to digestive enzymes and the mechanical action of the gut (hoverflies apparently do not need to grind pollen in order to extract nutrients, Gilbert 1981; Haslett 1983). Therefore, ingested pollen grains retain much of their external structure and so the original plant source can often be identified.

Hunt et al. (1991) devised a method for detecting pollen exines in the abdomens/gasters of dried, preserved insects. The body parts are cleaned and crushed then heated in a mixture of acetic anhydride and concentrated sulphuric acid. The mixture is then centrifuged and the exines eventually isolated and identified (Lewis et al. 1983; Hunt et al. 1991). This method has the advantage that evidence of pollen feeding can be sought from museum specimens, even very old ones.

The detection of pollen exines in the gut of a predator or parasitoid does not, however, constitute conclusive proof that the insect has been feeding directly at anthers. Pollen grains may fall or be blown from anthers and become trapped in nectar, honeydew or dew (Todd and Vansell 1942; Townes 1958; Hassan 1967; Sheldon and MacLeod 1971), while with predators such as carabid beetles, grains may enter the gut via the prey (Dawson 1965). Evidence is accumulating concerning the consumption of allochthonous pollen by predatory mites on crop plants, and the fitness benefits gained (van Rijn 2002). Nevertheless, it is possible that some natural enemies might inadvertently ingest appreciable amounts of pollen yet derive no fitness benefits whatsoever.

Sugars. Methods for detecting gut sugars in parasitoids and predators are cold anthrone tests and various forms of chromatography (Jervis et al. 1992; Heimpel and Jervis 2004; Heimpel et al. 2004). An important feature of all these methods is that they can distinguish between fructose and other sugars. Fructose is absent from insect hemolymph and is therefore a marker for ingested sugars in whole-insect preparations (van Handel 1984; Olson et al. 2000). Alternatively, insect guts can be dissected out and their contents analysed directly (Lewis and Domoney 1966).

Both anthrone tests and chromatography have been used on field-caught individuals of various species of biting flies (van Handel 1972; Yuval and Schlein 1986; Lewis and Domoney 1966; Burkett et al. 1999; Hunter and Ossowski 1999), yet the application of these methodologies to parasitoids and predators is in its infancy. In the parasitoid *Macrocentrus grandii* (Braconidae), both hot and cold anthrone tests (see below) were used to characterise trends in carbohydrate metabolism (Olson et al. 2000; Fadamiro and Heimpel 2001), and cold anthrone tests have been used to detect sugars in various parasitoid species (Lee and Heimpel 2003; Heimpel and Jervis 2004; Heimpel et al. 2004).

It is possible to determine whether individual parasitoids are close to starvation or relatively well-fed by quantifying levels of haemolymph sugars (presumably mainly trehalose) and glycogen with hot anthrone tests (Olson et al. 2000; Fadamiro and Heimpel 2001). Chromatographic methods have recently shown that a majority of *Cotesia glomerata* individuals captured in cabbage fields had fed upon honeydew sugars (Wäckers and Steppuhn 2003).

Analyses of gut sugars in field-caught parasitoids using the cold anthrone test and chromatography are beginning to reveal considerable differences in the proportion of sugar-starved parasitoids in the field. Although a small fraction of *M. grandii*, *Aphytis aonidiae* and *Trichogramma ostriniae* collected from corn fields interplanted with flowering buckwheat tested positive for gut sugars (not more than 20%), much higher levels of gut sugars were found for a number of parasitoids attacking cabbage pests, both in the vicinity of floral plantings and in areas that did not have obvious sugar sources (Lee and Heimpel 2003; Wäckers and Steppuhn 2003; Heimpel and Jervis 2004; Heimpel et al. 2004).

The hot anthrone test involves heating the reagent/test material mixture, so is inconvenient to perform outside the laboratory. The cold anthrone test, developed by van Handel (1972), is recommended for field use, as it is simple, convenient and rapid, allowing many individual insects to be tested.

Anthrone tests cannot, however, distinguish between sugars from nectar (floral or extrafloral), honeydew, or other rarer sources such as plant exudates and fruit juices (or sugar sprays), so they are of limited use, particularly with respect to providing information on the range of sugar sources exploited within a habitat. While chromatographic methods (GC, HPLC) can tell us much more about the specific kinds of sugar present in the gut, nectars rarely have 'signature' sugars (Heimpel and Jervis 2004). Nectars tend to have varying levels of sucrose, fructose and glucose (Percival 1961; van Handel et al. 1972; Harborne 1988), but these sugars are also present in the other sources just mentioned. Nevertheless, rhamnose can be used as a signature sugar for extrafloral nectar, and melibiose as a floral nectar marker. When the habitat in which a parasitoid is captured contains no honeydew-producing Homoptera, it may be possible to infer nectar feeding. However, even if honeydew is present, it typically contains its own signature sugars (Zoebelein 1955; Burkett et al. 1999), so we can infer that nectar feeding has occurred if honeydew signature sugars are absent but common sugars are present. Melezitose has often been used as a signature sugar for honeydew, but it also occurs occasionally in some floral nectars. There are several other alternative honeydew marker sugars; some are taxonomically very restricted in terms of their source insects, enabling identification of the homopteran producer (see Heimpel and Jervis 2004).

The spectrum of nitrogenous compounds present within nectar and honeydew tends to vary with source (Percival 1961; Maurizio 1975; Baker and Baker 1983). Therefore, in cases where a very narrow range of food sources is being exploited, it is possible to compare (using GC and HPLC) the chemical composition of the gut contents of parasitoids with that of potential food sources, and identify which nectars and honeydews are being fed upon.

A simpler, indirect method of determining whether natural enemies consume potential foods is to mark the latter with dyes or other markers such as rare elements, and examine the guts of the insects for the presence of the marker. For example, the commercially available insect food Eliminade™ incorporates a non-toxic food-dye to facilitate confirmation of its use by parasitoids (Stephen and Browne 2000). Freeman-Long et al. (1998) used the trace element rubidium (Rb) to mark various flowering plants near agricultural fields at three sites in central California. They then captured insects within the fields and examined them for the mark. Among the marked insects were individuals of three parasitoid species (*Hyposoter* sp., *Trichogramma* sp. and *Macrocentrus* sp.). Although this finding is consistent with the parasitoids feeding on marked plants, nectar may not be the only source. Rubidium would presumably have marked pollen as well as nectar in this study system, and would also have been passed on to herbivores of the rubidium-laced plants and ultimately on to parasitoids exploiting these marked herbivores (Payne and Wood 1984; Jackson et al. 1988; Hopper 1991; Hopper and Woolson 1991; Jackson 1991; Corbett and Rosenheim 1996). Therefore, in Freeman-Long et al.'s study, the mark could have been obtained by pollen feeding or, in the case of *Trichogramma* sp., host feeding (consumption of host blood or tissues, Jervis and Kidd 1986; Heimpel and Collier 1996). But as nectar appears to be a more universal food for parasitoids than is either pollen or hosts (Gilbert and Jervis 1998; Jervis 1998), the most likely source of the rubidium mark was nectar.

Mouthparts

Mouthpart structure can provide useful clues as to how entomophagous insects can be manipulated, in particular whether the species requires food, and what general types of non-host and non-prey food it habitually consumes (Allen 1929; Gilbert 1981; Gilbert and Jervis 1998; Jervis 1998). Vestigial mouthparts immediately identify a species as a non-feeder (Gilbert and Jervis 1998): not only will its females not respond behaviourally to food sources, but also their fecundity and longevity will not depend on food availability.

The morphology of functional mouthparts can provide a close approximation to the general type of food exploited. For example, in Tachinidae, possession of a greatly elongated proboscis is linked to exploitation of floral nectar, while a moderately long or a short proboscis is linked to feeding on a broader range of sugar-rich food types (floral nectar, honeydew, extrafloral nectaries) (Gilbert and Jervis 1998; based on field observation data of Allen 1929). There is also a correlation between proboscis length and diet among Syrphidae: as proboscis length increases, the flies are more often associated with flowers having deep corollae (i. e. more concealed nectar) and the proportion of nectar in the diet increases (Gilbert 1981).

Species in which the proboscis is insufficiently elongated are unlikely to be able to exploit deeply concealed nectar (Hickman and Wratten 1996; Jervis 1998; Gilbert and Jervis 1998) (unless that is, they can gain access to the nectar by virtue of being sufficiently small-bodied). To date, nectar 'theft' has been recorded in only one short-'tongued' parasitoid (Idris and Grafius 1997).

Some of the few parasitoid wasps that take pollen directly from anthers possess mouthpart specialisations for the purpose (modified palpal hairs, Jervis 1998). Some pollen-feeding predators also have such specialisations (Grinfel'd 1975; Gilbert 1981): the mouthparts of aphidophagous coccinellids mouthparts bear pollen brushes comprising short setae (these can be seen readily only in dissected insects). Other pollen-feeding predators such as lacewings lack any specialisations related to the consumption of such food (Carnard 2001).

Most parasitoid wasps, including those lacking mouthpart specialisations, require, to varying degrees, some form of sugar-rich food (honeydew, nectar) in order to achieve maximum realised fecundity (Gilbert and Jervis 1998; Heimpel and Jervis 2004). By the same token, Coccinellidae lack specialisations for feeding on sugar-rich foods, but the fact that the spraying of artificial honeydew onto crops can result in dramatic increases in coccinellid densities (Evans and Richards 1997) indicates that sugar limitation is important in influencing key life-history variables in these insects.

Body size

Body size is informative in several ways with regard to feeding requirements of natural enemies. First, a larger body size can physically restrict access to some floral nectar sources (Jervis et al. 1993; Wäckers et al. 1996) (see above and below), although it may facilitate access to others by enabling the insects to push past obstructing plant structures (Idris and Grafius 1997). Patt et al. (1997) assessed the compatibility of parasitoid wasp size and behaviour with real and artificial flowers of disparate architectures and thus varying nectar accessibility. The use of artificial flowers allowed the effects of floral scent to be controlled. The two small-bodied wasp species investigated, *Edovum puttleri* and *Pediobius foveolatus*, foraged efficiently on flowers with exposed nectaries. Only *P. foveolatus* could forage efficiently on flowers with partially obscured nectaries (certain umbellifers), and neither species could exploit nectar when it was contained in cup- or tube-shaped corollas, as their heads were wider than the floral apertures. Head width (which is strongly correlated with body size) was identified as an important constraint upon nectar exploitation as both species lacked an elongated proboscis.

Second, absolute metabolic rate increases with increasing body size in animals (Peters 1983; Schmidt-Nielsen 1984; Calder 1984), so a population of large-bodied natural enemies is likely to require a larger standing crop of energy-rich food than will an equivalent-sized population of small-bodied but otherwise biologically similar species.

Third, the ovigeny index of females (the proportion of the total lifetime potential egg complement that is mature and ready to lay upon emergence, see below) negatively correlates with body size across species (Jervis et al. 2003). The dependency on foods is likely to be greater in species having a lower (<1) index (Jervis et al. 2001) and thus greater body size. Pro-ovigenic (all eggs mature at emergence, ovigeny index = 1) species should also have minimal fat body reserves, as they are typically short-lived (Jervis et al. 2001, 2003) (see section on 'Ovigeny and fat body reserves').

Wings: aptery and brachyptery

Even if the adults of a parasitoid or a predator are shown to be food-limited, they may not be amenable to manipulation if they are physically incapable of travelling to or from the supplementary food source. If parasitoid wasps associated with important crop pests have reduced wings (e.g. some dryinids and bethylids), the area over which they forage for nectar, away from host patches, is presumably severely limited.

Anatomy

Characteristics of rectal pads, salivary glands and Malpighian tubules are hypothesised to be linked to food type (D.L.J. Quicke and M.A. Jervis, unpublished). These associations have not been tested but might be used as a predictor of general diet.

In insects, the rectal pads (1–20, usually six in number) serve to absorb water from the faeces (Chapman 1998). Species specialised in consuming deep-lying (concealed) nectar are predicted to show reduced rectal pad provision and a less extensive associated tracheal system than species exploiting exposed (extrafloral, floral) and semi-exposed (floral) nectar sources. Concealed nectar is more dilute on average than is exposed nectar, and is usually consumed in larger quantities (Gilbert and Jervis 1998; Jervis 1998), so reduction of faecal water loss will be less of a requirement. Reduction of rectal pads is most likely to be confined to species with an elongated proboscis, but it is possible that a cross-species continuum in organ reduction exists.

Unlike exposed nectar or other exposed sources of sugar-rich food that are consumed by morphologically unspecialised species, concealed nectar does not require dilution with saliva to enable ingestion (Gilbert and Jervis 1998). The saliva in this case is presumably used only to pre-digest oligosaccharides in the food. Therefore, species exploiting dilute nectar should have relatively smaller salivary glands (M.A. Jervis and D.L.J Quicke, unpublished). As with rectal pads, there may be a cross-species continuum in organ reduction.

In insects, the Malpighian tubules are responsible both for excretion of nitrogenous wastes and for water regulation (Chapman 1998). Parasitoid species feeding on nitrogen-rich materials such as pollen (and 'host feeders') should have a greater number of and longer Malpighian tubules than other species (D.L.J. Quicke and M.A. Jervis, unpublished).

Ovigeny and fat body reserves

Dissecting and counting eggs of very recently emerged females provides a useful clue to the dependency of a parasitoid species on foods. The ovigeny index is the initial egg load (the number of eggs that are mature at female emergence) divided by the total potential lifetime egg complement (estimated either from the total number of oöcytes present or from lifetime realised fecundity measured under the most favourable conditions, see Jervis et al. 2001). Females of pro-ovigenic species have entered the pupal stage with sufficient nutrient reserves to allow all the potential lifetime egg complement to mature by the time of emergence. Pro-ovigenic species, compared with strongly synovigenic species, are expected to have relatively little or no need for foods as adults (Gilbert and Jervis 1998). Mymarids appear to be exclusively pro-ovigenic (Jervis et al. 2001), and they are only occasionally observed feeding in the field (Jervis et al. 1993). However, up to 85% of the mymarid *Anagrus erythronurae* captured in vineyards were found to contain gut sugars (J.A. Rosenheim, pers. comm.; Heimpel and Jervis 2004).

Synovigeny (adult females continue to mature eggs after emergence) characterises the vast majority of parasitoid wasp species (Jervis et al. 2001; Ellers and Jervis 2003) and probably all

Table 5.1: Criteria for determining whether natural enemies are amenable to habitat manipulation involving supplementary food sources. An affirmative answer to a question indicates a higher likelihood of success in the programme.

Does the natural enemy demonstrate a need for sugar/pollen sources?
Feeding behaviour
Has the natural enemy been observed to feed on sugar sources (nectar, honeydew) or pollen in the field?
Do laboratory studies indicate that the natural enemy feeds on/has a requirement for sugars/honeydew/pollen?
Do natural enemies collected in the field show evidence of having fed on sugars or pollen?
Morphology
Does mouthpart structure suggest whether this natural enemy feeds or not?
Does the enemy possess a feeding-related specialisation for feeding on a particular food type, e.g. an elongated proboscis for exploiting concealed floral nectar?
Are the mouthparts unspecialised, suggesting the natural enemy might feed only from exposed sources? (If so, use floral sources with exposed nectar.)
Does the body size of the natural enemy facilitate or limit access to the insect's required floral food source? Is the natural enemy macropterous?
Anatomy
Are the rectal pads and the salivary glands reduced (in number and size, respectively), suggesting feeding on concealed nectar?
Do the number and length of the Malpighian tubules suggest feeding on nitrogen-rich materials such as pollen?
Does the natural enemy have small or large fat body reserves upon adult emergence?
Ovigeny
Does the natural enemy have a low ovigeny index (proportionately few eggs mature at female emergence)?
Will supplemental sugar/pollen provision in the field benefit natural enemies and pest control?
Functional response
Does the supplemental food source increase the per capita attack rate?
Do parasitoids have a greater egg load with supplemental foods?
Attraction and arrestment
Do visual and odour orientation studies show that the natural enemy readily responds to the food sources?
Can the insects travel easily between the supplemental food source and the crop?
If non-crop vegetation is established to provide supplemental food, does it become available before the crop and provide food when needed?
Numerical response
Does the natural enemy have more than one generation while pests are present?
When given supplemental food, do natural enemies show improved larval growth and development rate, and female fecundity and survival?
Does the food source lead to greater recruitment of natural enemies into the next generation?

predators. Synovigenic insects will, in general, be more dependent than pro-ovigenic species on foods for maintenance and further egg maturation. However, some synovigenic parasitoids and predators obtain most or all of the nutrients required through feeding on host or prey blood and/or tissues (Jervis and Kidd 1986; Heimpel and Collier 1996). Insects that host-feed may be less amenable to manipulation through the use of sugar-rich foods, although at least two species of *Aphytis* cannot achieve maximum longevity by host-feeding alone (Heimpel et al. 1994, 1997a). In *A. melinus*, host-feeding benefits both longevity and fecundity, but only if sugar meals are present as well (Heimpel et al. 1997b).

Species emerging with around half of their potential lifetime egg complement mature (ovigeny index = 0.5) should have the highest degree of allocation to fat body compared to either pro-ovigenic and extremely synovigenic species. This is because strictly pro-ovigenic species (ovigeny index = 1), are very short-lived (lifespan brevity makes the deposition of large reserves unnecessary) (Jervis et al. 2001) and extremely synovigenic species (ovigeny index = 0) are very long-lived and have the highest resource intake prospects (Boggs 1992, 1997). This hypothesis (M.A. Jervis and C.L. Boggs, submitted ms) on the pattern of fat body allocation has yet to be tested. If it holds, measurement of the fat body: abdominal biomass ratio, coupled with an examination of the ovaries (to measure ovigeny index) in newly emerged parasitoids and predators, may be a useful shortcut to assessing the dependency of females upon food.

Table 5.1 summarises specific questions for individual natural enemies with respect to the potential for manipulation using supplemental foods.

Alternative hosts and prey

Alternative host or prey species can serve as a reservoir of polyphagous natural enemies, maintaining them during periods of pest scarcity or absence, and allowing them to 'lay in wait' (Murdoch et al. 1985) until the pest becomes available. They form a ready source of natural enemy immigrants and minimise the population time-lag associated with the natural enemy locating and successfully colonising a newly established or re-established pest population (although too high an abundance of alternative prey/hosts during the critical stages of pest population development could impede transfer to the pest, and so lessen the enemy's impact. Hence the need for data, both on population abundance and on phenology). They can occur in areas such as field margin vegetation or can even be associated with another crop (Gilstrap 1988; Xu and Wu 1987; Corbett et al. 1991; Jiang et al. 1991), in which case it may be expedient to grow the two crops in very close proximity. Powell (1986) lists several cases where increased parasitism of a pest has been attributed to the presence of alternative hosts. Parasitoids of typhlocybine leafhopper eggs are among the best-known examples of natural enemies being manipulated through provision of alternatives (Doutt and Nakata 1973; Pickett et al. 1990; Yigit and Erkilic 1987; Murphy et al. 1996, 1998), although the alternative host's food plant was not adjacent to the crop in all these cases.

In the absence of information on host or prey range, the natural enemy may be identifiable as potentially polyphagous, based on the known habits of related species. Even the members of some typically polyphagous-tending taxonomic groups restrict their attacks to members of a particular prey or host taxon such as family or subfamily (Waloff and Jervis 1987). Some natural enemies, including a few parasitoids, attack insects from different orders, but still show a predictable degree of restriction in terms of the host's feeding niche (e.g. being confined to attacking insects in leaf-mines) (Lawton 1986).

Identifying alternative hosts and prey
To identify alternative hosts and prey, either the natural enemies can be tested with choices in laboratory experiments (parasitoids and predators should be presented with the pests on crop plants and the alternative host/prey on its food plant, which may itself be a crop, see above) or

they can be directly observed in the field. In both cases, preferences for particular prey can be measured, in addition to data being gathered on host and prey range (see Sadeghi and Gilbert 2000a; van Lenteren et al. 2003; Fellowes et al. 2004). The results of laboratory no-choice and even choice tests should be viewed with caution. In the former, the natural enemy may attack a herbivore species because its threshold for acceptance has been lowered by extreme hunger or a replete egg storage apparatus, and in the latter a natural enemy may ignore a less-preferred species when the encounter rate with the preferred one is above some threshold value (assuming recognition to be instantaneous). For practical advice on such matters, see Fellowes et al. (2004).

For parasitoids, possible alternative hosts are collected in the field and reared to determine whether the parasitoid uses the species, and to what extent it may be preferred/unpreferred compared to others, including the pest. Molecular techniques (such as those explored by Menalled et al., ch. 6 this volume) may be developed for detecting the immatures of some parasitoid species (see below), although a few parasitoids can be identified without recourse to such methods, by using larval morphology (Sunderland et al. 2004).

Identifying alternative hosts of parasitoids through observational studies (conducted in the field and/or the laboratory) presents additional challenges. First, a parasitoid should be recorded as actually depositing an egg during attack. This is confirmed by dissecting hosts shortly after exposure to parasitoids to locate eggs within the host body or by maintaining hosts until parasitism becomes detectable. Furthermore, although an insect species is accepted for oviposition, it does not necessarily follow that the species is suitable for successful parasitoid growth, development and survival. Under laboratory conditions some parasitoids will, if given no alternative, oviposit in unsuitable hosts, but subsequent monitoring of the parasitoids' progeny will show them not to complete development, for example they are killed by the host's physiological defences (Jervis et al. 2004). Such unsuitable hosts will act as a 'sink' for parasitoid eggs, contributing nothing to parasitoid recruitment (Hoogendoorn and Heimpel 2002; Heimpel et al. 2003) (see section 'Numerical responses').

Field-collecting of suspected prey is unlikely to be useful for establishing the trophic link between predator and alternative prey. It is very rare that injuries borne by cadavers of field-observed/collected herbivore species give information on the identity of the predator species responsible. Therefore, it is invariably necessary to collect predators from the field and to then identify the source of their gut contents. Reliable methods for establishing the identity of alternative prey from gut contents are gut dissection followed by visual examination of contents, and molecular (serology and DNA) techniques (electrophoresis has fallen out of favour, Sunderland et al. 2004). Gut dissection has been applied mainly to Carabidae, Staphylinidae and Coccinellidae which ingest the hard, indigestible parts of the prey (Sunderland and Vickerman 1980; Hengeveld 1980; Chiverton 1984; Luff 1987; Sunderland et al. 1987; Triltsch 1997; Sunderland et al. 2004; Hoogendoorn and Heimpel 2004). This method requires little equipment and is 'immediate'. The prey (and even plant remains) found in field-collected predators can be visually compared with the body parts of non-pest insects, other invertebrates and plant tissues. Identifying prey remains to species is sometimes possible if distinctive pieces of prey cuticle remain intact. Recognisable fragments generally found in the guts of carabid and staphylinid beetles include chaetae and skin of earthworms; cephalothoraces of spiders; claws, heads and/or antennae from Collembola; aphid siphunculi and claws; sclerotised cuticle, mandibles and legs from beetles; and the head and tarsal claws of some Diptera. Although the gut contents within a predator probably result from predation, carrion-feeding is also known in some species. A predator which acquires prey materials through secondary predation (feeding upon another predator species which itself contains prey items) can produce false positives (this problem also applies to the results of gut content studies involving other techniques, see Sunderland et al. 2004).

Many predators ingest prey materials in liquid form only, so molecular methods have to be applied in order to identify the prey source (Sunderland et al. 2004). The major attractiveness of serological techniques is their potential for extreme specificity due to biological recognition at the molecular level (although cross-reaction can still produce false positives in some cases) (Ragsdale et al. 1981; Greenstone and Hunt 1993; Hagler et al. 1995; Hagler and Naranjo 1997; Symondson et al. 1997, 1999; Hagler 1998; Agusti et al. 1999a; Sunderland et al. 2004). However, identifying alternative prey of a given predator would involve raising antibodies to each of the suspected prey species – a time-consuming and expensive process. Applying serological methods to detect and identify a parasitoid within a host is, by contrast, straightforward. Antibodies are raised to the immature stage of the parasitoid, and suspected alternative hosts collected in the field are screened for the presence of antigens. DNA techniques offer another method for performing gut contents analysis, and this technique has recently been developed for a few predator–prey systems (Agusti et al. 1999b, 2000; Zaidi et al. 1999; Chen et al. 2000; Hoogendoorn and Heimpel 2001, 2003). Similar analyses can also be used to detect immature parasitoids within host insects (Ratcliffe et al. 2002).

Even if an alternative host/prey species is identified and found to be suitable (i.e. the natural enemy is able to grow, develop and survive on that species), it may be unsuitable for use in the planned manipulation program because key offspring fitness parameters are seriously constrained (see sections 'Ease of transfer' and 'Dispersal studies').

Ease of transfer

Having (a) identified what appears to be a suitable alternative host or prey species, (b) established that the alternative species can be situated in close proximity to the crop and (c) established that the predators or parasitoids have sufficient powers of dispersal to be able to travel to and from the crop, it is necessary to determine that insects from the 'reservoir' are capable of successfully exploiting the pest.

Consideration needs to be given to the effects candidate alternative host or prey species have on parasitoid and predator fitness, in terms of key life-history variables such as body size, ovigeny index, fecundity, longevity and development rate, as these may vary significantly with host/prey species (Jervis et al. 2004, but see Sadeghi and Gilbert 2000b). For example, the success of natural enemy transfer to the pest may be constrained by a low fecundity (which will affect the functional response) associated with the alternative species, although such a problem may be offset by a host-/prey-related female oviposition preference (Sadeghi and Gilbert 2000b). Success of transfer may also be constrained by a smaller body size associated with the alternative species (see section 'Dispersal studies').

The possibility exists that the natural enemies being tested are so genetically different from those associated with the pest, that transfer of parasitism or predation is either difficult or impossible to achieve. This may be because laboratory cultures are genetically insufficiently diverse (made more likely by the practice of founding cultures with single, mated females) or because under field conditions the natural enemy populations associated with the alternative host/prey are already genetically different from those associated with the pest. Powell and Wright (1988) therefore recommended that a series of transfer trials be performed, each time using a parasitoid population taken from a different field locality. The application of genetic characterisation studies (Atanassova et al. 1998) would be a valuable complement to transfer studies by revealing underlying genetic differences between populations.

The ability of natural enemy populations to alternate between pest and alternative hosts/prey, as opposed to one-way transfer, should be assessed. Transfer and exchange can be impeded due to conditioning effects. For example, in a Y-tube olfactometer, the generalist aphid parasitoid *Aphidius colemani* shows a preference for the host plant complex (the host together

Table 5.2: Criteria for determining whether natural enemies are amenable to habitat manipulation involving alternative hosts and prey. An affirmative answer to a question indicates a higher likelihood of success in a programme.

Does the natural enemy demonstrate a need for alternative hosts/prey?
Are closely related species of the natural enemy in question known to be polyphagous?
Based on collecting and rearing a variety of hosts from the field, does the parasitoid develop on/in alternative hosts?
Does the parasitoid lay an egg into/onto, and successfully develop from, an alternative host?
Does the predator consume other types of prey in laboratory feeding tests? Do field-collected predators have alternative prey materials in their gut, based on dissections, serological and molecular techniques?
Will providing alternative hosts/prey in the field benefit natural enemies and pest control?
Can alternative hosts/prey be provided in proximity to the crop?
Are natural enemies capable of travelling between the alternative hosts/prey and the crop?
Under laboratory conditions, can natural enemies readily transfer from the alternative to the pest species?

with its food plant) on which it had been reared, even though the same aphid host species was involved. This preference appeared to be induced by chemical cues encountered on the 'mummy' casing at the time of female emergence (Storeck et al. 2000). Storeck et al. (2000) showed that the initial preference could be altered by subsequent foraging experiences (i.e. with a different host plant complex). In the field (e.g. the field margin vegetation) the likelihood of such experiences occurring may be very low, so reducing the probability of transfer to the pest. However, if the availability of the alternative host is suddenly reduced (e.g. by mowing vegetation or harvesting a crop) this would stimulate more widespread foraging, thereby increasing the likelihood of encounters with a different host and thus the likelihood of transfer occurring.

Table 5.2 summarises specific questions for individual natural enemies with respect to the potential for manipulation using alternative hosts and prey.

Physical refugia

Adults of many natural enemy species occupy physical refugia. In agroecosystems, physical refugia take various forms including crop residues, vegetation in field boundaries and crevices within and beneath the bark of living and dead trees. The importance of physical refugia in determining the abundance of predators and parasitoids has been shown in various settings, including cereal fields (Gilstrap 1988; Thomas et al. 1991), sugar cane (Joshi and Sharma 1989; Mohyuddinn 1991) and rice (Shepard et al. 1989). Refugia may protect natural enemies from mortality caused by weather (Luff 1965), desiccation (Van Driesche and Bellows 1996), insecticides (Lee et al. 2001), predation (Cottrell and Yeargan 1999), parasitism and pathogens.

Identifying the need for physical refugia

Determining whether natural enemies require physical refugia usually involves examining potential refugium sites for natural enemies during overwintering/summer aestivation and breeding periods. Places to examine include grassy field margins, hedgerows, nearby natural areas, weedy areas, cover crops and even farm buildings and associated paraphernalia. Predators can be collected from vegetation using vacuum nets such as the D-Vac. Tullgren funnel extraction can be applied to leaf litter, and flotation to soil cores. Radiolabelling of pre-migratory predators can be used to establish the eventual destination site of the insects (Sunderland et al. 2004). The results of such survey work can indicate which, among a variety of habitats, harbour the highest densities of a particular natural enemy (Dennis et al. 1994). Laboratory, semi-field and field experiments can reveal whether between-habitat variation in natural enemy numbers is the result of differences in survival rates and/or active selection of the habitat (Dennis et al. 1994).

The phenology and breeding patterns of natural enemies may affect their dependence on refugia. For example, predatory carabid beetles that overwinter as adults possibly benefit from refugia more than do species that overwinter as larvae (Lee and Landis 2002). Adult beetles are mobile before entering winter diapause and may actively select a burrowing site. Examination of refugia during winter consistently reveals very high densities of adult beetles (Sotherton 1984, 1985; Lys and Nentwig 1994). However, beetle species that overwinter as larvae depend on females selecting an appropriate oviposition site. The oviposition preferences should be tested to determine importance of refugia to the species.

Applications of refugia
Provision of refugia can in many cases be achieved by leaving some crop areas relatively unmanaged (e.g. as perennial field borders), by using cover crops (Altieri and Letourneau 1982; Hassall et al. 1992; Nentwig 1998; Thomas et al. 1991; Weiser 2001) and by conserving non-crop areas such as hedgerows bordering fields.

Using knowledge of a predator's or parasitoid's natural refugia, artificial refugia can be created. These have included wrappings of plant debris or cotton around tree trunks, piles of leaf and grass litter placed at the bases of trunks (Deng et al. 1988), bands of burlap and aluminium around trunks (Tamaki and Halfhill 1968), plastic or wooden chambers placed in and around crop fields (Gillaspy 1971; Lawson et al. 1961; McEwen and Sengonca 2001), polyethylene bags (Heirbaut and van Damme 1992) and empty cans (Schonbeck 1988). Adult lacewings (Chrysopidae) are known to overwinter individually and en masse in places such as unheated parts of buildings. Artificial chambers placed in the field in autumn in Europe have been shown to attract adults, with wooden chambers attracting more insects than plastic ones (Sengonca and Frings 1989; McEwen 1998). McEwen and Sengonca (2001) reviewed methods for optimising the attractiveness of chambers.

Artificial refugium design and placement has sometimes been based on trial and error or even on the investigator's intuition. While this may be successful in some cases, we advocate the following approaches.

- Characterise the known natural refugium in terms of measurements taken of temperature, relative humidity, light intensity, colour and other abiotic variables such as wind direction (see Unwin and Corbet 1991 for microclimate measurement techniques), then base artificial refugium construction and spatial arrangement (including aspect) on the results obtained.
- In the absence of information on the natural refugium: determine the natural enemy's thermal, light, humidity and substratum preferences (see references in Fraenkel and Gunn 1961 and Carthy 1971 for protocols), abiotic factor (especially humidity and temperature), tolerance ranges (see Bartlett 1962, Loveridge 1968, Edney 1971 and Jervis et al. 2004 for protocols), then base artificial refugium design and placement on the results obtained (microclimate measurements taken of refugium prototypes may be useful here).

Refugium provision strategies can range from augmenting the numbers of existing refugia to providing substitute refugia to replace those eliminated through agricultural practices. Refugia

Table 5.3: Criteria for determining whether natural enemies are amenable to habitat manipulation using refugia. An affirmative answer to a question indicates a higher likelihood of success in a program.

Does the natural enemy demonstrate a need for a refugium?
Do natural, weedy sites or cover crops, or other sites, harbour higher densities of overwintering or summer-aestivating natural enemies?
Do natural enemies oviposit and breed in non-crop habitats?
Do natural enemies actively select specific habitats for overwintering?
Does the phenology of the natural enemy enable it to use the refugia?
Do natural enemies readily commute between refugia and crop?
Are the artificial refugia attractive to natural enemies, and are they easy to deploy?
Are the physical characteristics of the refugia, e.g. temperature and humidity, suitable for the natural enemy in terms of survival; in particular, are the survival prospects of natural enemies improved in refugia?
Can the refugia be manipulated to encourage natural enemies to attack pests in the crop?

may need to be deliberately removed or destroyed the next season either as part of the biological control strategy (see below) or for some other reason; for example, the refugium may be judged to physically interfere with a mechanical farming practice such as early-season fertiliser or insecticide application. Destruction of refugia such as alternative breeding sites, when done at the appropriate time, can improve rather than impede pest control: natural enemies are forced to disperse and forage for prey on the crop.

Table 5.3 summarises specific questions for individual natural enemies with respect to the potential for manipulation using physical refugia.

Investigating the effects of resource provision on natural enemy searching efficiency

Parasitoid and predator functional responses

Food provision

Supplemental food provision may alter the shape and size of the functional response curve through its effect on foraging time allocation, egg production and female survival (Hassell 1978; Kidd and Jervis 1989; Fellowes et al. 2004; Jervis et al. 2004). Parasitoids and predators exhibit both a series of age-specific functional responses, and an overall lifetime response (Bellows 1985). The parasitism inflicted by each individual female parasitoid on a host population over her lifetime will be influenced by variations in age-specific fecundity and lifespan, both of which can alter with the availability and quality of energy-rich foods. Without an increase in lifespan, an increase in mean age-specific realised fecundity may not be sufficient to effect a significant reduction in host numbers. Lifetime functional response experiments, with food both present and absent, may therefore provide valuable insights into the potential efficacy of the food in a manipulation program. For protocols, see Bellows (1985), Sahragard et al. (1991) and Fellowes et al. (2004).

Lifespan needs to be considered in relation to mortality factors unrelated to nutrition. For example, if non-starvation mortality (e.g. that inflicted by predators) of female parasitoids in the field is such that average lifespan is reduced to that of starved insects (Heimpel et al. 1997b;

Rosenheim 1998), supplemental food provision will have little if any impact on lifespan, and therefore on the lifetime functional response, in the field. For a protocol aimed at investigating field mortality of parasitoids or predators, see Heimpel et al. (1997a).

The shape and size of the predator's functional response will vary with predator stage because larger predators are likely to move more rapidly, detect prey from greater distances and achieve a higher proportion of successful attacks; also, handling time is shorter for later predator stages (Dixon 1959; Wratten 1973; Glen 1975; Hassell 1978; Fellowes et al. 2004). For any one stage both attack rate and handling time are likely affected by hunger/satiation thresholds characteristic of that stage (Hassell 1978) and, as with parasitoids, there will be a series of age-specific functional responses and also a lifetime response. Supplying non-prey foods might result in a decreased per capita predation rate of a particular stage because predators might obtain substantial nutrient requirements for growth and development, egg production and survival from the supplement. This mechanism may explain why within-generation predation of olive moth eggs did not increase when artificial honeydew was sprayed in an olive orchard (Liber and Niccoli 1988; McEwen et al. 1993b). A negative, supplemental food-related change in the functional response has been demonstrated by van Rijn et al. (2002) working with a mite-thrips (*Neoseiulus cucumeris–Frankliniella occidentalis*) system. The functional response plateau lowered significantly when pollen of *Typha latifolia* was provided as a supplemental food. A similar effect is suggested by the results of Cottrell and Yeargan's (1998) study on the coccinellid *Coleomegilla maculata*. However, Hazzard et al. (1991) detected no change in the functional response of *Coleomegilla maculata* (Coccinellidae) when corn pollen was given as supplement to the Colorado beetle egg prey (use of a more nutritious type of pollen might have effected a change).

The functional response is essentially a within-patch phenomenon. Searching efficiency across patches – the more realistic scenario – is, however, sensitive not only to the patch-specific searching ability of the natural enemy but also the extent to which foraging is spatially non-random (Hassell 1982). To take account of this Hassell (1978, 1982) proposed a model for overall searching efficiency that takes account of both of these factors. For practical advice on how this model can be parameterised, see Kidd and Jervis (2004). The model does not incorporate foraging for non-host foods, but can nevertheless be used to compare the searching efficiency of the natural enemy with and without food.

Alternative hosts and prey

A parasitoid's or predator's functional response is known to vary with the host/prey species when the latter is presented alone, and it can also be also influenced by the abundance of the alternative host or prey species when it is spatially coincident with the main prey (e.g. the two victim species occur in the same laboratory arena) (Hassell 1978; Fellowes et al. 2004). In *Aphidius uzbekistanicus* and *Coccinella septempunctata* a sigmoid functional response was recorded when the unpreferred host and prey species was provided, compared with a type 2 (a continuously decelerating curve) when the preferred species was provided (Dransfield 1979; Hassell et al. 1977.)

Preference may vary with the relative abundance of two prey types, in which case if the predator or parasitoid eats or oviposits in disproportionately more of the more abundant type it is said to display (positive) switching behaviour (Murdoch 1969). Switching behaviour has aroused the interest of students of population dynamics because it results in a type 3 functional response to each of the two prey species (Hassell et al. 1977). Such a functional response in a discrete, natural enemy–two prey species interaction operating over several generations, can act as a powerful stabilising mechanism (Hassell and Comins 1978). Switching behaviour in parasitoids has been demonstrated in only a few parasitoids, including *Aphidius ervi* and *Praon pequodorum* (see Chow and Mackauer 1991). For advice on the design of experiments aimed at detecting switching behaviour (and at identifying the underlying behavioural mechanism), see Fellowes et al. (2004). The resulting data can be analysed using Murdoch's (1969) or other models (see Sherratt and Harvey 1993).

The spatial arrangement that typically pertains in manipulation programs involves the alternative host/prey species and the pest being situated on different plants that are some distance apart. We would expect switching behaviour to be generally less likely to occur the greater the separation between patches of the different host/prey species (this effect will be most pronounced in natural enemy species that have a small trivial range, *sensu* Southwood 1977, i.e. they forage over a smaller area, e.g. because of small body size, see Southwood 1978). We therefore regard investigation of switching behaviour to be of low research priority in most manipulation programs.

Investigating the effects of resource provision on natural enemy abundance in the crop

Attraction and arrestment behaviour

An appreciation of the stimuli and the responses involved in food and alternative host/prey location behaviour can be important in understanding how manipulation of natural enemy immigration and retention may be achieved (Tables 5.1 and 5.2). Attractant stimuli typically operate early in the searching sequence, eliciting orientation to areas that either contain or are likely to contain the hosts/prey/food. For many species, volatile chemicals are of primary importance in attraction (Shahjahan 1974; Lewis and Takasu 1990; Cortesero et al. 2000; Fellowes et al. 2004), although visual stimuli may also play an important role (Wäckers 1994). Parasitoids may, like other flower-visitors, have odour or colour preferences in relation to the range of flowering plants they exploit (for protocols, see Dafni 1992, Wäckers 1994). Arrestant stimuli typically operate later in the searching sequence, eliciting a reduction in the distance or area covered per unit time by parasitoids moving within such areas (i.e. result in reduced emigration, so contributing to retention) (Waage 1978; Fellowes et al. 2004). For many species these stimuli are chemical and are detected upon contact (Waage 1978).

Key contributions to be made by behavioural work include establishing whether, and to what degree, the resource encourages immigration into the crop. Responses to olfactory stimuli can be evaluated using olfactometers of various kinds (static-air, airflow, i.e. Y-tube, four-arm Petersson) and wind-tunnels (Dafni 1992; Kielty et al. 1996; Takasu and Lewis 1996; Ballal and Singh 1999; Jang et al. 2000; Raymond et al. 2000; Le Ru and Makaya-Makosso 2001; Fellowes et al. 2004). Wind-tunnels have the greatest potential for approximating field conditions. Where a natural enemy has been shown to locate the resource by anemotaxis, spatial arrangements of crop plantings and other resources (e.g. nectar sources, artificial honeydew) can take account of wind direction, increase the probability of colonisation and thus maximise the number of predators or parasitoid immigrants.

Attraction can also be studied under field conditions, and plants ranked according to their attractiveness, but investigators might find it difficult to control various significant confounding factors (Fellowes et al. 2004).

One of the aims of spraying a crop with an artificial honeydew or other substitute food is to increase immigration and/or reduce emigration (Butler and Ritchie 1971; Hagen et al. 1971, 1976; Carlson and Chiang 1973; Ben Saad and Bishop 1976; Nichols and Neel 1977; Hagley and Simpson 1981; Liber and Niccoli 1988; Evans and Swallow 1993; McEwen et al. 1993a; Mensah

1996, 1997; Evans and Richards 1997; Jacob and Evans 1998; Stephen and Browne 2000). Olfactometry and other behavioural techniques could prove very useful in comparing and optimising the attractant effects of honeydew mixtures (van Emden and Hagen 1976; Dean and Satasook 1983; McEwen et al. 1993b; Fellowes et al. 2004).

Once it has been shown that a parasitoid or predator can respond positively to the volatiles emanating from a plant or honeydew, the technique of coupled gas chromatography-mass spectrometry/electroantennography (coupled GC-MS/EAG) can be used to identify the specific chemicals that mediate attraction in the insects (Khan et al. 1997). Once the semiochemicals in a plant or other resource (e.g. natural honeydew) have been identified, known or potential resource-providing plant species, and artificial honeydew mixtures, can be screened, using GCMS, for their relative potential as attractors of natural enemies.

While a supplemental food (or source thereof) may be shown to be attractive, additional experiments should be conducted to determine whether the efficiency of food location is not significantly impeded under field conditions, as result of interference from odours emanating from other vegetation, including the pest-bearing crop. Odours from such sources may disrupt olfactory responses to food sources (Shahjahan and Streams 1973).

Natural honeydew is an arrestant, potentially contributing to population retention, for lacewings (McEwen et al. 1993b), coccinellids (Carter and Dixon 1984; van den Meiracker et al. 1990), syrphids (Budenberg and Powell 1992) and parasitoids (Ayal 1987; Budenberg 1990), and artificial honeydews are likely to have a similar effect, via the handling-time effect of feeding and/or through an effect on searching movements. Fellowes et al. (2004) provide guidance on designing experiments that can shed light on arrestment responses to artificial and natural honeydews. When planning manipulation using artificial honeydew sprays, consideration needs to be given to the possibility that too 'powerful' a formulation (in terms of arrestment), or a weak spatial association between spray deposits and host/prey patches, may confound the attempted manipulation by constraining the natural enemy's searching efficiency.

Van Rijn et al. (2002) investigated how the spatial distribution of supplemental pollen influenced the biological control of thrips in greenhouses. The local supply of pollen on otherwise pollen-free cucumber plants increased the densities of the predatory mites and suppressed growth of the herbivore population despite thrips also feeding on the pollen. A parameterised predator–prey model revealed how a uniform supply of alternative food enhanced pest population growth rate and escape from predator control. Conversely, a spatially restricted food supply caused predators to aggregate; this deterred prey due to the increased predation risk and thus the predators monopolised the food. Whether the predators aggregate as a result of increased attraction, increased arrestment, or both is not known.

State-dependent behaviour

Parasitoid and predator behaviour is state-dependent, varying with the number of mature eggs present in the ovaries at a given time (Jervis and Kidd 1986; Heimpel and Collier 1996; Jervis et al. 1996; Heimpel et al. 1996; Heimpel and Rosenheim 1998) as well as with nutritional state (Sabelis 1992; Wäckers 1994; Heimpel and Rosenheim 1995; Heimpel and Collier 1996; Hickman et al. 2001). In parasitoids, for example, physiological state influences responsiveness to plant odours, the choice between food- and host-containing patches, responsiveness to contact chemicals, patch time allocation, oviposition rate and possibly flight activity (Wäckers 1994; Jervis et al. 1996; Sirot and Bernstein 1996; Lewis et al. 1998). Natural enemy behaviour is also influenced by experience (e.g. Lewis et al. 1990; Vet et al. 1995; Fellowes et al. 2004), and possibly also age (but see Heimpel and Rosenheim 1995).

Dispersal studies

Investigations of natural enemy movement – by which we mean immigration into and emigration out of the crop (or the crop-associated habitat, e.g. in the case of some physical refugia) – are vital in studies of resource use in relation to habitat manipulation (Corbett 1998; MacLeod 1999) (Tables 5.1–3). While a natural or artificial resource may improve natural enemy survival, growth, development and reproduction, it does not necessarily guarantee that the resource, when placed in the vicinity of the crop, will be an effective source of immigrant parasitoids or predators. For example, incorporation of knotweed in alfalfa fields attracted more predators in the general area but did not cause an increase in predator densities in alfalfa, the economic crop (Bugg et al. 1987).

By the same token, an identified food source, natural or artificial, may be placed in or near to the crop, but the parasitoids or predators may not use the food. Proof is required that the natural enemies can travel between refuge and crop, between alternative host/prey and pest populations, or between food and pests, in sufficient numbers to improve pest suppression. Methods for monitoring parasitoid and predator movements are reviewed in Sunderland et al. (2004). Inexpensive techniques include marking coccinellids, carabids and other hard-bodied insects with paints (MacLeod 1999) and dusting smaller and soft-bodied insects such as parasitoids with fluorescent powder. However, fluorescent powder can be harmful to small-bodied parasitoids (Garcia-Salazar and Landis 1997). Other studies have involved labelling parasitoids with rubidium (Corbett and Rosenheim 1996) and immuno-markers (Hagler and Jackson 1998). One-way travel capability, between food source and crop, can also be assessed using the food materials themselves as markers: pollen grains located within the gut (e.g. see Wratten et al. 2003) or on the body surface, or signature sugars/dyes in nectar and honeydew present within the gut. Marking and tracking is reviewed by Lavandero et al. (ch. 7 this volume).

Theoretical studies can help predict whether adding beneficial vegetational structure for natural enemies of known dispersal powers will improve pest control. A simulation model by Corbett and Plant (1993) described the spatial distribution of natural enemies after the addition of strips of vegetation intersecting a field. Generally, if natural enemies used the vegetation before the crop germinated, the vegetation was a 'source' of natural enemies. Natural enemies would be available to exert pressure on pests once the crop germinated. However, if the vegetation and crop germinated simultaneously, the vegetation strip might serve as a 'sink' of natural enemies by reducing their activity in the economically important crop (see Mensah and Sequeira, ch. 12 this volume). Also, by incorporating the diffusion rate of the natural enemy, the model predicted the scale at which beneficial effects of the resource will be observed in the field. To what extent this model applies to real natural enemies is unclear, but with appropriate modification the model would be a useful basis for investigations aimed at assessing the importance of dispersal rate, coupled with plant phenology, in influencing outcome of manipulation programs.

Knowing the absolute distance a natural enemy can travel, and how this is affected by physiological state, is important. Evidence of commuting behaviour between refuge and crop (this would constitute migration, see Southwood 1962) or between food and crop (this would constitute foraging by the natural enemy within its trivial range, see Southwood 1977, 1978) should also be sought.

Body size can influence dispersal by natural enemies in two ways: directly, through its biomechanical effect on flight performance (Dudley 2000) and indirectly through its effect on the size of the fat body which fuels both somatic maintenance and, in some insects, locomotion (Chapman 1998). Ellers et al. (1998) showed that in the parasitoid *Asobara tabida* (Braconidae), the greater the distance travelled from a central release point, the more the fat reserves were depleted. (The quantity of lipids in the fat body of individual insects can be measured using the ether extraction method of Ellers (1996) or the vanillin reaction (Olson et al. 2000).

Furthermore, larger females had larger fat reserves (see also Rivero and West 2002), and dispersed over greater distances than smaller females (Ellers et al. 1998). Note that a decline in fat reserves was recorded independently of female age (Ellers et al. 1998) (lipid reserves have been shown to decline with age in several parasitoid wasp species, see Jervis and Kidd 1986, Casas et al. 2003) carbohydrates, not fats, are used by Hymenoptera as a 'flight fuel' (Casas et al. 2003) and the same is likely to apply to the fuelling of ambulation. This points to egg production as the most likely cause of the fat body decline. At least as far as the body size–fat body effect was concerned, there is support for this: lifetime reproductive success (measured as the egg load of recaptured females plus the numbers of eggs they were estimated to have laid) increased with body size in *A. tabida*. Conservation biological control investigators should be mindful of the body size–dispersal distance relationship when planning a manipulation program: a nutritionally sub-optimal alternative host or prey species will produce smaller adult progeny (see Jervis et al. 2004) which will be inferior dispersers (see above).

Ellers et al's (1998) study was of dispersal within *A. tabida*'s trivial range. Body size and fat body size have been measured in a migratory range context (migration to overwintering refugia) for *Coccinella septempunctata* (Zhou et al. 1995).

Last, the physical attributes of companion plants can significantly influence the number of predators dispersing to crop plants (Cottrell and Yeargan 1999).

Numerical responses

Supplemental foods

An increase in the numerical response (the rate of recruitment into the subsequent generation) of the natural enemy population is an important consideration where the pest is present in the crop for longer than the duration of one natural enemy generation. Van Rijn et al. (2002) point out that the effect of provision of supplemental foods on the natural enemy's numerical response has nevertheless been ignored by some authors. The numerical response depends largely on three components: development rate, larval survival and the realised fecundity of females (Beddington et al. 1976; Hassell 1978). Food supplements have the potential to enhance some or all of these. For example, McEwen et al. (1993c, 1996) found diet-related variation in development and survival in the larvae of *Chrysoperla carnea*. Larvae given artificial honeydew, as opposed to water, were more likely both to complete development to the second larval moult (there are three larval instars) at low egg densities and to survive to adult eclosion. Pollen feeding increases development rate and survival to the adult stage in some phytoseiid mites (McMurtry and Scriven 1964; Osakabe 1988). Food supplements can enable immature predators to compensate for the energetic costs of maintenance metabolism (Beddington et al. 1976; Jervis et al. 2004) and achieve a higher adult weight (McEwen et al. 1993c), a higher egg load upon eclosion and a larger store of energy reserves for use in somatic maintenance.

Food provision may also affect female fecundity by altering the threshold prey density at which eggs are laid by anautogenous predators (Beddington et al. 1976; Jervis et al. 2004). Anautogenous insects cannot produce mature eggs without first ingesting a certain quantity of an appropriate food, due to insufficient resource carry-over from the larval stage. Anautogeny is found in Coccinellidae and Chrysopidae (Dixon 1959; Beddington et al. 1976; McEwen et al. 1996). *Chrysoperla carnea* displays plasticity in its dependence on food for initial egg production: it can be either anautogenous or, when given a non-prey food supplement during larval life, autogenous, capable of producing mature eggs before ingesting a meal as an adult (McEwen et al. 1996). If the food supplement contains adequate nutrients to allow initial egg production, then it should lower the prey density threshold for oviposition. Food provision can also increase age-specific and lifetime fecundity (Jervis et al. 2004).

Protocols for measuring the effects that natural and artificial supplementary foods have on the main components of the numerical response can be derived from van Lenteren et al. (1987), Idris and Grafius (1995), Gilbert and Jervis (1998), Crum et al. (1998), Evans (2000), Limburg and Rosenheim (2001) and Jervis et al. (2004). Negative correlations (indicative of physiological trade-offs) between key life-history variables can occur: supplemental food provision may increase one variable (e.g. fecundity) at the expense of another (e.g. egg size, which will influence larval survival) (see Crum et al. 1998). Thus it is advisable to study a range of life-history variables to obtain insights into the potential benefits of supplemental or substitute food provision (caution should be exercised in using any single life-history trait as a proxy measure of fitness, see the review by Roitberg et al. 2001).

Alternative foods may be investigated as potential substitutes for prey during periods when the latter are absent or very scarce, for example during the period following harvesting and reseeding/planting of the crop (see Crum et al. 1998 for protocols applied to predators). They may serve only as a stop-gap resource for some predator species (Gurr et al. 2004), enabling survival and a high search rate over a short time period (e.g. Limburg and Rosenheim 2001), whereas for others they may allow a significant amount of development, reproduction and survival to take place (Smith 1961, 1965; Kiman and Yeargan 1985; Cottrell and Yeargan 1998; Crum et al. 1998; van Rijn and Tanigoshi 1999). Provision of substitute foods may be more effective during some phases of the life-cycle than others (Crum et al. 1998).

Even when a supplement or substitute is found to be highly effective in promoting development, reproduction and survival, it still needs to be tested for its effect on the numerical response under field conditions. Obvious problems such as rainfall and humidity extremes aside, artificial resources can become ineffective due to bacterial or fungal contamination. Microorganisms may reduce the performance-enhancing value of the artificial food, and may even make the food repellent. For example, survival of adult *Microplitis croceipes* was greatly reduced when given honey-water that was contaminated with bacteria (Sikorowski et al. 1992).

Van Rijn et al. (2002) studied supplemental food effects under glasshouse conditions, removing immigration and retention effects, and showed that pollen provision can greatly improve the control of thrips by predatory mites even though pollen was fed upon by the prey as well as by the predator. Improved control results from an increase in the predators' numerical response to pollen and thrip density. The numerical response outweighs the negative effects of a reduction in the functional response (see above) and the accelerated population growth of the pest due to its feeding on pollen.

Provision of alternative hosts and prey

Among the range of possible population processes resulting from the provision of alternative hosts/prey are apparent competition and apparent mutualism (see Holt 1977). In the former case, the natural enemy increases in abundance (numerical response) on the alternative herbivore and, assuming the natural enemy readily transfers to the pest, the latter will decrease in the abundance – a result that is in line with the goal of pest management through habitat manipulation, although the precise level of the economic threshold would determine the degree to which the program has been successful. If, however, there is only a weak tendency for the natural enemy to transfer to the pest, and the alternative's abundance remains sufficiently high, the parasitoids can become egg-depleted (the alternative host acting as a 'sink' for the parasitoids' eggs, see Heimpel et al. 2003) or the predators can become satiated, so limiting the impact the enemy might have on the pest. ('Apparent mutualism' applies here because the pest indirectly benefits from the presence of the alternative herbivore species.) The pest is likely to suffer much less mortality than anticipated by the biological control practitioner so, depending on the precise level of the economic threshold, the manipulation program may be a failure. Guidance on the design and interpretation of manipulation experiments for studying apparent competition and mutualism can be derived from Müller and Godfray (1997), Morris et al. (2001) and Heimpel et al. (2003).

Conclusions

Our purpose in writing this chapter was to outline how natural enemy life-history may affect the degree to which natural enemies may respond to habitat manipulation that is designed to improve biological control. The three habitat manipulations discussed are food supplementation, provision of alternative hosts and provision of physical refugia. Clearly, a wide range of life-histories will mediate the effectiveness of these and other methods of conservation biological control. Designing habitat manipulation programs that best take advantage of natural enemy life-history traits should lead to more effective and predictable science of conservation biological control (Barbosa 1998; Pickett and Bugg 1998; Landis et al. 2000). Our approach can be summarised and illustrated by posing a series of questions for each of the manipulations discussed above. The effectiveness of supplementary food sources is centred on two broad questions (Table 5.1). First, does the natural enemy demonstrate a need for the supplemental food? Second, is the natural enemy likely to benefit from the supplemental food source? For the provision of alternative hosts or prey, questions focus on whether the resource is limiting for the natural enemy, and whether it will actually be used by the natural enemy in the field (Table 5.2). Finally, for the incorporation of physical refugia in the field, questions centre on the usefulness of diverse habitats and the ability of natural enemies to commute between the refugium and the pest habitat (Table 5.3). These are apparently simple questions, but as discussed in this chapter, each involves a suite of biological complexities. Accordingly, entomologists can, irrespective of their particular research specialism, make a valuable contribution to knowledge of the life-history traits that are potentially important in conservation biological control. Even evolutionary biologists have a key role to play by discovering negative and positive correlations between life-history traits.

Acknowledgements

We are very grateful to two anonymous reviewers for their constructive comments on the manuscript, and to Neil Kidd for useful discussions.

References

Agusti, N., Aramburu, J. and Gabarra, R. (1999a). Immunological detection of *Helicoverpa armigera* (Lepidoptera: Noctuidae) ingested by heteropteran predators: time-related decay and effect of meal size on detection period. *Annals of the Entomological Society of America* 92: 56–62.

Agusti, N., De Vicente, M.C. and Gabarra, R. (1999b). Development of sequence amplified characterized region (SCAR) markers of *Helicoverpa armigera*: a new polymerase chain reaction-based technique for predator gut analysis. *Molecular Ecology* 8: 1467–1474.

Agusti, N., De Vicente, M.C. and Gabarra, R. (2000). Developing SCAR markers to study predation on *Trialeurodes vaporariorum*. *Insect Molecular Biology* 9: 263–268.

Allen, H.W. (1929). An annotated list of the Tachinidae of Mississippi. *Annals of the Entomological Society of America* 22: 676–690.

Altieri, M.A. and Letourneau, D.K. (1982). Vegetation management and biological control in agroecosystems. *Crop Protection* 1: 405–430.

Andow, D. and Risch, S.J. (1985). Predation in diversified agroecosystems: relations between a coccinellid predator *Coleomegilla maculata* and its food. *Journal of Applied Ecology* 22: 357–372.

Atanassova, P., Brookes, C.P., Loxdale, H.D. and Powell, W. (1998). Electrophoretic study of five aphid parasitoid species of the genus *Aphidius* (Hymenoptera: Braconidae), including evidence for reproductively isolated sympatric populations and a cryptic species. *Bulletin of Entomological Research* 88: 3–13.

Avidov, Z., Balshin, M. and Gerson, U. (1970). Studies on *Aphytis coheni*, a parasite of the California red scale, *Aonidiella aurantii*, in Israel. *Entomophaga* 15: 191–207.

Ayal, Y. (1987). The foraging strategy of *Diaeretiella rapae*. I. The concept of the elementary unit of foraging. *Journal of Animal Ecology* 56: 1057–1068.

Baggen, L.R. and Gurr, G.M. (1998). The influence of food on *Copidosoma koehleri* (Hymenoptera: Encyrtidae) and the use of flowering plants as a habitat management tool to enhance biological control of potato moth, *Phthorimaea operculella* (Lepidoptera: Gelechiidae). *Biological Control* 11: 9–17.

Baggen, L.R., Gurr, G.M. and Meats, A. (1999). Flowers in tri-trophic systems: mechanisms allowing selective exploitation by insect natural enemies for conservation biological control. *Entomologia Experimentalis et Applicata* 91: 155–161.

Baker, H.G. and Baker, I. (1983). A brief historical review of the chemistry of nectar. In *The Biology of Nectaries* (B. Bentley and T. Elias, eds), pp. 126–152. Columbia University Press, New York.

Ballal, C.R. and Singh, S.P. (1999). Host plant-mediated orientational and ovipositional behaviour of three species of chrysopids (Neuroptera: Chrysopidae). *Biological Control* 16: 47–53.

Barbosa, P. (1998). *Conservation Biological Control*. Academic Press, New York. 396 pages.

Bartlett, B.R. (1962). The ingestion of dry sugars by adult entomophagous insects and the use of this feeding habit for measuring the moisture needs of parasites. *Journal of Economic Entomology* 55: 749–753.

Beddington, J.R., Hassell, M.P. and Lawton, J.H. (1976). The components of arthropod predation. II. The predator rate of increase. *Journal of Animal Ecology* 45: 165–185.

Bellows, T.S. (1985). Effects of host and parasitoid age on search behaviour and oviposition (Hymenoptera: Pteromalidae). *Researches on Population Ecology* 27: 65–76.

Ben Saad, A.A. and Bishop, G.W. (1976). Attraction of insects of potato plants through use of artificial honeydews and aphid juice. *Entomophaga* 21: 49–57.

Boggs, C.L. (1992). Resource allocation: exploring connections between foraging and life history. *Functional Ecology* 6: 508–518.

Boggs, C.L. (1997). Reproductive allocation from reserves and income in butterfly species with differing diets. *Ecology* 78: 181–191.

Budenberg, W.J. (1990). Honeydew as a contact kairomone for aphid parasitoids. *Entomologia Experimentalis et Applicata* 55: 139–148.

Budenberg, W.J. and Powell, W. (1992). The role of honeydew as an ovipositional stimulant for two species of syrphids. *Entomologia Experimentalis et Applicata* 64: 57–61.

Bugg, R.L., Ehler, L.E. and Wilson, L.T. (1987). Effect of common knotweed (*Polygonum aviculare*) on abundance and efficiency of insect predators of crop pests. *Hilgardia* 55: 1–52.

Burkett, D.A., Kline, D.L and Carlson, D.A. (1999). Sugar meal composition of five north central Florida mosquito species (Diptera: Culicidae) as determined by gas chromatography. *Journal of Medical Entomology* 36: 462–467.

Butler, G.D. and Ritchie, P.L. (1971). Feed wheat and the abundance and fecundity of *Chrysoperla carnea*. *Journal of Economic Entomology* 64: 933–934.

Calder, W.A. (1984). *Size, Function and Life History*. Harvard University Press, Cambridge. 431 pages.

Carlson, R.E. and Chiang, H.C. (1973). Reduction of an *Ostrinia nubilalis* population by predatory insects attracted by sucrose sprays. *Entomophaga* 18: 205–211.

Carnard, M. (2001). Natural food and feeding habits of lacewings. In *Lacewings in the Crop Environment* (P. McEwen, T.R. New and A.E. Whittington, eds), pp. 116–129. Cambridge University Press, Cambridge.

Carter, M.C. and Dixon, A.F.G. (1984). Honeydew: an arrestant stimulus for coccinellids. *Ecological Entomology* 9: 383–387.

Carthy, J.D. (1971). *An Introduction to the Behaviour of Invertebrates*. Hafner, New York. 380 pages.

Casas, J., Driessen, G., Mandon, N., Wileaard, S., Desouhant, E., van Alphen, J., Lapchin, L., Rivero, A., Christides, J.P. and Bernstein, C. (2003). Energy dynamics in a parasitoid foraging in the wild. *Journal of Animal Ecology* 72: 691–697.

Chapman, R.F. (1998). *The Insects: Structure and Function*. Cambridge University Press, Cambridge. 770 pages.

Chen, Y., Giles, K.L., Payton, M.E. and Greenstone, M.H. (2000). Identifying key cereal aphid predators by molecular gut analysis. *Molecular Ecology* 9: 1887–1898.

Chiverton, P.A. (1984). Pitfall-trap catches of the carabid beetle *Pterostichus melanarius*, in relation to gut contents and prey densities, in insecticide treated and untreated spring barley. *Entomologia Experimentalis et Applicata* 36: 23–30.

Chow, A. and Mackauer, M. (1991). Patterns of host selection by four species of aphidiid (Hymenoptera) parasitoids: influence of host switching. *Ecological Entomology* 16: 403–410.

Clapham, A.R., Tutin, T.G. and Moore, D.M. (1989). *Flora of the British Isles*. Cambridge University Press, Cambridge. 688 pages.

Collins, F.L. and Johnson, S.J. (1985). Reproductive response of caged adult velvetbean caterpillar and soybean looper to the presence of weeds. *Agriculture, Ecosystems and Environment* 14: 139–149.

Corbett, A. (1998). The importance of movement in the response of natural enemies to habitat manipulation. In *Enhancing Biological Control* (C.H. Pickett and R.L. Bugg, eds), pp. 25–48. University of California Press, Berkeley.

Corbett, A. and Plant, R.E. (1993). Role of movement in response of natural enemies to agroecosystem diversification: a theoretical evaluation. *Environmental Entomology* 22: 519–531.

Corbett, A. and Rosenheim, J.A. (1996). Impact of a natural enemy overwintering refuge and its interaction with the surrounding landscape. *Ecological Entomology* 21: 155–164.

Corbett, A., Leigh, T.F. and Wilson, L.T. (1991). Interplanting alfalfa as a source of *Metaseiulus occidentalis* (Acari: Phytoseiidae) for managing spider mites in cotton. *Biological Control* 1: 188–196.

Cortesero, A.M., Stapel, J.O. and Lewis, W.J. (2000). Understanding and manipulating plant attributes to enhance biological control. *Biological Control* 17: 35–49.

Cottrell, T.E. and Yeargan, K.V. (1998). Influence of native weed, *Acalypha ostryaefolia* (Euphorbiaceae) on *Coleomegilla maculata* (Coleoptera: Coccinellidae) population density, predation and cannibalism in sweet corn. *Environmental Entomology* 27: 1375–1385.

Cottrell, T.E. and Yeargan, K.V. (1999). Factors influencing dispersal of larval *Coleomegilla maculata* from the weed *Acalypha ostryaefolia* to sweet corn. *Entomologia Experimentalis et Applicata* 90: 313–322.

Cowgill, S.E., Wratten, S.D. and Sotherton, N.W. (1993). The effect of weeds on the numbers of hoverfly (Diptera: Syrphidae) adults and the distribution and composition on their eggs in winter wheat. *Annals of Applied Biology* 123: 499–514.

Crum, D.A., Weiser, L.A.. and Stamp, N.E. (1998). Effects of prey scarcity and plant material as a dietary supplement on an insect predator. *Oikos* 81: 549–557.

Dafni, A. (1992) *Pollination Ecology: A Practical Approach*, IRL Press, Oxford.

Dawson, N. (1965). A comparative study of the ecology of eight species of fenland Carabidae (Coleoptera). *Journal of Animal Ecology* 34: 299–314.

Dean, G.L. and Satasook, C. (1983). Response of *Chrysoperla carnea* (Stephens) (Neuroptera: Chrysopidae) to some potential attractants. *Bulletin of Entomological Research* 73: 619–624.

Deng, X., Zheng, Z.Q., Zhang, N.X. and Jia, X.F. (1988). Methods of increasing winter-survival of *Metaseiulus occidentalis* (Acari: Phytoseiidae) in northwest China. *Chinese Journal of Biological Control* 4: 97–101.

Dennis, P., Thomas, M.B. and Sotherton, N.W. (1994). Structural features of field boundaries which influence the overwintering densities of beneficial arthropod predators. *Journal of Applied Ecology* 31: 361–370.

Deyrup, M.A. (1988). Pollen-feeding in *Poecilognathus punctipennis* (Diptera: Bombyliidae). *Florida Entomologist* 71: 597–604.

Dixon, A.F.G. (1959). An experimental study of the searching behaviour of the predatory coccinellid beetle *Adalia decempunctata* (L.). *Journal of Animal Ecology* 28: 259–281.

Doutt, R.L. and Nakata, J. (1973). The *Rubus* leafhopper and its egg parasitoid: an endemic biotic system useful in grape-pest management. *Environmental Entomology* 2: 381–386.

Downes, W.L. and Dahlem, G.A. (1987). Keys to the evolution of Diptera: role of Homoptera. *Environmental Entomology* 16: 847–854.

Dransfield, R.D. (1979). Aspects of host-parasitoid interactions of two aphid parasitoids, *Aphidius urticae* (Haliday) and *Aphidius uzbekistanicus* (Luzhetski) (Hymenoptera, Aphidiidae). *Ecological Entomology* 4: 307–316.

Drumtra, D.E.W. and Stephen, F.M. (1999). Incidence of wildflower visitation by hymenopterous parasitoids of southern pine beetle, *Dendroctonus frontalis* Zimmermann. *Journal of Entomological Science* 34: 484–488.

Dudley, R. (2000). *The Biomechanics of Insect Flight*. Princeton University Press, Princeton, NJ. 476 pages.

Edney, E.B. (1971). Some aspects of water balance in tenebrionid beetles and a thysanuran from the Namib Desert of South Africa. *Physiological Zoology* 44: 61–76.

Ellers, J. (1996). Fat and eggs: an alternative method to measure the trade-off between survival and reproduction in insect parasitoids. *Netherlands Journal of Zoology* 46: 227–235.

Ellers, J. and Jervis, M. (2003). Body size and the timing of egg production in parasitoid wasps. *Oikos*. 102: 164–172

Ellers, J., van Alphen, J.J.M. and Sevenster, J.G. (1998). A field study of size–fitness relationships in the parasitoid *Asobara tabida*. *Journal of Animal Ecology* 67: 318–324.

Elliott, N.C., Simmons, G.A. and Sapio, F.J. (1987). Honeydew and wildflowers as food for the parasites *Glypta fumiferanae* (Hymenoptera: Ichneumonidae) and *Apanteles fumiferanae* (Hymenoptera: Braconidae). *Journal of the Kansas Entomological Society* 60 (1): 25–29.

Erdtmann, G. (1969). *Handbook of Palynology*. Munksgaard, Copenhagen. 580 pages.

Evans, E.W. (1993). Indirect interactions among phytophagous insects: aphids, honeydew and natural enemies. In *Individuals, Populations and Patterns in Ecology* (A.D. Watt, S. Leather, N.J. Mills and K.F.A. Walters, eds), pp. 287–298. Intercept Press, Andover.

Evans, E.W. (2000). Egg production in response to combined alternative foods by the predator *Coccinella transversalis*. *Entomologia Experimentalis et Applicata* 94: 141–147.

Evans, E.W. and England, S. (1993). Indirect interactions in biological control of insects: pests and natural enemies in alfalfa. *Ecological Applications* 6: 920–930.

Evans, E.W. and Richards, D.R. (1997). Managing the dispersal of ladybird beetles (Col.: Coccinellidae): use of artificial honeydew to manipulate spatial distributions. *Entomophaga* 42: 93–102.

Evans, E.W and Swallow, J.G. (1993). Numerical responses of natural enemies to artificial honeydew in Utah alfalfa. *Environmental Entomology* 22: 1392–1401.

Fadamiro, H.Y and Heimpel, G.E. (2001). Effects of partial sugar deprivation on lifespan and carbohydrate mobilization in the parasitoid *Macrocentrus grandii* (Hymenoptera: Braconidae). *Annals of the Entomological Society* 94: 909–916.

Faegri, K. and Iversen, J. (1989) (4th edn). *A Textbook of Pollen Analysis* (revised by K. Faegri, P.E. Kaland and K. Krzywinski). John Wiley & Sons, Chichester. 328 pages.

Feinsinger, P. and Swarm, L.A. (1978). How common are ant-repellent nectars? *Biotropica* 10: 238–239.

Fellowes, M.D.E., van Alphen, J.J.M. and Jervis, M.A. (2004). Foraging behaviour. In *Insects as Natural Enemies: A Practical Perspective* (M.A. Jervis, ed.). Kluwer Academic, Dordrecht. In press.

Fraenkel, G.S. and Gunn, D.L. (1961). *The Orientation of Animals: Kineses, Taxes and Compass Reactions*. Dover Publications, New York.

Freeman-Long, R., Corbett, A., Lamb, C., Reberg-Horton, C., Chandler, J. and Stimmann, M. (1988). Beneficial insects move from flowering plants to nearby crops. *California Agriculture* September–October: 23–26.

Garcia-Salazar, C. and Landis, D.A. (1997). Marking *Trichogramma brassicae* Bezdenko (Hymenoptera: Trichogrammatidae) with a fluorescent marker dust and its effect on survival and flight behavior. *Journal of Economic Entomology* 90: 1546–1550.

Gilbert, F.S. (1981). Foraging ecology of hoverflies: morphology of the mouthparts in relation to feeding on nectar and pollen in some common urban species. *Ecological Entomology* 6: 245–262.

Gilbert, F.S. (1986). *Hoverflies, Naturalists' Handbooks No. 5*. Cambridge University Press, Cambridge.

Gilbert, F.S. and Jervis, M.A. (1998). Functional, evolutionary and ecological aspects of feeding-related mouthpart specializations in parasitoid flies. *Biological Journal of the Linnaean Society* 63: 495–535.

Gillaspy, J.E. (1971). Papernest wasps (*Polistes*): observations and study methods. *Annals of the Entomological Society of America* 64: 1357–1361.

Gilstrap, F.E. (1988). Sorghum-corn-Johnsongrass and Banks grass mite: a model for biological control in field crops. In *The Entomology of Indigenous and Naturalized Systems in Agriculture* (M.K. Harris and C.E. Rogers, eds), pp. 141–159. Westview Press, Boulder.

Glen, D.M. (1975). Searching behaviour and prey-density requirements of *Blepharidopterus angulatus* (Fall.) (Heteroptera: Miridae) as a predator of the lime aphid, *Eucallipterus tiliae* (L.), and leafhopper, *Alnetoidea alneti* (Dahlbom). *Journal of Animal Ecology* 44: 115–134.

Greenstone, M.H. and Hunt, J.H. (1993). Determination of prey antigen half-life in *Polistes metricus* using a monoclonal antibody-based immunodot assay. *Entomologia Experimentalis et Applicata* 68: 1–7.

Grinfel'd, E.K. (1975). Anthophily in beetles (Coleoptera) and a critical evaluation of the cantharophilous hypothesis. *Entomological Review* 54: 18–22.

Gurr, G.M., van Emden, H.F. and Wratten, S.D. (1998). Habitat manipulation and natural enemy efficiency: implications for the control of pests. In *Conservation Biological Control* (P. Barbosa, ed.), pp. 155–183. Academic Press, San Diego.

Gurr, G.M., Wratten, S.D., Tylianakis, J., Kean, J. and Keller, M. (2004). Providing plant foods for insect natural enemies in farming systems: balancing practicalities and theory. In *Plant-provided Food and Plant-carnivore Mutualism* (F. Wäckers, P. van Rijn and J. Bruin, eds). Cambridge University Press, Cambridge. In press.

Györfi, J. (1945). Beobachtungen uber die ernahrung der schlupfwespenimagos. *Erdészeti Kisérletek* 45: 100–112.

Haber, W.A, Frankie, G.W., Baker, H.G., Baker, I. and Koptur, S. (1981). Ants like flower nectar. *Biotropica* 13: 211–214.

Hagen, K.S., Sawall, E.F. Jr and Tassan, R.L. (1971). In *Proceedings Tall Timber Conference Ecological Animal Control Habitat Management*, pp. 59–81. Tall Timbers Research Station, Tallahassee.

Hagen, K.S., Bombosch, S. and McMurty, J.A. (1976). The biology and impact of predators. In *Theory and Practice of Biological Control* (C. Huffaker and P.S. Messenger, eds), pp. 93–142. Academic Press, New York.

Hagler, J.R. (1998). Variation in the efficacy of several predator gut content immunoassays. *Biological Control* 12: 25–32.

Hagler, J.R. and Jackson, C.G. (1998). An immunomarking technique for labeling minute parasitoids. *Environmental Entomology* 27: 1010–1016.

Hagler, J.R. and Naranjo, S.E. (1997). Measuring the sensitivity of an indirect predator gut content ELISA: detectability of prey remains in relation to predator species, temperature, time, and meal size. *Biological Control* 9: 112–119.

Hagler, J.R., Buchmann, S.L. and Hagler, D.A. (1995). A simple method to quantify dot blots for predator gut analyses. *Journal of Entomological Science* 30: 95–98.

Hagley, E.A.C. and Barber, D.R. (1992). Effect of food sources on the longevity and fecundity of *Pholetesor ornigis* (Weed) (Hymenoptera: Braconidae). *Canadian Entomologist* 124: 341–346.

Hagley, E.A.C. and Simpson, C.M. (1981). Effect of food sprays on numbers of predators in an apple orchard. *Canadian Entomologist* 113: 75–77.

Harborne, J.D. (1988). *Introduction to Ecological Biochemistry*. Academic Press, London. 356 pages.

Haslett, J.R. (1983). A photographic account of pollen digestion by adult hoverflies. *Physiological Entomology* 8: 167–171.

Haslett, J.R. (1989). Interpreting patterns of resource utilization: randomness and selectivity in pollen-feeding by adult hoverflies. *Oecologia* 78: 433–442.

Haslett, J.R. and Entwistle, P.F. (1980). Further notes on *Eriozona syrphoides* (Fall.) (Dipt., Syrphidae) in Hafren Forest, mid-Wales. *Entomologist's Monthly Magazine* 116: 36.

Hassall, M., Hawthorne, A., Maudsley, M., White, P. and Cardwell, C. (1992). Effects of headland management on invertebrate communities in cereal fields. *Agriculture, Ecosystems and Environment* 40: 155–178.

Hassan, E. (1967). Untersuchungen uber die bedeutung der kraut- und strauchschicht al nahrungsquelle für imagines entophager Hymenopteren. *Zeitschrift für Angewandte Entomologie* 60: 238–265.

Hassell, M.P. (1978). *The Dynamics of Arthropod Predator-Prey Systems*. Princeton University Press, Princeton. 237 pages.

Hassell, M.P. (1982). What is searching efficiency? *Annals of Applied Biology* 101: 170–175.

Hassell, M.P., and Comins, H.N. (1978). Sigmoid functional responses and population stability. *Theoretical Population Biology* 14: 62–67.

Hassell, M.P., Lawton, J.H. and Beddington, J.R. (1977). Sigmoid functional responses by invertebrate predators and parasitoids. *Journal of Animal Ecology* 46: 249–262.

Hausmann, C., Wäckers, F.L. and Dorn, S. (2003) Establishing preferences for nectar constituents in the parasitoid *Cotesia glomerata*. *Entomologia Experimentalis et Applicata* (in press.)

Hazzard, R.V., Ferro, D.N, Van Driesche, R.G. and Tuttle, A.F. (1991). Mortality of eggs of Colorado potato beetle (Coleoptera: Chrysomelidae) from predation by *Coleomegilla maculata* (Coleoptera: Coccinellidae). *Environmental Entomology* 20: 841–848.

Heimpel, G.E. and Collier, T.R. (1996). The evolution of host-feeding behaviour in insect parasitoids. *Biological Reviews* 71: 373–400.

Heimpel, G.E. and Jervis, M.A. (2004). An evaluation of the hypothesis that floral nectar improves biological control by parasitoids. In *Plant-provided Food and Plant–carnivore Mutualism* (F. Wäckers, P. van Rijn and J. Bruin, eds). Cambridge University Press, Cambridge. In press.

Heimpel, G.E. and Rosenheim, J.A. (1995). Dynamic host feeding by the parasitoid *Aphytis melinus*: the balance between current and future reproduction. *Journal of Animal Ecology* 64: 153–167.

Heimpel, G.E. and Rosenheim, J.A. (1998). Egg limitation in insect parasitoids: a review of the evidence and a case study. *Biological Control* 11: 160–168.

Heimpel, G.E., Rosenheim, J.A. and Adams, J.M. (1994). Behavioral ecology of host feeding in *Aphytis* parasitoids. Proceedings of the 7th European Workshop on Insect Parasitoids. *Norwegian Journal of Agricultural Sciences Supplement* 16: 101–115.

Heimpel, G.E., Rosenheim, J.A. and Mangel, M. (1996). Egg limitation, host quality, and dynamic behavior in a parasitoid wasp. *Ecology* 77: 2410–2420.

Heimpel, G.E., Rosenheim, J.A. and Mangel, M. (1997a). Predation on adult *Aphytis* parasitoids in the field. *Oecologia* 110: 346–352.

Heimpel, G.E., Rosenheim, J.A. and Kattari, D. (1997b). Adult feeding and lifetime reproductive success in the parasitoid *Aphytis melinus*. *Entomologia Experimentalis et Applicata* 83: 305–315.

Heimpel, G.E., Neuhauser, C. and Hoogendoorn, M. (2003). Effects of parasitoid fecundity and host resistance on indirect interactions in host-parasitoid population dynamics. *Ecology Letters* 6: 556–566.

Heimpel, G.E., Lee, J.C., Wu, Z., Weiser, L., Wäckers, F. and Jervis, M.A. (2004). Gut sugar analysis in field-caught parasitoids: adapting methods originally developed for biting flies. *International Journal of Pest Management* (in press.)

Heirbaut, M. and van Damme, P. (1992). The use of artificial nests to establish colonies of the black cocoa ant (*Dolichoderus thoracicus* Smith) used for biological control of *Helopeltis theobromae* Mill in Malaysia. *Mededlingen van de Faculteit Landbouwwetenschappen* 57: 533–542.

Hemptinne, J.L. and Desprets, A. (1986). Pollen as a spring food for *Adalia bipunctata*. In *Ecology of Aphidophaga* (I. Hodek, ed.), pp. 29–35. Academia, Dordrecht.

Hengeveld, R. (1980). Polyphagy, oligophagy and food specialization in ground beetles (Coleoptera, Carabidae). *Netherlands Journal of Zoology* 30: 564–584.

Henneberry, T.J., Forlow Jech, L. and de la Torre, T. (2002). Effects of cotton plant water stress on *Bemisia tabaci* strain B (Homoptera: Aleyrodidae) honeydew production. *Southwestern Entomologist* 27: 117–133.

Hickey, M. and King, C.J. (1981). *100 Families of Flowering Plants*. Cambridge University Press, Cambridge. 619 pages.

Hickman, J.M. and Wratten, S.D. (1996). Use of *Phacelia tanacetifolia* (Hydrophyllaceae) as a pollen source to enhance hoverfly (Diptera: Syrphidae) populations in cereal fields. *Journal of Economic Entomology* 89: 832–840.

Hickman, J.M., Wratten, S.D., Jepson, P.C. and Frampton, C.M. (2001). Effect of hunger on yellow water trap catches of hoverfly (Diptera: Syrphidae) adults. *Agricultural and Forest Entomology* 3: 35–40.

Hocking, H. (1967). The influence of food on longevity and oviposition in *Rhyssa persuasoria* (L.) (Hymenoptera: Ichneumonidae). *Journal of the Australian Entomological Society* 6: 83–88.

Hodek, I. (1971). *Biology of Coccinellidae*. W. Junk, The Hague/Czechoslovakian Academy of Sciences.

Holloway, B.A. (1976). Pollen-feeding in hover-flies (Diptera: Syrphidae). *New Zealand Journal of Zoology* 3: 339–350.

Holt, R.D. (1977). Predation, apparent competition, and the structure of prey communities. *Theoretical Population Ecology* 12: 197–229.

Hoogendoorn, M. and Heimpel, G.E. (2001). PCR–based gut content analysis of insect predators: using ribosomal ITS–1 fragments from prey to estimate predation frequency. *Molecular Ecology* 10: 2059–2067.

Hoogendoorn, M. and Heimpel, G.E. (2002). Indirect interactions between an introduced and a native ladybird beetle species mediated by a shared parasitoid. *Biological Control* 25: 224–230.

Hoogendoorn, M. and Heimpel, G.E. (2003). PCR–based gut content analysis of insect predators: using ribosomal ITS–1 fragments from prey to estimate predation frequency. In *Proceedings of the 1st International Symposium on Biological Control of Arthropods* (R.G. Van Driesche, ed.), pp. 91–97. United States Dept of Agriculture, Forest Service, Morgantown, WV, FHTET–(2003)–05.

Hoogendoorn, M. and Heimpel, G.E. (2004). Competitive interactions between an exotic and a native ladybeetle: a field cage study. *Entomologia Experimentalis et Applicata.* In press.

Hopper, K.R. (1991). Ecological applications of elemental labelling: analysis of dispersal, density, mortality and feeding. *Southwestern Entomologist Supplement* 14: 71–83.

Hopper, K.R. and Woolson, E.A. (1991). Labelling a parasitic wasp, *Microplitis croceipes* (Hymenoptera: Braconidae) with trace elements for mark-recapture studies. *Annals of the Entomological Society of America* 84: 255–262.

Hunt, J.H., Brow, P.A., Sago, K.M. and Kerker, J.A. (1991). Vespid wasps eat pollen. *Journal of Kansas Entomological Society* 64: 127–130.

Hunter, F.F. and Ossowski, A.M (1999). Honeydew sugars in wild-caught female horse flies (Diptera: Tabanidae). *Journal of Medical Entomology* 36: 896–899.

Idoine, K. and Ferro, D.N. (1988). Aphid honeydew as a carbohydrate source for *Edovum puttleri* (Hymenoptera: Eulophidae). *Environmental Entomology* 17: 941–944.

Idris, A.B. and Grafius, E. (1995). Wildflowers as nectar sources for *Diadegma insulare* (Hymenoptera: Ichneumonidae), a parasitoid of diamondback moth (Lepidoptera: Yponomeutidae). *Environmental Entomology* 24: 1726–1735.

Idris, A.B. and Grafius, E. (1997). Nectar-collecting behavior of *Diadegma insulare* (Hymenoptera: Ichneumonidae), a parasitoid of diamondback moth (Lepidoptera: Plutellidae). *Environmental Entomology* 26: 114–120.

Jackson, C.G. (1991). Elemental markers for entomophagous insects. *Southwestern Entomology Supplement* 14: 65–70.

Jackson, C.G., Cohen, A.C. and Verdugo, C.L. (1988). Labeling *Anaphes ovijentatus* (Hymenoptera: Mymaridae) an egg parasite of *Lygus* (Hemiptera: Miridae), with rubidium. *Annals of the Entomological Society of America* 81: 919–922.

Jacob, H.S. and Evans, E.W. (1998). Effects of sugar spray and aphid honeydew on field populations of the parasitoid *Bathyplectes curculionis* (Thomson) (Hymenoptera: Ichneumonidae). *Environmental Entomology* 27: 1563–1568.

Jang, E.B., Messing, R.H., Klungness, L.M. and Carvalho, L.A. (2000). Flight tunnel responses of *Diachasmimorpha longicaudata* (Ashmead) (Hymenoptera: Braconidae) to olfactory and visual stimuli. *Journal of Insect Behavior* 13: 525–538.

Jervis, M.A. (1990). Predation of *Lissonota coracinus* (Gmelin) (Hymenoptera: Ichneumonidae) by *Dolichonabis limbatus* (Dahlbom) (Hemiptera: Nabidae). *Entomologist's Gazette* 41: 231–233.

Jervis, M.A. (1998). Functional and evolutionary aspects of mouthpart structure in parasitoid wasps. *Biological Journal of the Linnaean Society* 63: 461–493.

Jervis, M. A. and Kidd, N.A.C. (1986). Host-feeding strategies in hymenopteran parasitoids. *Biological Reviews* 61: 395–434.

Jervis, M.A., Kidd, N.A.C. and Walton, M. (1992). A review of methods for determining dietary range in adult parasitoids. *Entomophaga* 37: 565–574.

Jervis, M.A., Kidd, N.A.C., Fitton, M.G., Huddleston, T. and Dawah, H.A. (1993). Flower-visiting by hymenopteran parasitoids. *Journal of Natural History* 27: 67–105.

Jervis, M.A., Kidd, N.A.C. and Heimpel, G.E. (1996). Parasitoid adult feeding and biological control – a review. *Biocontrol News and Information* 17: 1N–22N.

Jervis, M.A, Heimpel, G.E., Ferns, P.N., Harvey, J.A. and Kidd, N.A.C. (2001). Life-history strategies in parasitoid wasps: a comparative analysis of 'ovigeny.' *Journal of Animal Ecology* 70: 442–458.

Jervis, M.A., Ferns, P. and Heimpel, G.E. (2003). Body size and the timing of egg production in parasitoid wasps: a comparative analysis. *Functional Ecology* 17: 375–383.

Jervis, M.A., Copland, M.J.W. and Harvey, J.A. (2004). The life cycle. In *Insects as Natural Enemies: A Practical Perspective* (M.A. Jervis, ed.). Kluwer Academic, Dordrecht. In press.

Jiang, J.Q., Zhang, X.D. and Gu, D.X. (1991). A bionomical study of *Scirpophaga praelata* (Lep. Pyralidae), a 'bridging host' of *Tetrastichus schoenobii* (Hym.: Eulophidae). *Chinese Journal of Biological Control* 7: 13–15.

Joshi, R.K. and Sharma, S.K. (1989). Augmentation and conservation of *Epiricania melanoleuca* Fletcher, for the population management of sugarcane leafhopper, *Pyrilla perpusilla* Walker, under arid conditions of Rajasthan. *Indian Sugar* 39: 625–628.

Kevan, P.G. (1973). Parasitoid wasps as flower visitors in the Canadian high arctic. *Anzeiger Schädlungskunde* 46: 3–7.

Khan, Z.R., Ampong-Nyarko, K., Chiliswa, P., Hassanali, A., Kimani, S., Lwande, W., Overholt, W.A., Pickett, J.A., Smart, L.E., Wadhams, L.J. and Woodcock, C.M. (1997). Intercropping increases parasitism of pests. *Nature* 388: 631–632.

Kidd, N.A.C. and Jervis, M.A. (1989). The effects of host-feeding and oviposition by parasitoids in relation to host stage. *Researches on Population Ecology* 33: 13–28.

Kidd, N.A.C. and Jervis, M.A. (2004). Population dynamics. In *Insects as Natural Enemies: A Practical Perspective* (M.A. Jervis, ed.). Kluwer Academic, Dordrecht. In press.

Kielty, J.P., Allen Williams L.J., Underwood, N. and Eastwood, E.A. (1996). Behavioural responses of three species of ground beetle (Coleoptera: Carabidae) to olfactory cues associated with prey and habitat. *Journal of Insect Behavior* 9: 237–250.

Kiman, Z.B. and Yeargan, K.V. (1985). Development and reproduction of the predator *Orius insidiosus* (Hemiptera: Anthocoridae) reared on diets of selected plant material and arthropod prey. *Annals of the Entomological Society of America* 78: 464–467.

Klingauf, F.A. (1987). Feeding, adaptation and excretion. In *Aphids: Their Biology, Natural Enemies and Control*, vol. 2A (A.K. Minks and P Harrewin, eds), pp. 225–253. Elsevier Press, Amsterdam. 450 pages.

Landis, D.A., Wratten, S.D. and Gurr, G.M. (2000). Habitat management to conserve natural enemies of arthropod pests in agriculture. *Annual Review of Entomology* 45: 175–201.

Lawson, F.R., Rabb, R.L., Guthrie, F.E. and Bowery, T.G. (1961). Studies of an integrated control system for hornworms on tobacco. *Journal of Economic Entomology* 54: 93–97.

Lawton, J.H. (1986). The effect of parasitoids on phytophagous insect communities. In *Insect Parasitoids* (J. Waage and D. Greathead, eds), pp. 265–287. Academic Press, London.

Lee, J.C. and Heimpel, G.E. (2003). Nectar availability and parasitoid sugar feeding. In *Proceedings of the 1st International Symposium on Biological Control of Arthropods* (R.G. Van Driesche, ed.), pp. 220–225. United States Dept of Agriculture, Forest Service, Morgantown, WV, FTET–(2003)–05.

Lee, J.C. and Landis, D.A. (2002). Non-crop habitat management for carabid beetles. In *Carabids and Agriculture* (J. Holland, ed.), pp. 279–303. Intercept, Andover.

Lee, J.C., Menalled, F.D. and Landis, D.A. (2001). Refuge habitats modify insecticide disturbance on carabid beetle communities. *Journal of Applied Ecology* 38: 472–483.

Leereveld, H. (1982). Anthecological relations between reputedly anemophilous flowers and syrphid flies III. Worldwide survey of crop and intestine contents of certain anthophilous syrphid flies *Melanostoma* and *Platycheirus* and some related taxa, ecologic aspect. *Tijdschrift voor Entomologie* 125: 25–35.

Leius, K. (1960). Attractiveness of different foods and flowers to the adults of some hymenopterous parasites. *Canadian Entomologist* 92: 369–376.

Leius, K. (1961). Influence of food on fecundity and longevity of adults of *Itoplectis conquisitor* (Say) (Hymenoptera: Ichneumonidae). *Canadian Entomologist* 93: 771–780.

Leius, K. (1963). Effects of pollens on fecundity and longevity of adult *Scambus buolianae* (Htg.) (Hymenoptera: Ichneumonidae). *Canadian Entomologist* 95: 202–207.

Le Ru, B. and Makaya-Makosso, J.P. (2001). Prey habitat location by the cassava mealybug predator *Exochomus flaviventris*: olfactory responses to odor of plant, mealybug, plant-mealybug complex and plant-mealybug-natural enemy complex. *Journal of Insect Behavior* 14: 557–572.

Lewis, D.J. and Domoney, C.R. (1966). Sugar meals in Phlebotominae and Simuliidae (Diptera). *Proceedings of the Royal Entomological Society of London*, A. 41: 175–179.

Lewis, W.J. and Takasu, K. (1990). Use of learned odours by a parasitic wasp in accordance with host and food needs. *Nature* 348: 635–636.

Lewis, W.H., Vinay, P. and Zenger, V.E. (1983). *Airborne and Allergenic Pollen of North America*. Johns Hopkins University Press, Baltimore. 254 pages.

Lewis, W.J., Vet, L.E.M., Tumlinson, J.H., van Lenteren, J.C. and Papaj, D.R. (1990). Variations in parasitoid foraging behavior: essential element of a sound biological control theory. *Environmental Entomology* 19: 1183–1193.

Lewis, W.J., Stapel, J.O., Cortesero, A.M. and Takasu, K. (1998). Understanding how parasitoids balance food and host needs: importance to biological control. *Biological Control* 11: 175–183.

Liber, H. and Niccoli, A. (1988). Observations on the effectiveness of an attractant food spray in increasing chrysopid predation on *Prays oleae* (Bern.) eggs. *Redia* 71: 467–482.

Limburg, D.D. and Rosenheim, J.A. (2001). Extrafloral nectar consumption and its influence on survival and development of an omnivorous predator, larval *Chrysoperla plorabunda* (Neuroptera: Chrysopidae). *Environmental Entomology* 30: 595–604.

Loveridge, J.P. (1968). The control of water loss in *Locusta migratoria migratorioides* R., and F. I. Cuticular water loss. *Journal of Experimental Biology* 49: 1–13.

Luff, M.L. (1965). The morphology and microclimate of *Dactylis glomerata* tussocks. *Journal of Ecology* 53: 771–787.

Luff, M.L. (1987). Biology of phytophagous ground beetles in agriculture. *Agricultural Zoology Reviews* 2: 237–278.

Lys, J.A. and Nentwig, W. (1994). Improvement of the overwintering sites for Carabidae, Staphylinidae and Araneae by strip-management in a cereal field. *Pedobiologia* 38: 238–242.

MacLeod, A. (1999). Attraction and retention of *Episyrphus balteatus* DeGeer (Dipera: Syrphidae) at an arable field margin with rich and poor flora resource. *Agriculture, Ecosystems and Environment* 73: 237–244.

Majerus, M.E.N. (1994). *Ladybirds*. Harper Collins, London. 367 pages.

Maurizio, A. (1975). How bees make honey. In *Honey: A Comprehensive Survey* (E. Crane, ed.), pp. 77–105. Heinemann, London.

McEwen, P.K. (1998). Overwintering chambers for green lacewings (*Chrysoperla carnea*): effect of chemical attractant, material and size. *Journal of Neuropterology* 1: 17–21.

McEwen, P.K. and Sengonca, Ç. (2001). Artificial overwintering chambers for *Chrysoperla carnea* and their application in pest control. In *Lacewings in the Crop Environment* (P. McEwen, T.R. New and A.E. Whittington, eds), pp. 487–491. Cambridge University Press, Cambridge.

McEwen, P., Jervis, M.A. and Kidd, N.A.C. (1993a). The effect on olive moth (*Prays oleae*) population levels, of applying artificial food to olive trees. *Proceedings A.N.P.P. 3rd International Conference on Pests in Agriculture*, Montpellier, December (1993): 361–368.

McEwen, P.K., Clow, S., Jervis, M.A. and Kidd, N.A.C. (1993b). Alteration in searching behaviour of adult female green lacewings (*Chrysoperla carnea*) (Neur: Chrysopidae) following contact with honeydew of the black scale (*Saissetia oleae*) (Hom: Coccidae) and solutions containing L–tryptophan. *Entomophaga* 38: 347–354.

McEwen, P., Jervis, M.A. and Kidd, N.A.C. (1993c). Influence of artificial honeydew on larval development and survival in *Chrysoperla carnea* (Neur.: Chrysopidae). *Entomophaga* 38: 241–244.

McEwen, P.K., Jervis, M.A. and Kidd, N.A.C. (1996). The influence of an artificial food supplement on larval and adult performance in the green lacewing *Chrysoperla carnea* (Stephens). *International Journal of Pest Management* 42: 25–27.

McMurtry, J.A. and Scriven, G.T. (1964). Studies on the feeding, reproduction and development of *Amblyseius hibisci* (Acarina: Phytoseiidae) in various food substances. *Annals of the Entomological Society of America* 57: 649–655.

Mensah, R.K. (1996). Suppression of *Helicoverpa* spp. (Lepidoptera: Noctuidae) oviposition by use of the natural enemy food supplement Envirofeast®. *Australian Journal of Entomology* 35: 323–329.

Mensah, R.K. (1997). Local density responses of predatory insects of *Helicoverpa* spp. to a newly developed food supplement 'Envirofeast' in commercial cotton in Australia. *International Journal of Pest Management* 43: 221–225.

Mohyuddinn, A.I. (1991). Utilization of natural enemies for the control of insect pests of sugar-cane. *Insect Science and its Application* 12: 19–26.

Moore, P.D., Webb, J.A. and Collinson, M.E. (1991). *Pollen Analysis*. Blackwell, Oxford. 216 pages.

Morris, R.J., Müller, C.B. and Godfray, H.C.J. (2001). Field experiments testing for apparent competition between primary parasitoids mediated by secondary parasitoids. *Journal of Animal Ecology* 70: 301–309.

Müller, C.B. and Godfray, H.C.J. (1997). Apparent competition between two aphid species. *Journal of Animal Ecology* 66: 57–64.

Murdoch, W.W. (1969). Switching in generalist predators: experiments on predator specificity and stability of prey populations. *Ecological Monographs* 39: 335–354.

Murdoch, W.W., Chesson, J. and Chesson, P.L. (1985). Biological control theory and practice. *American Naturalist* 125: 344–366.

Murphy, B.C., Rosenheim, J.A. and Granett, J. (1996). Habitat diversification for improving biological control: abundance of *Anagrus epos* (Hymenoptera: Mymaridae) in grape vineyards. *Environmental Entomology* 25: 495–504.

Murphy, B.C., Rosenheim, J.A., Dowell, R.V. and Granett, J. (1998). Habitat diversification tactic for improving biological control: parasitism of the western grape leafhopper. *Entomologia Experimentalis et Applicata* 87: 225–235.

Nentwig, W. (1988). Augmentation of beneficial arthropods by strip-management I: Succession of predaceous arthropods and long-term change in the ratio of phytophagous and predaceous arthropods in a meadow. *Oecologia* 76: 597–606.

Nichols, P.R. and Neel, W.W. (1977). The use of wheat as a supplemental food for *Coleomegilla maculata* (DeGeer) (Coleoptera: Coccinellidae) in the field. *Southwest Entomologist* 2: 102–105.

Olson, D.M., Fadamiro, H., Lundgren, J.G. and Heimpel, G.E. (2000). Effects of sugar feeding on carbohydrate and lipid metabolism in a parasitoid wasp. *Physiological Entomology* 25: 17–26.

Osakabe, M. (1988). Relationships between food substances and developmental success in *Amblyseius sojaensis* Ehara (Acarina: Phytoseiidae). *Applied Entomology and Zoology* 23: 45–51.

Patt, J.M., Hamilton, G.C. and Lashomb, J.H. (1997). Foraging success of parasitoid wasps on flowers: interplay of floral architecture and searching behavior. *Entomologia Experimentalis et Applicata* 83: 21–30.

Payne, J.A. and Wood, B.W. (1984). Rubidium as a marking agent for the hickory shuckworm (Lepidoptera: Tortricidae). *Environmental Entomology* 13: 1519–1521.

Percival, M.S. (1961). Types of nectar in angiosperms. *New Phytology* 60: 235–281.

Peters, R.H. (1983). *The Ecological Implications of Body Size*. Cambridge University Press, New York. 329 pages.

Pickett, C.H. and Bugg, R.L. (1998). *Enhancing Biological Control*. University of California Press, Berkeley. 422 pages.

Pickett, C.H., Wilson, L.T. and Flaherty, D.L. (1990). The role of refuges in crop protection, with reference to plantings of French prune trees in a grape agroecosystem. In *Monitoring and Integrated Management of Arthropod Pests of Small Fruit* Crops (N.J. Bostanian, L.T. Wilson and T.J. Dennehy, eds), pp. 151–165. Intercept, Andover.

Powell, W. (1986). Enhancing parasitoid activity in crops. In *Insect Parasitoids* (J. Waage and D. Greathead, eds), pp. 319–340. Academic Press, London.

Powell, W. and Wright, A.F. (1988). The abilities of the aphid parasitoid *Aphidius ervi* Haliday and *A. rhopalosiphi* De Stefani Perez (Hymenoptera: Braconidae) to transfer between different known host species and the implication for the use of alternative hosts in pest control strategies. *Bulletin of Entomological Research* 78: 682–693.

Ragsdale, D.W., Larson, A.D. and Newsom, L.D. (1981). Quantitative assessment of the predators of *Nezara viridula* eggs and nymphs within a soybean agroecosystem using an ELISA. *Environmental Entomology* 10: 402–405.

Ratcliffe, S., Robertson, H.M., Jones, C.J., Bollero, G.A. and Weinzierl, R.A. (2002). Assessment of parasitism of house fly and stable fly (Diptera) pupae by pteromalid (Hymenoptera: Pteromalidae) parasitoids using a polymerase chain reaction assay. *Journal of Medical Entomology* 39 (1): 52–60.

Raymond, B., Darby, A.C. and Douglas, A.E. (2000). The olfactory responses of coccinellids to aphids on plants. *Entomologia Experimentalis et Applicata* 95: 113–117.

Reitsma, G. (1966). Pollen morphology of some European Roasaceae. *Acta Botanica Neerlandensis* 15: 290–307.

Rivero, A. and West, S.A. (2002). The physiological costs of being small in a parasitic wasp. *Evolutionary Ecology Research* 4: 407–420.

Roitberg, B.D., Boivin, G. and Vet, L.E.M. (2001). Fitness, parasitoids, and biological control: an opinion. *Canadian Entomologist* 133: 429–438.

Romeis, J. and Wäckers, F. (2000). Feeding responses by female *Pieris brassicae* butterflies to carbohydrates and amino acids. *Physiological Entomology* 25: 247–253.

Rosenheim, J.A. (1998). Higher order predators and the regulation of insect herbivore populations. *Annual Review of Entomology* 43: 421–448.

Sabelis, M.W. (1992). Predatory arthropods. In *Natural Enemies: The Population Biology of Predators, Parasites and Diseases* (M. Crawley, ed.), pp. 225–264. Blackwell Science Publishers, London.

Sadeghi, H. and Gilbert, F. (2000a). Oviposition preferences of aphidophagous hoverflies. *Ecological Entomology* 25: 91–100.

Sadeghi, H. and Gilbert, F. (2000b). Aphid suitability and its relationship to oviposition preference in predatory hoverflies. *Journal of Animal Ecology* 69: 771–784.

Sahragard, A., Jervis, M.A. and Kidd, N.A.C. (1991). Influence of host availability on rates of oviposition and host-feeding and on longevity in *Dicondylus indianus* Olmi (Hym., Dryinidae), a parasitoid of the rice brown planthopper, *Nilaparvata lugens* Stål (Hem., Delphacidae). *Journal of Applied Entomology* 112: 153–162.

Sawyer, R. (1981). *Pollen Identification for Beekeepers*. University College, Cardiff Press, Cardiff. 112 pages.

Sawyer, R. (1988). *Honey Identification*. Cardiff Academic Press, Cardiff. 115 pages.

Schmidt-Nielsen, K. (1984). *Scaling: Why is Animal Size so Important?* Cambridge University Press, New York. 241 pages.

Schonbeck, H. (1988). Biological control of aphids on wild cherry. *Allgemeine Forstzeitxchrift* 34: 944.

Sengonca, Ç. and Frings, B. (1989). Enhancement of the green lacewing *Chrysoperla carnea* (Stephens) by providing artificial facilities for hibernation. *Turkiye Entomoloji Dergisi* 13: 245–250.

Shahjahan, M. (1974). *Erigeron* flowers as a food and attractive odour source for *Peristenus pseudopallipes*, a braconid parasitoid of the tarnished plant bug. *Environmental Entomology* 3: 69–72.

Shahjahan, M. and Streams, F.A. (1973). Plant effects on host-finding by *Leiophron pseudopallipes* (Hymenoptera: Braconidae), a parasitoid of the tarnished plant bug. *Environmental Entomology* 2: 921–925.

Shea, K., Possingham, H.P., Murdoch, W.W. and Roush, R. (2002). Active adaptive management in insect pest and weed control: intervention with a plan for learning. *Ecological Applications* 12: 927–936.

Sheehan, W. (1986). Response by specialist and generalist natural enemies to agroecosystems diversification: a selective review. *Environmental Entomology* 15: 456–461.

Sheldon, J.K. and MacLeod, E.G. (1971). Studies on the biology of the Chrysopidae. I. the feeding behaviour of the adult of *Chrysopa carnea* (Neuroptera). *Psyche* 78: 107–121.

Shepard, M., Rapusas, H.R. and Estano, D.B. (1989). Using rice straw bundles to conserve beneficial arthropod communities in ricefields. *International Rice Research News* 14: 30–31.

Sherratt, T.N. and Harvey, I.F. (1993). Frequency-dependent food selection by arthropods: a review. *Biological Journal of the Linnaean Society* 48: 167–186.

Sikorowski, P.P., Powell, J.E. and Lawrence, A.M. (1992). Effects of bacterial contamination on development of *Microplitis croceipes* (Hym, Braconidae). *Entomophaga* 37: 475–481.

Sirot, E. and Bernstein, C. (1996). Time sharing between host searching and food searching in parasitoids: state-dependent optimal strategies. *Behavioral Ecology* 7: 189–194.

Smith, B.C. (1961). Results of rearing some coccinellid (Coleoptera: Coccinellidae) larvae on various pollens. *Proceedings of the Entomological Society of Ontario* 91: 270–271.

Smith, B.C. (1965). Growth and development of coccinellid larvae on dry foods (Coleoptera, Coccinellidae). *Canadian Entomologist* 97: 760–768.

Sotherton, N.W. (1984). The distribution and abundance of predatory arthropods overwintering on farmland. *Annals of Applied Biology* 105: 423–429.

Sotherton, N.W. (1985). The distribution and abundance of predatory Coleoptera overwintering in field boundaries. *Annals of Applied Biology* 106: 17–21.

Southwood, T.R.E. (1962). Migration of terrestrial arthropods in relation to habitat. *Biological Reviews* 37: 171–214.

Southwood, T.R.E. (1977). Habitat, the templet for ecological strategies. *Journal of Animal Ecology* 46: 337–365.

Southwood, T.R.E. (1978). The components of diversity. In *Diversity of Insect Faunas* (L.A. Mound and N. Waloff, eds), pp. 19–40. Blackwell Scientific Publications, Oxford.

Stelleman, P. and Meeuse, A.D.J. (1976). Anthecological relations between reputedly anemophilous flowers and syrphid flies. I. The possible role of syrphid flies as pollinators of *Plantago*. *Tijdschrift voor Entomologie* 119: 15–31.

Stephen, F.M. and Browne, L.E. (2000). Application of Eliminade™ parasitoid food to boles and crowns of pines (Pinaceae) infested with *Dendroctonus frontalis* (Coleoptera: Scolytidae). *Canadian Entomologist* 132: 983–985.

Stephens, M.J., France, C.M., Wratten, S.D. and Frampton, C. (1998). Enhancing biological control of leafrollers (Lepidoptera: Tortricidae) by sowing buckwheat (*Fagopyrum esculentum*) in an orchard. *Biocontrol Science and Technology* 8: 547–558.

Storeck, A., Poppy, G.M, van Emden, H.F. and Powell, W. (2000). The role of plant chemical cues in determining host preference in the generalist aphid parasitoid *Aphidius colemani*. *Entomologia Experimentalis et Applicata* 97: 41–46.

Sunderland, K.D. and Vickerman, G.P. (1980). Aphid feeding by some polyphagous predators in relation to aphid density in cereal fields. *Journal of Applied Ecology* 17: 389–396.

Sunderland, K.D., Crook, N.E., Stacey, D.L. and Fuller, B.T. (1987). A study of feeding by polyphagous predators on cereal aphids using ELISA and gut dissection. *Journal of Applied Ecology* 24: 907–933.

Sunderland, K.D., Symondson, W.O.C. and Powell, W. (2004). Populations and communities. In *Insects as Natural Enemies: A Practical Perspective* (M.A. Jervis, ed.). Kluwer Academic, Dordrecht.

Syme, P.D. (1975). The effect of flowers on the longevity and fecundity of two native parasites of the European pine shoot moth in Ontario. *Environmental Entomology* 4: 337–346.

Symondson, W.O.C., Erickson, M.L. and Liddell, J.E. (1997). Species-specific detection of predation by Coleoptera on the milacid slug *Tandonia budapestensis* (Mollusca: Pulmonata). *Biocontrol Science and Technology* 7: 457–465.

Symondson, W.O.C., Erickson, M.L. and Liddell, J.E. (1999). Development of a monoclonal antibody for the detection and quantification of predation on slugs within *Arion hortensis* agg. (Mollusca: Pulmonata). *Biological Control* 16: 274–282.

Takasu, K. and Lewis, W.J. (1996). The role of learning in adult food location by the larval parasitoid *Microplitis croceipes* (Hymenoptera: Braconidae). *Journal of Insect Behavior* 9: 265–281.

Tamaki, G. and Halfhill, J.E. (1968). Bands on peach trees as shelters for predators of the green peach aphid. *Journal of Economic Entomology* 61: 707–711.

Thomas, M.B., Wratten, S.D. and Sotherton, N.W. (1991). Creation of island habitats in farmland to manipulate populations of beneficial arthropods: predator densities and emigration. *Journal of Applied Ecology* 28: 906–917.

Todd, F.E. and Vansell, G.H. (1942). Pollen grains in nectar and honey. *Journal of Economic Entomology* 35: 728–731.

Townes, H. (1958). Some biological characteristics of the Ichneumonidae (Hymenoptera) in relation to biological control. *Journal of Economic Entomology* 51: 650–652.

Triltsch, H. (1997). *The Ladybug* Coccinella septempunctata *L. in the Winter Wheat-cereal Aphid Antagonist Complex*. Verlag Agrarokologie Bern, Hannover. 159 pages.

Unwin, D.M. and Corbet, S.A. (1991). *Insects, Plants and Microclimate*. Richmond Publications, Slough. 68 pages.

van den Meiracker, R.A.F., Hammond, W.N.O. and van Alphen, J.J.M. (1990). The role of kairomones in prey finding by *Diomus* sp., and *Exochomus* sp., two coccinellid predators of the cassava mealybug, *Phenacoccus manihoti*. *Entomologia Experimentalis et Applicata* 56: 209–217.

van der Goot, V.S. and Grabant, R.A.J. (1970). Some species of the genera *Melanostoma*, *Platycheirus* and *Pyrophaena* (Diptera, Syrphidae) and their relation to flowers. *Entomologische Berichten* 30: 135–143.

Van Driesche, R.G. and Bellows, T.S. Jr (1996). *Biological Control*. Chapman & Hall, New York. 447 pages.

van Emden, H.F. and Hagen, K.S. (1976). Olfactory reactions of the green lacewing, *Chrysopa carnea*, to typrophan and certain breakdown products. *Environmental Entomology* 5: 469–473.

van Handel, E. (1972). The detection of nectar in mosquitoes. *Mosquito News* 32: 458.

van Handel, E. (1984). Metabolism of nutrients in the adult mosquito. *Mosquito News* 44: 573–579.

van Handel, E., Haeger, J.S. and Hansen, C.W. (1972). The sugars of some Florida nectars. *American Journal of Botany* 59: 1030–1032.

van Lenteren, J.C., van Vianen, A., Gast, H.F. and Kortenhoff, A. (1987). The parasite–host relationship between *Encarsia formosa* Gahan (Hymenoptera: Aphelinidae) and *Trialeurodes vaporariorum* (Westwood) (Homoptera: Aleyrodidae). XVI. Food effects on ooegensis, life span and fecundity of *Encarsia formosa* and other hymenopterous parasites. *Zietschrift für Angewandte Entomologie* 103: 69–84.

van Lenteren, J.C., Babendreier, D., Bigler, F., Burgio, G., Hokkanen, H.M.T., Kuske, S., Loomans, A.J.M., Menzler-Hokkanen, I., van Rijn, P.C.J., Thomas, M.B., Tommasini, M.G. and Zeng, G.G. (2003). Environmental risk assessment of exotic natural enemies used in inundative biological control. *BioControl* 48: 3–38.

van Rijn, P.C.J. (2002). The impact of supplementary food on a prey–predator interaction. Ph.D. thesis, University of Amsterdam.

van Rijn, P.C.J. and Tanigoshi, L.K. (1999). The contribution of extrafloral nectar to survival and reproduction of the predatory mite *Iphiseius degenerans* on *Ricinus communis*. *Experimental Applied Acarology* 23: 281–296.

van Rijn, P.C.J., van Houten, Y.M. and Sabelis, M.W. (2002). How plants benefit from providing food to predators even when it is also edible to herbivores. *Ecology* 83: 2664–2679.

Vet, L.E.M., Lewis, W.J. and Carde, R.T. (1995). Parasitoid foraging and learning. In *Chemical Ecology of Insects* 2 (R.T. Carde and W.J. Bell, eds), pp. 65–101. Chapman & Hall, New York.

Waage, J.K. (1978). Arrestment responses of the parasitoid, *Nemeritis canescens*, to a contact chemical produced by its host, *Plodia interpunctella*. *Physiological Entomology* 3: 135–146.

Waage, J.K. (1990). Ecological theory and the selection of biological control agents. In *Critical Issues in Biological Control* (M. Mackauer, L.E. Ehler and J. Roland, eds), pp. 135–157. Intercept, Andover.

Wäckers, F.L. (1994). The effect of food deprivation on the innate visual and olfactory preferences in the parasitoid *Cotesia rubecula*. *Journal of Insect Physiology* 40: 641–649.

Wäckers, F. L. (1999). Gustatory response by the hymenopteran parasitoid *Cotesia glomerata* to a range of nectar and honeydew sugars. *Journal of Chemical Ecology* 25: 2863–2877.

Wäckers, F.L. (2000). Do oligosaccharides reduce the suitability of honeydew for predators and parasitoids? A further facet to the function of insect-synthesized honeydew sugars. *Oikos* 90: 197–201.

Wäckers, F.L. (2001). A comparison of nectar- and honeydew sugars with respect to their utilization by the hymenopteran parasitoid *Cotesia glomerata*. *Journal of Insect Physiology* 47: 1077–1084.

Wäckers, F.L. and Steppuhn, A. (2003). Characterizing nutritional state and food source use of parasitoids collected in fields with high and low nectar availability. *Proceedings of the IOBC/WPRS Study Group on Landscape Management for Functional Biodiversity*. Bologna, Italy.

Wäckers, F.L., Bjornsten, A. and Dorn, S. (1996). A comparison of flowering herbs with respect to their nectar accessibility for the parasitoid *Pimpla turionellae*. *Proceedings of Experimental and Applied Entomology* 7: 177–182.

Waloff, N. and Jervis, M.A. (1987). Communities of parasitoids associated with leafhoppers and planthoppers in Europe. *Advances in Ecological Research* 17: 281–402.

Weiser, L.A. (2001). Enhancing biotic mortality of the potato leafhopper *Empoasca fabae*. Ph.D. dissertation. Iowa State University, Ames.

Wratten, S.D. (1973). The effectiveness of the coccinellid beetle, *Adalia bipunctata* (L.) as a predator of the lime aphid, *Eucallipterus tiliae* L. *Journal of Animal Ecology* 42: 785–802.

Wratten, S.D., White, A.J., Bowie, M.H., Berry, N.A. and Weigmann, U. (1995). Phenology and ecology of hover flies (Diptera: Syrphidae) in New Zealand. *Environmental Entomology* 24: 595–600.

Wratten, S.D., Bowie, M.H., Hickman, J.M., Evans, A.M., Sedcole, J.R. and Tylianakis, J.M. (2003). Field boundaries as barriers to movement of hover flies (Diptera: Syrphidae) in cultivated land. *Oecologia* 134: 605–611.

Xu, F.Y. and Wu, D.W. (1987). Control of bamboo scale insects by intercropping rape in the bamboo forest to attract coccinellid beetles. *Chinese Journal of Biological Control* 5: 117–119.

Yigit, A. and Erkilic, L. (1987). Studies on egg parasitoids of grape leafhopper, *Arboridia adanae* Dlab. (Hom., Cicadellidae) and their effects in the region of South Anatolia. In *Tukiye I. Entomologi Kongresi Bildirileri*, 13–16 Ekim (1987), pp. 35–42. Ege Universitesi, Bornova, Izmir. Ege Universitesi/Ataturk Kultur Merkezi, Bornova/Ismir.

Yuval, B. and Schlein, Y. (1986). Leishmaniasis. In the Jordan Valley. III. Nocturnal activity of *Phlebotomus papatasi* (Diptera: Psychodidae) in relation to nutrition and ovarian development. *Journal of Medical Entomology* 23: 411–415.

Zaidi, R.H., Jaal, Z., Hawkes, N.J., Hemingway, J. and Symondson, W.O.C. (1999). Can multiple-copy sequences of prey DNA be detected amongst the gut contents of invertebrate predators? *Molecular Ecology* 8: 2081–2087.

Zhou, X., Honek, A., Powell, W. and Carter, N. (1995). Variations in body length, weight, fat content and survival in *Coccinella septempunctata* at different hibernation sites. *Entomologia Experimentalis et Applicata* 75: 99–107.

Zoebelein, G. (1955). Der honigtau als nahrung der insekten I. *Zietschrift für Angewandte Entomologie* 38: 369–416.

Chapter 6

Molecular techniques and habitat manipulation approaches for parasitoid conservation in annual cropping systems

Fabian D. Menalled, Juan M. Alvarez and Douglas A. Landis

Introduction

Although parasitoids represent a valuable and diverse group of natural enemies of agricultural pests, their success in annual cropping systems has been limited. Among the factors constraining the effectiveness of parasitoids as biological control agents are their small size and their high species diversity that could hinder correct identification. Further, the harsh environment created by agricultural practices reduces parasitoid survivorship. In this chapter we first introduce several molecular-based techniques that could aid the study of parasitoids. Second, we summarise case studies where molecular techniques have allowed entomologists to address difficult questions of parasitoid identification, biology and ecology. Then we discuss the impact that common agricultural practices have on parasitoids. Finally, we explain how combining habitat manipulation practices with molecular techniques can help advance biological control.

The need to manage insect pests has always been a key component in agricultural production systems. At the beginning of the twenty-first century, different strategies for pest management coexist and are sometime perceived as contradictory by the industry, farmers and consumers (Liebman and Davis 2000; Juna 2002). On one hand, low-input and organic farmers base their pest-management programs on the integration of a series of cultural, biological and ecological tactics aimed at reducing pest survivorship and/or impact. On the other hand, high-input conventional farmers rely on pesticides, mechanical control practices and transgenic crops to reduce pest impact. Both approaches to pest management are reflected in the current scientific literature.

In recent years, several researchers have evaluated the potential of ecological theory to foster the design of low-input and organic agricultural systems (Altieri 1995; Vandermeer 1995; Gliessman 1998; Ewel 1999; Gliessman 2000; Liebman et al. 2001). Specifically, Barbosa (1998a), Landis et al. (2000) and Pickett and Bugg (1998) have evaluated the potential of habitat manipulation and conservation biological control to promote natural enemies of agricultural pests. The goal of these practices is to manipulate the environment to enhance the survivorship and/or performance of natural enemies, as explored in detail in chapter 2.

At the same time as the growing interest in ecological approaches, there has been an expansion of research into use of transgenic organisms as a component of high-input conventional farming systems. The debate about ecological risks and benefits of using transgenic organisms to combat arthropod pests is vast (Trewavas 1999; Beringer 2000; Hails 2000; Wolfenbarger and

Phifer 2000; Cox 2002). Potential negative outcomes include evolution of resistant pests, impact on non-target insects, escape of transposable genes into wild relatives, cascading multitrophic level impacts, and changes in farm management practices that could reduce the survivorship of natural enemies. Potential benefits include high efficiency in pest population regulation, reduced environmental impact of pesticide applications, and increased yield. Such factors are explored in greater depth by Altieri et al. (ch. 2 this volume).

Contributions of molecular technologies to pest management go beyond the development and release of transgenic organisms. Among the areas that could benefit are an increased ability to identify pests and natural enemies, evaluation of failure or success in the establishment of natural enemies, study of dispersal patterns, and assessment of genetic divergence among and within populations. Despite these advantages, few field ecologists have fully incorporated molecular tools into their research (Snow and Parker 1998).

In this chapter, we explore potential benefits of linking molecular-based technologies with ecologically based habitat manipulation programs to enhance parasitoid survivorship in annual agroecosystems. We do that by first presenting a brief introduction to several molecular-based techniques that could aid entomologists to study parasitoids. Second, we present several case studies where molecular-based techniques allowed entomologists to address difficult questions of parasitoid identification, biology and ecology that cannot be solved with standard methods. Third, we explore how common agricultural management practices of annual crop systems such as tillage, insecticide application, nutrient management and harvest that influence parasitoid survivorship. Finally, we describe how combining habitat manipulation practices with molecular-based techniques can help advance ecologically based biological control.

Molecular techniques and parasitoids

Parasitoids constitute a ubiquitous and diverse group of insects with enormous ecological and economic importance, particularly as biological control agents of agricultural pests (Greathead 1986; Quicke 1997). However, the small size of many parasitoids and their high species diversity have hampered their correct identification and selection for pest-management programs. Molecular methods can increase our ability to characterise parasitoid biology, genetic diversity and ecological requirements. This in turn allows an unambiguous identification of parasitoid species and an improved selection of the best genotype to be released as a biological control agent (Ives and Hochberg 2000; Liu et al. 2000). Molecular methods can also help field entomologists better characterise the parasitoid community already present in agricultural systems, understand host associations and assess the evolution of parasitoid behavioural traits that affect host selection. These factors could play important roles in the success of parasitoids as biological control agents. The following is a brief introduction to some of the molecular tools used to solve questions related to parasitoid ecology, biology and population genetics. For a detailed description of molecular methods we recommend Symondson and Hemingway (1997), Loxdale and Lushai (1998) and Parker et al. (1998).

Enzyme electrophoresis

Allozymes are variant proteins produced by allelic forms of the same locus that can be separated by electrophoresis. The analysis of variation in allozymes (a reflection of changes in their gene codes) was the earliest molecular method successfully used to determine genetic variation among insect species and for taxa identification (Berlocher 1979; Gonzalez et al. 1979; May 1992). Loxdale and den Hollander (1989) and Menken and Ulenberg (1987) provide a detailed summary of cases in which allozyme electrophoresis has been used in agricultural entomology. In the specific case of parasitoids, this technique allowed successful discrimination between several species of *Trichogramma* (Hymenoptera: Trichogrammatidae), a genus of minute parasitoids of great importance in biological control (Symondson and Hemingway 1997). More recently, the use of allozymes has been supplemented with techniques generating direct or indirect estimates of nucleic acid variation.

Although enzyme electrophoresis is inexpensive and relatively easy to conduct, it investigates only some of the variation in the most conserved class of DNA (the slowly evolving coding DNA), underestimating the amount of genetic variation in the non-coding DNA. This non-coding DNA, also called 'junk', 'parasitic' or 'selfish' DNA, may constitute 30–90% of the insect genome (Hoy 1994). In addition, insects in the order Hymenoptera (most of the effective parasitoids used in agriculture) have exceptionally low allozyme variability (Graur 1985).

DNA molecular markers and nucleotide polymorphisms

The recent advent of molecular methods to directly investigate the DNA molecule has increased accuracy and resolution in genome analysis. DNA molecules are constructed of monomeric units called nucleotides consisting of a purine or pyrimidine base, a pentose and a phosphoric acid group. Molecular methods allow direct evaluation of genomes, determining the proportion of nucleotide sites differing between two or more DNA sequences (nucleotide polymorphisms). Genomes can also be compared indirectly by assessing the bands produced in an electrophoresis gel by DNA fragments obtained from known individuals (DNA molecular markers) (Alvarez 2000).

The number of molecular techniques available for entomological studies has greatly expanded since the advent of the polymerase chain reaction (PCR). Kary Mullins conceived the PCR method in 1983, receiving the Nobel Prize in chemistry for the invention. PCR is an in vitro method for amplifying a target region of DNA by means of enzymes that catalyse the formation of DNA from deoxyribonucleoside triphosphates, using single-stranded DNA as a template. The presence of conserved regions in DNA sequences such as the mitochondrial DNA and nuclear ribosomal DNA makes it possible to amplify fragments from organisms for which there is no specific sequence information available (Kocher et al. 1989). Although the part of the genome to be used will depend on the level of resolution needed by the researcher, several mitochondrial (COI, COII, 16S) and nuclear (ITS1, ITS2, 28S, D2, D3 and EF-1) regions have proved useful for discriminating parasitoid species (Alvarez 2000).

The number of PCR-based techniques is expanding every year and they have been successfully used in ecological and entomological studies. Among the different areas that have benefited from PCR-based techniques are systematics, population genetics and insecticide resistance assessment (Loxdale and Lushai 1998; Parker et al. 1998). PCR-based techniques offer several advantages, including the possibility of working with extremely small insects, such as many of the effective parasitoids used in biological control. They are also unaffected by the life stage of insects and can potentially be used with stored, dry or old material. Currently, the main disadvantage of using molecular methods is the cost involved.

Some of the PCR-based techniques commonly used are restriction fragment length polymorphism (RFLP) analysis, microsatellite analysis, single-strand conformation polymorphisms (SSCPs), random amplified polymorphic DNA (RAPD) and amplified fragment length polymorphism (AFLP) fingerprinting (Hedrick 1992; Hughes and Queller 1993; Mueller et al. 1996; Symondson and Hemingway 1997). Table 6.1 presents a comparison of different molecular methods in terms of their sensitivity, cost, efficiency, level of discrimination and application.

The selection of any given molecular technique depends on the type of problem to be solved, the cost of the technique, its ease of use and the sample size to be analysed. The section of the genome to be investigated depends on the level of variation that the researcher needs to resolve. Among the different PCR-based techniques, RAPD-PCR has been very popular in entomological

Table 6.1: Molecular tools for assessing variation in DNA fragments or PCR products and qualitative assessment of the cost, effectiveness, levels of discrimination, type of data obtained and applications of the different tools.

Attribute	PCR-RFLPs	Microsatellites	SSCPs	RAPD-PCR	AFLP
Sensitivity	Moderate	High	Moderate–high	Moderate	High
Cost	Low	Moderate	Low	Low	Moderate
Efficiency	High	High	High	High	High
Level of discrimination/ type of data (gene frequencies or base pair changes)	Single nucleotides in nuclear and mt-DNA/gene frequency data	Differences in individuals and populations in nuclear DNA/neither gene frequency nor base pair changes	Single nucleotide in nuclear DNA/gene frequency data	Differences in single nucleotides in nuclear DNA/gene frequency data	Differences in single nucleotides in nuclear DNA/gene frequency
Application	Diversity, geographic variation, hybrid zones and species boundaries, phylogeny	Mating systems, diversity, parentage, relatedness	Parentage, relatedness	Hybrid zones and species boundaries, biotype and subspecies identification	Subspecies identification

Adapted from Dowling et al. (1996).

studies perhaps because the technique does not require prior knowledge of DNA sequence, provides a rapid way of identifying genetic markers, is inexpensive and is easy to develop (Williams et al. 1990; Alvarez 2000). However, RAPD markers have been criticised as a tool for molecular identification of species because their results can have poor reproducibility (Ellsworth et al. 1993). Reineke et al. (1999) recommended amplifying the DNA in a second reaction in order to assess the reproducibility of the banding patterns produced by the RAPD primers used in a test.

Examples of molecular techniques used in biological control

Cryptic species identification

Unequivocal taxonomical identification of pests and natural enemies is important in biological control. Although evaluation of morphological features is the predominant tool to distinguish insect species, many parasitic groups are visually undistinguishable (cryptic species) despite their high genetic diversity (Pinto and Stouthamer 1994; Atanassova et al. 1998; Stouthamer et al. 2000; Alvarez and Hoy 2002). This genetic diversity, in turn, could be important to the success of parasitoids as biological control agents (Goldson et al. 1997).

Beginning in 1994, two *Ageniaspis* (Hymenoptera: Encyrtidae) populations, one from Australia and one from Taiwan, were introduced into Florida as part of a classical biological control project against the Asian citrus leafminer *Phyllocnistis citrella* (Lepidoptera: Gracillariidae) (Hoy and Nguyen 1997). Although taxonomic specialists identified the two populations as belonging to the same species, *Ageniaspis citricola*, there were clear biological and physiological differences between them (Yoder and Hoy 1998; Hoy et al. 2000). By 1996, the Australian population of *A. citricola* had colonised most of Florida's 344 250 ha of citrus groves and parasitism of citrus leafminer pupae was found to be as high as 99% in some sites (Hoy and Nguyen 1997). Alvarez (2000) and Alvarez and Hoy (2002) collected *Ageniaspis* individuals in 10 field sites at six Florida counties and took advantage of RAPDs to show no evidence that the Taiwan population had established in Florida by the end of 1999. After these results were obtained, only Australian *A. citricola* was sent to Caribbean and Latin American countries for biological control of the citrus leafminer, to ensure establishment.

Differentiation between exotic and indigenous parasitoids

In biological control programs where exotic strains of parasitoids are introduced into a new area, it is not always possible to distinguish between imported and native strains using morphological features. Molecular methods have been used to differentiate strains of indigenous and exotic biological control agents prior to their release. For example, many strains of exotic parasitic Hymenoptera that are investigated in the US to control the Russian wheat aphid, *Diuraphis noxia* (Homoptera: Aphididae), are indistinguishable from the indigenous ones. Narang et al. (1994) took advantage of RAPD-PCR to identify and characterise four species of hymenopteran parasitoid introduced for the control of the Russian wheat aphid and develop an easy-to-use dichotomous key for parasitoid identification. Zhu and Greenstone (1999), using the resulting PCR product in a gel of the amplified ribosomal ITS2 region, were able to distinguish three species and two strains of *Aphelinus* (Hymenoptera: Aphelinidae) endoparasitoids that had been released against the Russian wheat aphid. In South Africa, Prinsloo et al. (2002) used specific primers for both the ribosomal ITS2 and mitochondrial 16s DNA sequences of several morphologically similar parasitoid species and strains of *Aphelinus*, digested them with a restriction enzyme and showed in an agarose gel electrophoresis the differences between strains. These procedures allowed a reliable separation of *Aphelinus* species and confirmation that the exotic aphid parasitoid *A. hordei* had successfully spread in wheat fields, suggesting it could be effective at controlling the Russian wheat aphid.

Parasitoid rearing

Contamination problems are recurrent in laboratory colonies of parasitoids, especially when rearing small, morphologically similar species (Dourojeanni 1990; Fernando and Walter 1997). In many cases, the biological control worker is not aware of the contamination but molecular methods can be used to monitor the taxonomic quality of parasitoid colonies or to assist in selecting desirable traits prior to release. For example, Rosen and DeBach (1973) cited that cultures of *Encarsia perniciosi* (Hymenoptera: Aphelinidae) imported from Germany to control the San Jose scale, *Quadraspidiotus perniciosus* (Homoptera: Diaspididae) were contaminated with a related but ineffective species, *E. fasciata*. It is estimated that about 4 million *E. fasciata* were released, but never became established.

There is fairly extensive prior work on the use of molecular markers for identification and characterisation of *Trichogramma* species and strains. For example, Chang et al. (2001), Laurent et al. (1998), Pinto et al. (2002), Silva et al. (1999) and Vanlerberghe-Masutti (1994) found specific markers for several morphologically similar *Trichogramma* strains and species with the use of RAPD, ITS2 sequences and RFLP analyses. These DNA markers could potentially be used in rearing laboratories for rapid species identification.

Agricultural practices and parasitoids

Modern agricultural practices create a harsh environment for parasitoids (Hassell 1986; LaSalle 1993). High-input conventional annual cropping systems usually depend on large amounts of off-farm chemical and mechanical inputs such as insecticide and herbicide applications, primary and secondary tillage, fertilisation, cultivation and harvest. From an ecological point of view, these management practices represent disturbances that disrupt the characteristics of the ecosystem, the community or the population through a rapid change in resources, substrate availability or physical environment (Pickett and White 1985).

On a temporal dimension, annual crop fields can be characterised as ephemeral and predictable habitats where early successional plants (the crops) are established as short-lived monocultures. These plants have been selected to have low chemical and mechanical defences and are readily attacked by herbivores (Kogan and Latin 1993; Obrycki et al. 1997). Spatially, the current tendency of removing non-crop habitats has severely reduced and fragmented native habitats that could provide resources for parasitoids including overwintering sites, pollen and nectar, suitable environmental microsites and alternative hosts. In this way, industrialised agriculture generates a mosaic of small and isolated late-successional patches with high species diversity embedded in a matrix of extensive early-successional agricultural fields with low species diversity (Auclair 1976; Pogue and Schnell 2001) (Table 6.2). Schmidt et al. (ch. 4 this volume) explores landscape-level factors in more depth.

The joint impact of disturbances, temporal predictability, resource allocation and habitat fragmentation has been identified as a limiting factor conditioning the abundance and effectiveness of parasitoids in annual agroecosystems (Kruess and Tscharntke 1994; Price and Waldbauer 1994; Tscharntke and Kruess 1999; Tscharntke 2000; Menalled et al. 2003). These unique biotic and abiotic characteristics, coupled with the relatively low cost and high effectiveness of pesticide applications, has discouraged the use of natural enemies, including parasitoids, as a tool to manage arthropod pests in annual cropping systems (Smith et al. 1997; Barbosa 1998b).

Despite the increased reliance on off-farm chemical inputs to manage pests, there is a growing body of theoretical work and empirical evidence suggesting that the high-input conventional farming model may not be sustainable and could pose long-term environmental and

Table 6.2: Comparison between annual crops and non-crop habitats with reference to disturbance regimes, vegetation structure and parasitoid community size.

Disturbance regime	Annual crops	Non-crop habitats
Frequency	High, several times a year	Low, one event every several years
Magnitude	High	Low, moderate or high
Area	Extensive areas	In general, reduced areas
Turnover rate	High: several times a year	Low: once every several years
Spatial distribution of elements	Homogeneous	Heterogeneous
Predictability	High	Low
Examples of disturbances	Tillage, cultivation, pesticides etc.	Fire, windstorm, flood, landslide etc.
Successional stage	Early	Mid- to late-successional
Plant type	Herbaceous, in general annuals	Shrubs and trees
Vegetational structure	Simple: one stratum and life form	Complex: several strata and life forms
Plant species diversity	Low	High
Parasitoid community size	Small	Large
Presence of shelter, food or alternate host for parasitoids	Low	High

health risks (Matson et al. 1997; Liebman et al. 2001). In response to rising public support for environmentally sound agricultural practices, there has been a growing interest in the development of ecologically based pest-management systems (National Research Council 1996). In this context, several studies evaluated the importance of ecological theory to foster the abundance and/or effectiveness of natural enemies including parasitoids (Landis and Menalled 1998; Letourneau 1998; Altieri and Nicholls 1999; Hawkins and Cornell 1999; Hochberg and Ives 2000; Altieri et al., ch. 2 this volume).

Improving parasitoid survivorship through habitat manipulation and molecular biology

Habitat manipulation, defined as a series of environment manipulations to provide natural enemies with the necessary resources to improve their effectiveness at combating agricultural pests, emerges as a unifying theme to mitigate the negative impact of agricultural practices (Landis et al. 2000). The underlying concept of habitat manipulation is that provision of supplementary and complementary food, microclimate modification and existence of refuge habitats in close association with crop fields might help natural enemies to cope with the detrimental impact of agricultural practices. For example, many species of adult parasitoids use wildflowers and aphid honeydew as food resources that are not provided in agricultural fields in which weeds are controlled (Jervis et al. 1993; Idris and Grafius 1995; Dyer and Landis 1996; Jervis et al., ch. 5 this volume). Practices such as no tillage and conservation tillage, cover cropping, crop residue conservation, intercropping and establishment of herbaceous strips in close spatial association with crop fields represent viable within-crop field approaches to provide the necessary resources to enhance parasitoid survival (Van Driesche and Bellows 1996; Khan et al. 1997; Tscharntke 2000).

As explained in the following sections, molecular tools can be integrated with an in-depth knowledge of insect biological and ecological requirements to help entomologists improve habitat manipulation practices for parasitoid conservation. The integration of these tools and knowledge could be done at different spatial scales.

Within-crop field activities and molecular-based techniques

Pesticide applications can be the most important factors reducing parasitoid survivorship within crop fields. In the short term, pesticides can kill large numbers of parasitoids, which may lead to an outbreak of secondary pests. In the long term, repeated pesticide applications may select resistant biotypes and influence parasitoid population dynamics (Tolstova and Atanov 1982). Ruberson et al. (1998) described different approaches to integrate pesticide applications with natural enemies. Among the recommended tactics are periodic scouting of crop fields, use of selective pesticides and sublethal doses, and spatial and temporal separation of pesticides and natural enemies.

Molecular technologies can help biological control practitioners to integrate pesticide use with natural enemies through the selection of pesticide-resistant strains of natural enemies. Although this approach appears to be more successful with predators than with parasitoids, recent progress in parasitoid selection programs could reverse this situation (Johnson and Tabashnik 1994). For example, RAPD-PCR has proved to be a useful technique to distinguish between two populations of the walnut aphid parasitoid, *Trioxys pallidus* (Hymenoptera: Aphidiidae), differing in resistance to pesticides (Edwards and Hoy 1995). Although the experiments were performed in the laboratory, the authors suggest that RAPD-PCR could be used to determine the fate of pesticide-resistant parasitoids released in the field.

Conserving parasitoids at the farm and landscape level

While practices at the within-field level might enhance parasitoid survivorship, agricultural practices occurring at larger scales might negate such conservation efforts. Several studies have shown that field size, crop rotation and presence of non-crop habitats play a critical role in determining within-field parasitoid abundance. For example, Landis and Haas (1992) determined that parasitism of the European cornborer, *Ostrinia nubilalis* (Lepidoptera: Pyralidae) by its larval parasitoid, *Eriborus terebrans* (Hymenoptera: Ichneumonidae) was significantly higher at the borders of maize (also known as corn) fields than in field interiors. They further determined that the greatest levels of parasitism were observed at wooded field edges. In accordance, it has been observed that access to plant nectar, aphid honeydew or sugar improved *E. terebrans* survival in crop and non-crop habitats (Dyer and Landis 1996; Landis and Marino 1999a). These studies suggested that the absence of food resources in large maize fields, combined with high temperatures before canopy closure, was responsible for the low abundance of parasitoids observed in field interiors (Dyer and Landis 1997).

Similarly, Langer (2001) determined that short rotation coppice hedges and clover/grass grazing areas spread among crop fields forming a series of annual, biannual and semi-perennial habitats correlated with an increase in parasitism of the pest aphid *Sitobion avenae* (Homoptera: Aphididae). These and other studies (e.g. Höller 1990; Duelli 1997; Lee et al. 2001) highlight the importance of semi-natural habitats established in close association with crop fields for natural enemy conservation. Ideally, agricultural landscapes should contain a series of non-crop habitats interspersed with crop fields to provide shelter and food for parasitoids. Mid- and late-successional habitats provide parasitoids with adult food resources, alternative hosts, overwintering sites and shelter for adverse conditions (Marino and Landis 1996; Landis and Menalled 1998; Menalled et al. 1999; Schmidt et al., ch. 4 this volume).

Because parasitoids use semiochemicals in host location, some of which emanate from plants (De Moraes et al. 1998; Khan and Pickett, ch.10 this volume), non-crop vegetation interspersed with crop fields can also affect parasitoid behavior. Understanding changes in parasitoid behaviour to plant assemblage composition is a key element in the development of habitat practices aimed at increasing vegetational diversity of agricultural systems (Landis et al. 2000). Yet due to their small size, vagility and the inherent difficulty of mark and recapture experiments, very little is known about parasitoid dispersal behaviour (Hastings 2000).

Molecular-based techniques may aid entomologists to assess the impact of agricultural landscape structure on parasitoid population and community structure (Loxdale and Lushai 1998). Vaughn and Antolin (1998) used RAPD-PCR markers visualised by SSCP analysis to study the population structure of *Diaretiella rapae* (Hymenoptera: Braconidae), a parasitoid of several aphid species. The authors found that larger genetic variation occurred between fields separated by short distances than between areas separated by longer distances. They argued that the reduced genetic exchange among subpopulations indicates that released *D. rapae* would not disperse between fields.

In another study, Althoff and Thompson (2001) used the mtDNA cytochrome oxidase 1 (CO1) gene sequence and nuclear rDNA RFLPs to compare patterns of host search behaviour among six populations of *Agathis* n.sp. (Braconidae: Agathidinae) located in a relatively large geographic area in south-eastern Washington, USA. The results showed no isolation by distance, suggesting long-distance dispersal among populations. The authors argued that phenotypic differences in host–parasitoid interactions such as time allocated to searching, ovipositor length and place of searching appear to be driven by local plant characteristics. Although this study was not developed within the context of annual cropping systems, it provides a framework that could be used to foster the understanding of how local, mid- and large-scale agricultural landscape affect parasitoid abundance, searching behaviour and distribution.

Conclusion: integrating molecular techniques with field studies to improve habitat manipulation

Despite the large potential of biological control in sustainable agriculture, the commercial development and success of this approach has been extremely limited (Frey 2001). This is particularly true in annual cropping systems where disturbances, resource paucity and habitat simplification reduce natural enemy survivorship. More successful biological control may be possible if agricultural entomologists take advantage of newly developed concepts and techniques deriving from a wide variety of areas ranging from molecular-based technologies to landscape ecology.

Molecular technologies are among the fastest-growing sectors in modern agriculture with the potential of transforming farming systems (National Research Council 1996; Ives and Hochberg 2000). To date, much of the debate on this issue has been on the benefits and risks of releasing transgenic organisms (see Letourneau and Burrows 2002 for a review). As discussed in this chapter, these newly developed technologies can also increase the adoption and success of biological control agents through greater taxonomic accuracy, higher precision in insect identification and a better understanding of the genetic and population structure of natural enemy populations.

There is a rich collection of evidence that ecological diversity plays a key role in the resilience of farm systems (Collins and Qualset 1999; Brookfield 2001). More specifically, increasing diversity appears to be a cornerstone of habitat manipulation to enhance parasitoid survivorship (Roland 1993; Kruess and Tscharntke 1994; Roland and Taylor 1997; Cappuccino et al. 1998; Landis and Menalled 1998; Landis and Marino 1999a, b; Thies and Tscharntke 1999; Kruess and Tscharntke 2000; Tscharntke 2000). Encouraging diversity per se, however, might not be enough, and sometimes it may even have negative consequences (van Emden 1990). Instead, it is necessary to identify the key elements of diversity needed for the particular set of parasitoid species under consideration (Landis et al. 2000) in what Gurr et al. (2004) consider 'directed' approaches to the use of diversity to support biological control. Undoubtedly, correct insect identification is crucial. Parasitoids and alternative hosts must be accurately identified before any existing information on their environmental needs could be used in the selection of within-field, farm- and landscape-level practices that will enhance their survivorship and effectiveness. Molecular techniques provide a useful tool to further the accuracy and ease of parasitoid and host identification.

Although 'no other aspect of applied biology is more intimately dependent on sound taxonomy than is biological control of pests' (Rosen 1986), tools for accurate parasitoid identification are not well developed. This is particularly true among cryptic species or biological races with different environmental requirements and biological attributes. In the past, taxonomic research relied mostly on morphological characteristics. Indeed, the advent of phase-contrast and scanning electron microscopes allowed taxonomists to separate closely related parasitoid species on the basis of minor morphological variations (Rosen and DeBach 1973; Rosen 1986). More recently, DNA molecular markers have allowed parasitoid specialists to base their classification on genetic variations. Molecular-based techniques can also be used to more accurately quantify gene flow among various agricultural landscapes and habitat fragments, and to select the best-adapted genotype for a given environmental scenario. Other avenues for future research include the use of molecular techniques to modify plants in a way that could improve the efficacy of natural enemies, for example the selection of plants with more accessible sources of food for parasitoid adults.

Combining molecular technologies with ecological concepts of habitat manipulation appears a promising approach to improve parasitoids' efficiency and survivorship in agricultural systems. Still, most biological control professionals 'are broadly trained in entomology and ecology, but rarely in genetics and are thus unaware of the genetic techniques available' (Hoy

1986). Also, few geneticists 'are trained in biological control and most do not understand the need of such applied projects' (Hoy 1986). More than 18 years later, these statements are still valid. It is still necessary to have close collaboration among taxonomists, molecular biologists, ecologists and entomologists at all phases of a biological control program. Hopefully, this chapter will stimulate future collaborative studies.

Acknowledgements

We thank John Pleasants, Paul Marino and John Obrycki for their helpful comments and suggestions.

References

Althoff, D.M. and Thompson, J.N. (2001). Geographic structure in the searching behavior of a specialist parasitoid: combining molecular and behavioral approaches. *Journal of Evolutionary Biology* 14: 406–417.

Altieri, M.A. (1995). *Agroecology: The Science of Sustainable Agriculture*. Westview Press, Boulder, Colorado. 433 pages.

Altieri, M.A. and Nicholls, C.I. (1999). Biodiversity, ecosystem function, and insect pest management in agricultural systems. In *Biodiversity in Agroecosystems* (W.W. Collins and C.O. Qualset, eds), pp. 69–84. CRC Press, Boca Raton, Florida.

Alvarez, J.M. (2000). Use of molecular tools for discriminating between two populations of the citrus leafminer parasitoid *Ageniaspis* (Hymenoptera: Encyrtidae). Ph.D. dissertation. Department of Entomology and Nematology, University of Florida, Gainesville, Florida. 77 pages.

Alvarez, J.M. and Hoy, M.A. (2002). Evaluation of the ribosomal ITS2 DNA sequences in separating closely related populations of the parasitoid *Ageniaspis* (Hymenoptera: Encyrtidae). *Annals of the Entomological Society of America*, 95: 250–257.

Atanassova, P., Brookes C.P., Loxdale H.D. and Powell, W. (1998). Electrophoretic study of five aphid parasitoid species of the genus *Aphidis* (Hymenoptera: Braconidae), including evidence for reproductively isolated sympatric populations and cryptic species. *Bulletin of Entomological Research* 88: 3–13.

Auclair, A.U. (1976). Ecological factors in the development of intensive-management ecosystems in the midwestern United States. *Ecology* 57: 431–434.

Barbosa, P. (1998a). *Conservation Biological Control*. Academic Press, San Diego. 416 pages.

Barbosa, P. (1998b). Agroecosystems and conservation biological control. In *Conservation Biological Control* (P. Barbosa, ed.), pp. 39–54. Academic Press, San Diego.

Beringer, J.E. (2000). Releasing genetically modified organisms: will any harm outweigh any advantage? *Journal of Applied Ecology* 37: 207–214.

Berlocher, S.H. (1979). Biochemical approaches to strain, race, and species discrimination. In *Genetics in Relation to Insect Management* (M.A. Hoy and J.J. McKelvey, eds), pp. 137–144. Rockefeller Foundation, New York.

Brookfield, H. (2001). *Exploring Agrodiversity*. Columbia University Press, New York. 608 pages.

Cappuccino, N., Lavertu, D., Bergeron, Y. and Régnière, J. (1998). Spruce budworm impact, abundance and parasitism rate in a patchy landscape. *Oecologia* 114: 236–242.

Chang, S.C., Hu, N.T., Hsin, C.Y. and Sun, C.N. (2001). Characterization of differences between two *Trichogramma* wasps by molecular markers. *Biological Control* 21: 75–78.

Collins, W.W. and Qualset, C.O. (1999). *Biodiversity in Agroecosystems*. CRC Press, Boca Raton, Florida. 352 pages.

Cox, T.S. (2002). The mirage of genetic engineering. *American Journal of Alternative Agriculture* 17: 41–43.

De Moraes, C.M., Lewis, W.J., Pare, P.W., Alborn, H.T. and Tumlinson, J.H. (1998). Herbivore-infested plants selectively attract parasitoids. *Nature* 393: 570–573.

Dourojeanni, M.J. (1990). Entomology and biodiversity conservation in Latin America. *American Entomologist* 36: 88–93.

Dowling, T.E., Moritz, G., Palmer, J.D. and Rieseber, L.H. (1996). Nucleic acids III: analysis of fragments and restriction sites. In *Molecular Systematics* (D.M. Hillis, C. Moritz and B.K. Mable, eds), pp. 249–320. Sinauer Associates, Sunderland, Massachusetts.

Duelli, P. (1997). Biodiversity evaluation in agricultural landscapes: an approach at two different scales. *Agriculture, Ecosystems and Environment* 62: 81–91.

Dyer, L.E. and Landis, D.A. (1996). Effect of habitat, temperature, and sugar availability on longevity of *Eriborus terebrans* (Hymenoptera: Ichneumonidae). *Environmental Entomology* 25: 1192–1201.

Dyer, L.E. and Landis, D.A. (1997). Influence of non-crop habitats on the distribution of *Eriborus terebrans* (Hymenoptera: Ichneumonidae) in corn fields. *Environmental Entomology* 26: 924–932.

Edwards, O.R. and Hoy, M.A. (1995). Monitoring laboratory and field biotypes of the walnut parasite, *Trioxys pallidus,* in a population cages using RAPD-PCR. *Biocontrol Science and Technology* 5: 313–327.

Ellsworth, D.L., Rittenhouse, K.D. and Honeycutt, R.L. (1993). Artifactual variation in randomly amplified polymorphic DNA banding patterns. *BioTechniques* 14: 214–217.

Ewel, J.J. (1999). Natural systems as models for the design of sustainable systems of land use. *Agroforestry Systems* 45: 1–21.

Fernando, L.C.P. and Walter, G.H. (1997). Species status of two host-associated populations of *Aphytis lingnanensis* (Hymenoptera: Aphelinidae) in citrus. *Bulletin of Entomological Research* 87: 137–144.

Frey, P.M. (2001). Biocontrol agents in the age of molecular biology. *Trends in Biotechnology* 19: 432–433.

Gliessman, S.R. (1998). *Agroecology: Ecological Processes in Sustainable Agriculture.* Ann Arbor Press, Ann Arbor, Michigan. 357 pages.

Gliessman, S.R. (2000). *Agroecosystems Sustainability, Developing Practical Strategies.* CRC Press, Boca Raton, Florida. 210 pages.

Goldson, S.L., Phillips, C.B., McNeill, M.R. and Barlow, N.D. (1997). The potential of parasitoid strains in biological control: observations to date on *Microctonus* spp. intraspecific variation in New Zealand. *Agriculture, Ecosystems and Environment* 64: 115–124.

Gonzalez, D., Gordh, G., Thompson, S.N. and Adler, J. (1979). Biotype discrimination and its importance to biological control. In *Genetics in Relation to Insect Management: A Rockefeller Foundation Conference* (M.A. Hoy and J.J. McKelvey, eds), pp. 129–136. 31 March –5 April 1978, Bellagio, Italy. Rockefeller Foundation, New York.

Graur, D. (1985). Gene diversity in Hymenoptera. *Evolution* 39: 190–199.

Greathead, D.J. (1986). Parasitoids in classical biological control. In *Insect Parasitoids. 13th Symposium of the Royal Entomological Society of London* (J.K. Waage and D.J. Greathead, eds), pp. 289–318. 18–19 September 1985. Academic Press, London.

Gurr, G.M., Wratten, S.D., Tylianakis, J., Kean, J. and Keller, M. (2004). Providing plant foods for insect natural enemies in farming systems: balancing practicalities and theory. In *Plant-derived Food and Plant–carnivore Mutualism* (F.L. Wackers, P.C.J. van Rijn and J. Bruin, eds). Cambridge University Press, Cambridge. In press.

Hails, R.S. (2000). Genetically modified plants – the debate continues. *Trends in Ecology and Systematics* 15: 14–18.

Hassell, M.P. (1986). Parasitoids and population regulation. In *Insect Parasitoids. 13th Symposium of the Royal Entomological Society of London* (J.K. Waage and D.J. Greathead, eds), pp. 201–224. 18–19 September 1985. Academic Press, London.

Hastings, A. (2000). Parasitoid spread: lessons from invasion biology. In *Parasitoid Population Biology* (M.E. Hochberg and A.R. Ives, eds), pp. 70–82. Princeton University Press, Princeton, New Jersey.

Hawkins, B.A. and Cornell, H.V. (1999). *Theoretical Approaches to Biological Control.* Cambridge University Press, Cambridge. 412 pages.

Hedrick, P. (1992). Shooting the RAPD's. *Nature* 355: 679–680.

Hochberg, M.E. and Ives, A.R. (2000). *Parasitoid Population Biology.* Princeton University Press, Princeton, New Jersey. 366 pages.

Höller, C. (1990). Overwintering and hymenopterus parasitism in autumn of the cereal aphid *Sitobion avenae* (F.) in northern FR Germany. *Journal of Applied Entomology* 109: 21–28.

Hoy, M.A. (1986). Use of genetic improvement in biological control. *Agriculture, Ecosystems and Environment* 15: 109–119.

Hoy, M.A. (1994). *Insect Molecular Genetics: An Introduction to Principles and Applications*. Academic Press, San Diego. 540 pages.

Hoy, M.A. and Nguyen, R. (1997). Classical biological control of the citrus leafminer *Phyllocnistis citrella* Stainton (Lepidoptera: Gracillariidae): theory, practice, art, and science. *Tropical Lepidoptera* 8: 1–19.

Hoy, M.A., Jeyaprakash, A., Morakote, R., Lo, P.K.C. and Nguyen, R. (2000). Genomic analysis of two populations of *Ageniaspis citricola* (Hymenoptera: Encyrtidae) suggests that a cryptic species may exist. *Biological Control* 17: 1–10.

Hughes, C.R. and Queller, D.C. (1993). Detection of highly polymorphic microsatellite loci in a species with little allozyme polymorphism. *Molecular Ecology* 2: 131–137.

Idris, A.B. and Grafius, E. (1995).Wildflowers as nectar sources for *Diadegma insulare* (Hymenoptera: Ichneumonidae), a parasitoid of diamondback moth (Lepidoptera: Yponomeutidae). *Environmental Entomology* 24: 1726–1735.

Ives, A.R. and Hochberg, M.E. (2000). Conclusions: debating parasitoid population biology over the next twenty years. In *Parasitoid Population Biology* (M.E. Hochberg, and A.R. Ives, eds), pp. 278–303. Princeton University Press, Princeton, New Jersey.

Jervis, M.A., Kidd, M.A.C., Fitton, M.G., Huddleston, T. and Dawah, H.J.A. (1993). Flower-visiting by hymenopteran parasitoids. *Journal of Natural History* 27: 67–105.

Johnson, M.W. and Tabashnik, B.E. (1994). Laboratory selection for pesticide resistance in natural enemies. In *Applications of Genetics to Arthropods of Biological Control Significance* (S.K. Narang, A.C. Bartlett and R.M. Faust, eds), pp. 91–105. CRC Press, Boca Raton, Florida.

Juna, C. (2002). Sustaining tropical agriculture. *Nature* 415: 960–961.

Khan, Z.R., Ampong-Nyarko, K., Chiliswa, P., Hassanali, A., Kimani, S., Lwande, W., Overholt, W.A., Pickett, J.A., Smart, L.E., Wadhams, L.J. and Woodcock, C.M. (1997). Intercropping increases parasitism of pests. *Nature* 388: 631–632.

Kocher, T.D., Thomas, W.K., Meyer, A., Edwards, S.V., Pääbo, S., Villablanca, F.X. and Wilson, A.C. (1989). Dynamics of mitochondrial DNA evolution in animals: amplification and sequencing with conserved primers. *Proceedings of the National Academy of Sciences of the United States of America* 86: 6196–6200.

Kogan, M. and Latin, J.E. (1993). Insect conservation and pest management. *Biodiversity and Conservation* 2: 242–257.

Kruess, A. and Tscharntke, T. (1994). Habitat fragmentation, species loss, and biological control. *Science* 264: 1581–1584.

Kruess, A. and Tscharntke, T. (2000). Species richness and parasitism in a fragmented landscape: experiments and field studies with insects and *Vicia sepium*. *Oecologia* 122: 129–137.

Landis, D.A. and Haas, M.J. (1992). Influence of landscape structure on abundance and within-field distribution of *Ostrinia nubilalis* Hübner (Lepidoptera: Pyralidae) larval parasitoids in Michigan. *Environmental Entomology* 21: 409–416.

Landis, D.A. and Marino, P.C. (1999a). Landscape structure and extra-field processes: impact on management of pests and beneficials. In *Handbook of Pest Management* (J. Ruberson, ed.), pp. 79–104. Marcel Dekker, New York.

Landis, D.A. and Marino, P.C. (1999b). Conserving parasitoid communities of native pests: implications for agricultural landscape structure. In *Biological Control of Native or Indigenous Pests* (L. Charlet and G. Brewer, eds), pp. 38–51. Thomas Say Publications in Entomology. Entomological Society of America, Lanham, Maryland.

Landis, D.A. and Menalled, F.D. (1998). Ecological considerations in conservation of parasitoids in agricultural landscapes. In *Conservation Biological Control* (P. Barbosa, ed.), pp. 101–121. Academic Press, San Diego.

Landis, D.A., Wratten, S.D. and Gurr, G.M. (2000). Habitat management to conserve natural enemies of arthropod pests in agriculture. *Annual Review of Entomology* 45: 175–210.

Langer, V. (2001). The potential of leys and short rotation coppice hedges as reservoirs for parasitoids of cereal aphids in organic agriculture. *Agriculture, Ecosystems and Environment* 87: 81–92.

LaSalle, J. (1993). Parasitic Hymenoptera, biological control and biodiversity. In *Hymenoptera and Biodiversity* (J. LaSalle and I.D. Gaud, eds), pp. 197–216. CAB International, Wallingford.

Laurent, V., Wajnberg, E., Mangin, B., Schiex, T., Gaspin, C. and Vanlerberghe-Masutti, F. (1998). A composite genetic map of the parasitoid wasp *Trichogramma brassicae* based on RAPD markers. *Genetics* 150: 275–282.

Lee, J., Menalled, F.D. and Landis, D.A. (2001). Refuge habitats modify impact of insecticide disturbance on carabid beetle communities. *Journal of Applied Ecology* 38: 472–483.

Letourneau, D.K. (1998). Conservation biology: lessons for conserving natural enemies. In *Conservation Biological Control* (P. Barbosa, ed.), pp. 9–38. Academic Press, San Diego.

Letourneau, D.K. and Burrows, B.E. (2002). *Genetically Engineered Organisms. Assessing Environmental and Human Health Effects*. CRC Press, Boca Raton, Florida. 438 pages.

Liebman, M. and Davis, A.S. (2000). Integration of soil, crop and weed management in low-external-input farming systems. *Weed Research* 40: 27–47.

Liebman, M., Mohler, C.L. and Staver, C.P. (2001). *Ecological Management of Agricultural Weeds*. Cambridge University Press, Cambridge. 532 pages.

Liu, J., Poinar, G.O. Jr and Berry, R.E. (2000). Control of insect pests with entomopathogenic nematodes: the impact of molecular biology and phylogenetic reconstruction. *Annual Review of Entomology* 45: 287–306.

Loxdale, H.D. and den Hollander, J. (1989). *Electrophoretic Studies on Agricultural Pests*. Systematic Association Special vol. 39. Clarendon Press, New York. 497 pages.

Loxdale, H.D. and Lushai, G. (1998). Molecular markers in entomology. *Bulletin of Entomological Research* 88: 577–600.

Marino, P.C. and Landis, D.A. (1996). Effect of landscape structure on parasitoid diversity in agroecosystems. *Ecological Applications* 6: 276–284.

Matson, P.A., Parton, W.J., Power, G. and Swift, M.J. (1997). Agricultural intensification and ecosystem properties. *Science* 277: 504–509.

May, B. (1992). Starch gel electrophoresis of allozymes. In *Molecular Genetic Analysis of Populations: A Practical Approach* (A.R. Hoelzel, ed.), pp. 1–28. Oxford University Press, Oxford.

Menalled, F.D., Marino, P.C., Gage, S. and Landis, D.A. (1999). Does agricultural landscape structure affect parasitism and parasitoid diversity? *Ecological Applications* 9: 634–641.

Menalled, F.D., Costamagna, A.C., Marino, P.C. and Landis, D.A. (2003).Temporal variation in the response of parasitoids to agricultural landscape structure. *Agriculture, Ecosystems and Environment* 96: 29–35.

Menken, S.B.J. and Ulenberg, S.A. (1987). Biochemical characters in agricultural entomology. *Agricultural Zoology Reviews* 2: 305–360.

Mueller, U.G., Lipari, S.E. and Milgroom, M.G. (1996). Amplified fragment length polymorphism (AFLP) fingerprinting of symbiotic fungi cultured by the fungus-growing ant *Cyphomyrmex minutus*. *Molecular Ecology* 5: 119–122.

Narang, S.K., Leopold, R.A., Krueger, C.M. and DeVault, J.D. (1994). Dichotomous RAPD-PCR key for identification of four species of parasitic hymenoptera. In *Applications of Genetics to Arthropods of Biological Control Significance* (S.K. Narang, A.C. Bartlett and R.M. Faust, eds), pp. 53–67. CRC Press, Boca Raton, Florida.

National Research Council (1996). *Ecologically Based Pest Management. New Solutions for a New Century*. Board on Agriculture, Committee on Pest and Pathogen Control Agents and Enhanced Cycles and Natural Processes. National Academy Press, Washington, DC. 144 pages.

Obrycki, J.J., Lewis, L.C. and Orr, D.B. (1997). Augmentative releases of entomophagous species in annual cropping systems. *Biological Control* 10: 30–36.

Parker, P.G., Snow, A.S., Schug, M.D., Booton G.C. and Fuerst, P.A. (1998). What molecules can tell us about populations: choosing and using a molecular marker. *Ecology* 79: 361–382.

Pickett, C.H. and Bugg, R.L. (1998). *Enhancing Biological Control. Habitat Management to Promote Natural Enemies of Agricultural Pests*. University of California Press, Berkeley, California. 422 pages.

Pickett, S.T.A. and White, P.S. (1985).*The Ecology of Natural Disturbance and Patch Dynamics*. Academic Press, San Diego. 472 pages.

Pinto, J.D. and Stouthamer, R. (1994). Systematics of the Trichogrammatidae with emphasis on *Trichogramma*. In *Biological Control with Egg Parasitoids* (E. Wajnberg and S.A. Hassan, eds), pp. 1–36. CAB International, Wallingford.

Pinto, J.D., Koopmanschap, A.B., Platner, G.R. and Stouthamer, R. (2002). The North American *Trichogramma* (Hymenoptera: Trichogrammatidae) parasitizing certain Tortricidae (Lepidoptera) on apple and pear, with ITS2 DNA characterizations and description of a new species. *Biological Control* 21: 134–142.

Pogue, D.W. and Schnell, G.D. (2001). Effects of agriculture on habitat complexity in a prairie-forest ecotone in the Southern Great Plains of North America. *Agriculture, Ecosystems and Environment* 87: 287–298.

Price, P.W. and Waldbauer, G.P. (1994). Ecological aspects of pest management. In *Introduction to Pest Management* (R.L. Metcalf and W.H. Luckmann, eds), pp. 33–65. Wiley-Interscience, New York.

Prinsloo, G., Chen, Y., Giles, K.L. and Greenstone, M.H. (2002). Release and recovery in South Africa of the exotic aphid parasitoid *Aphelinus hordei* verified by the polymerase chain reaction. *BioControl* 47: 127–136.

Quicke, D.L.J. (1997). *Parasitic Wasps*. Chapman & Hall, London. 470 pages.

Reineke, A., Karlovsky, P. and Zebitz, C.P.W. (1999). Suppression of randomly primed polymerase chain reaction products (random amplified polymorphic DNA) in heterozygous diploids. *Molecular Ecology* 8: 1449–1455.

Roland, J. (1993). Large-scale forest fragmentation increases the duration of tent caterpillar outbreak. *Oecologia* 93: 25–30.

Roland, J. and Taylor, P.D. (1997). Insect parasitoid species respond to forest structure at different spatial scales. *Nature* 386: 710–713.

Rosen, D. (1986). The role of taxonomy in effective biological control programs. *Agriculture, Ecosystems and Environment* 15: 121–129.

Rosen, D. and DeBach, P. (1973). Systematics, morphology, and biological control. *Entomophaga* 18: 215–222.

Ruberson, J.R., Nemoto, H. and Hirose, Y. (1998). Pesticides and conservation of natural enemies in pest management. In *Conservation Biological Control* (P. Barbosa, ed.), pp. 207–220. Academic Press, San Diego.

Silva, I.M.M.S., Honda, J., Kan, F. van, Hu, J., Neto, L., Pintureau, B. and Stouthamer, R. (1999). Molecular differentiation of five *Trichogramma* species occurring in Portugal. *Biological Control* 16: 177–184.

Smith, J.W., Wiedenmann, R.N. and Gilstrap, F.E. (1997). Challenges and opportunities for biological control in ephemeral crop habitats: an overview. *Biological Control* 10: 2–3.

Snow, A.A. and Parker, P.G. (1998). Molecular markers for population biology. *Ecology* 79: 359–360.

Stouthamer, R., Jochemsen, P., Platner, G.R. and Pinto, J.D. (2000). Crossing incompatibility between *Trichogramma minutum* and *T. platneri* (Hymenoptera: Trichogrammatidae): implications for application in biological control. *Environmental Entomology* 29: 832–837.

Symondson, W.O.C. and Hemingway, J. (1997). Biochemical and molecular techniques. In *Methods in Ecological and Agricultural Entomology* (D.R. Dent and M.P. Walton, eds), pp. 293–350. CAB International, Wallingford.

Thies, C. and Tscharntke, T. (1999). Landscape structure and biological control in agroecosystems. *Science* 285: 893–895.

Tolstova, Y.S. and Atanov, N.M. (1982). Action of chemical substances for plant protection on the arthropod fauna of the orchard. I. Long-term action of pesticides in agroecoenoses. *Entomological Review* 61: 1–14.

Trewavas, A. (1999). Much food, many problems. *Nature* 402: 231–232.

Tscharntke, T. (2000). Parasitoid populations in the agricultural landscape. In *Parasitoid Population Biology* (M.E. Hochberg and A.R. Ives, eds), pp. 235–253. Princeton University Press, Princeton, New Jersey.

Tscharntke, T. and Kruess, A. (1999). Habitat fragmentation and biological control. In *Theoretical Approaches to Biological Control* (B.A. Hawkins and H.W. Cornell, eds), pp. 190–205. Cambridge University Press, Cambridge.

Vandermeer, J.H. (1995).The ecological basis of alternative agriculture. *Annual Review of Ecology and Systematics* 26: 201–224.

Van Driesche, R.G. and Bellows, T.S. (1996). *Biological Control*. Chapman & Hall, New York. 539 pages.

van Emden, H.F. (1990). Plant diversity and natural enemy efficiency in agroecosystems. In *Critical Issues in Biological Control* (M. Mackauer, L.E. Ehler and J. Roland, eds), pp. 63–80. Intercept, Andover, Massachusetts.

Vanlerberghe-Masutti, F. (1994). Molecular identification and phylogeny of parasitic wasp species (Hymenoptera: Trichogrammatidae) by mitochondrial DNA RFLP and RAPD markers. *Insect Molecular Biology* 3: 229–237.

Vaughn, T.Y.T. and Antolin, M.F. (1998). Population genetics of an opportunistic parasitoid in an agricultural landscape. *Heredity* 80: 152–162.

Williams, J.G.K., Kubelik, A.R., Livak, K.J., Rafalski, J.A. and Tingey, S.V. (1990). DNA polymorphisms amplified by arbitrary primers are useful as genetic markers. *Nucleic Acids Research* 18: 6531–6535.

Wolfenbarger, L.L. and Phifer, P.R. (2000). The ecological risks and benefits of genetically engineered plants. *Science* 290: 2088–2093.

Yoder, J.A. and Hoy, M.A. (1998). Differences in water relations among the citrus leafminer and two different populations of its parasitoid inhabiting the same apparent microhabitat. *Entomologia Experimentalis et Applicata* 89: 169–173.

Zhu, Y.C. and Greenstone, M.H. (1999). Polymerase chain reaction techniques for distinguishing three species and two strains of *Aphelinus* (Hymenoptera: Aphelinidae) from *Diuraphis noxia* and *Schizaphis graminum* (Homoptera: Aphididae). *Annals of the Entomological Society of America* 92: 71–79.

Chapter 7

Marking and tracking techniques for insect predators and parasitoids in ecological engineering

Blas I. Lavandero, Steve D. Wratten, James Hagler and Jason Tylianakis

Introduction

Ecologically based research into biological control of natural enemy populations usually requires an understanding of population dynamics (Barbosa and Wratten 1998; Gurr et al. 2000), dispersal into the crop from a particular resource or habitat patch (Thomas et al. 1992) or immigration into a region. In order to investigate these processes, assessment of natural enemy movement is essential. Consequently, different methods for marking and tracking individuals in the field have been developed. Such techniques should be easy to use, cost-effective, environmentally safe and persist without affecting the animal's behaviour (Southwood 1978; Hagler and Jackson 2001).

Marking and tracking arthropods in habitat manipulation for pest management

Knowledge of the dynamics of insect movement is crucial in habitat manipulation, in which top-down and bottom-up effects can be understood and enhanced. Through provision of pollen, nectar and shelter, population densities and fitness of beneficial arthropods can be enhanced, thereby facilitating top-down reduction of pest populations in a manner consistent with Root's (1973) 'enemies' hypothesis. However, plant resources added to provide nectar and/or pollen may influence the pest's host-plant finding, so may therefore also generate bottom-up effects, supporting the 'resource concentration' hypothesis (Root 1973). It is now widely accepted that top-down and bottom-up forces operate in concert (Oksanen 1991; Leibold 1996), as communities are structured by both plant and predator diversity and abundance.

When resource subsidies (Kean et al. 2003; Gurr et al. 2004; Tylianakis et al. 2004) such as nectar or pollen are added to highly modified habitats, such as agricultural crops, the spatial scale over which the natural enemies move after having access to the resources is largely unknown (Landis et al. 2000; Schellhorn et al. 2000), and the use of efficient markers can therefore help resolve these questions. Experiments showing that the use of flowers enhances parasitoid activity and parasitism rates have been carried out (see Landis et al. 2000; Tylianakis et al. 2004). Even parasitoid sex ratios favouring a higher proportion of females has been observed following the provision of floral resources (Berndt et al. 2002). However, none of these studies addressed spatial questions, the answers to which should help growers determine how to place floral resources within the landscape. The role of different artificial or naturally occurring refuges to enhance top-down effects can also be a critical question, which can be addressed only by the use of marking or tracking techniques (see section 'Dispersal in relation to refugia').

When studying the effects of intercropping and trap cropping, intercrop dispersal of the pest, pest or natural enemy recolonisation patterns, or any study involving population and meta-populations dynamics, some means of marking or tracking insects is required.

Need for marking and tracking arthropods in biological control

The estimation of population size and dispersal are key measures required to understand the dynamics of pests and their natural enemies in integrated pest management (IPM) systems (Dent 2000). This information can act as a basis for decision-making in terms of when and how to intervene with pesticides in relation to economic thresholds and whether natural enemies are exerting sufficient control. Also, the ability to forecast pest outbreaks can depend on knowledge of pest and natural enemies' birth, death, development and dispersal rates. Absolute population estimates require some form of marking or tracking in order to assess total population numbers (Krebs 1985; Southwood and Henderson 2000). The dynamics of host–parasitoid interactions may be influenced by spatial patchiness, interactions with other species and meta-population structure (Hassell 2000). Understanding this structure will require knowledge of migration and recruitment of subpopulations, usually involving some marking or tracking technique. There are also several examples where marking techniques have been used to study metapopulation dynamics in predators (Thomas et al. 1997; Garcia et al. 2000; Tuda and Shima 2002; Wratten et al. 2003) and parasitoids (Lei et al. 1999; Weisser 2000).

Marking and tracking techniques can determine the spatial structure of the population, therefore influencing the area to manage (Thomas 2001). Understanding the seasonal movement of natural enemies, and the quantification of the role of sinks or sources of predators and parasitoids in agroecosystems, can inform decisions on refuge placement and deployment of nectar and pollen as resource subsidies for natural enemies, as well as determining the optimal time for their enhancement (Gurr et al. 2003; Gurr et al. 2004; Wratten et al. 2004).

Dispersal in relation to refugia

Marking and tracking techniques can be used to evaluate refuges as possible sources or sinks of natural enemies following disturbance. Refugia can be classified into four types: natural, artificial, strip- or modified-harvest and within-crop refugia.

Natural refugia

Natural refugia are important because they can be abundant and, if not mismanaged, are likely to harbour high populations of natural enemy species and guilds typical of a region or location. These kinds of refugia can be an important source of natural enemies and of alternative prey/hosts for them (Griffiths and Wratten 1979; Thies and Tscharntke 1999). They can also provide nesting sites and shelter for vertebrates (see ch. 13), many species of which are declining on western European farmland (Krebs et al. 1999; Brickle et al. 2000).

Artificial refugia

Artificial refugia can include habitats such as 'beetle banks' (Wratten et al. 1990; Thomas et al. 1991; Thomas et al. 1992; MacLeod et al. 2004), hedges and multicrop systems where crop areas are planted with non-crop species or are untreated with herbicide. Refugia of this type can comprise a range of natural or introduced plant species harbouring a high diversity of natural enemies (Thomas et al. 2002), although the invertebrate community structure can differ between region and country.

For example, in New Zealand, beetle banks have been established but farmland carabid (ground beetle) species do not feature strongly in the community. Spider species dominate, so perhaps in New Zealand these refugia should more appropriately be called 'spider strips'. However, recent work in New Zealand (a country with an extremely modified, European-oriented agricultural system) has shown that the predominant spider community in refugia in agriculture is dominated by native and endemic species, which do not move far from these semi-permanent habitats (McLachlan and Wratten 2003). Studies using rubidium (as RbCl) as a marker demonstrated that refuge strips supported a large population of early-season parasitoids which later migrated into adjacent crops. For example, this marker made it possible to assess movement of parasitoids of *Bemisia tabaci* between refuge sites and the adjacent crop, also showing the importance of appropriate plant selection for refuges (Pickett et al. 2004).

Refugia involving strip or modified harvests
Unharvested or undisturbed areas in a crop can act as refugia for natural enemies. In a typical agroecosystem, pesticide use, harvesting and other practices such as burning can affect size, distribution and dispersal of insect parasitoid and predator populations (Hossain et al. 2001; Hossain et al. 2002). This in turn may affect recolonisation of the crop by natural enemies in the next season. For example, the use of a fluorescent dye to mark a parasitoid (*Diadegma semiclausum*) of the diamondback moth (*Plutella xylostella*) made it possible to determine the dispersal range of the parasitoid and therefore the percentage of crop required to be left unharvested, in order to allow recolonisation of future crops (Schellhorn and Silberbauer 2002; Schellhorn et al. 2004). In the same way, other marking and tracking systems could be used to study these events in other agroecosystems.

Within-crop refugia
Areas where weeds have not been completely removed can act as refugia for natural enemies. The type of refugium could have different implications for pest management, by increasing or aiding recolonisation by natural enemies.

Within-crop refuges can also act as barriers, limiting the dispersal of natural enemies into a field. For example, *Phacelia tanacetifolia* pollen was used as an internal marker (see section 'Internal markers') to show the ways in which field boundaries act as barriers to movement of hoverflies (Wratten et al. 2003). Such barriers could change the spatial population dynamics of the predatory hoverflies by influencing their movement between fields, and have an impact on any IPM program that uses hoverflies to manage aphid populations (Lövei et al. 1992).

Studies on introduced biological control agents
The need to evaluate the quality and performance of commercially purchased predators for augmentative biological control is another area where tracking and marking will probably gain importance (Heimpel et al. 2004). Releases are made and often a general success or failure is recorded, but frequently no attempt is made to determine the reasons behind the often-low success rate. This is also the case for all attempts at 'classical' biological control of arthropods (Gurr and Wratten 2000; Gurr et al. 2000). Understanding the fate of released individuals is an essential first step towards knowledge-based biological control, and marking techniques may provide this information. The use of protein markers, for example, can facilitate the quality-control monitoring of commercially purchased predators (such as *Hippodamia convergens*) and parasitoids for augmentative biological control (Hagler et al. 2002; Hagler and Naranjo 2004).

Marking and tracking techniques for biological control

Marking and tracking techniques can often provide information that cannot be obtained through other means. This information is critical when population (including meta-populations) estimates and/or dispersal have to be addressed (Krebs 1985; Southwood and Henderson 2000), both of which are usually key components of biological control programs (Van Driesche and Bellows 1996; Gurr et al. 2000; Landis et al. 2000; Gurr et al. 2003). The use of a particular marking or tracking technique will of course depend on the system, and the objectives of the study. This section briefly reviews a sample from a broad spectrum of techniques that are worthy of further research in habitat manipulation and conservation biological control.

Marking techniques

Marking techniques are broadly classified into two groups:

1 external markers, where the mark is applied, abraded or cut into the insect's surface and where individual insect recognition usually occurs through visual methods, following recapture;
2 internal markers, where recognition of markers involves visual, chemical or biochemical methods, following dissection or digestion of the recaptured insects.

External markers

Pollen. As a self-marking technique, pollen can be used to analyse floral resource use and subsequently determine the potential of a particular plant species or family to provide resource subsidies to adult parasitoids and predators. Information on movement of natural enemies between areas of floral resource subsidies and hosts/prey can be determined easily, depending on the system. Pollen contains pollenium, a very enduring protein (Faegri and Iversen 1975; Kearns and Inouye 1993), facilitating its use as a marker.

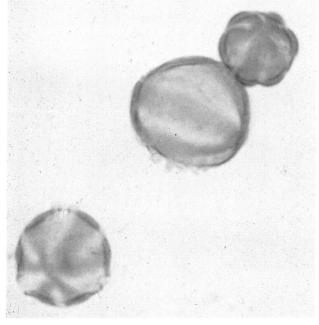

Figure 7.1: *Phacelia* pollen.

Many plants that depend on insect pollination have pollen grains that adhere to the insect's body (see section 'Internal markers', below). Pollen can be distinguished at the species level, allowing determination of the flower-visiting history of a captured insect. To determine migratory pattern, the pollen must come from a geographically remote source (Hagler and Jackson 2001), such as specific and uncommon flowering plants that are added to the crop system to provide marker pollen or to provide resource subsidies (Wratten et al. 2003).

Pollen can be identified using light microscopy (either normal light or phase-contrast) (Figure 7.1) and scanning electron microscopy (SEM) (Turnock et al. 1978; Bryant et al. 1991; DeGrandi-Hoffman et al. 1992). For light microscopy, the pollen can be removed from the insect, washed and mounted on a slide. Mounting can be achieved through acetolysis, glycerin slides, Hoyer's medium or, for permanent preparations, polyvinyl/lactophenol (Kearns and Inouye 1993).

To prepare insects for SEM examination, specimens are normally dried then sputter-coated with gold (Silberbauer et al. 2004). For accurate identification of the pollen to family and species level, the magnification used is approximately 1000× (see Silberbauer et al. 2004). Although SEM can be more accurate, preparation for analysis of pollen can be time-consuming and expensive (Turnock et al. 1978) therefore light microscopy is a better alternative whenever possible (see section 'Internal markers').

Abrasion and branding. Manual abrasion with drills, pins or scalpels can be used to permanently mark the surface of an insect. These kinds of marks can be made easily when the insect's exoskeleton is sufficiently sclerotised (Thomas 1995). Size is an important constraint on the use of these marks and the process can be time-consuming. Because of these drawbacks, abrasion marking is impractical for marking a large number of insects. Although not possible in all insects, these marks have been used successfully to study carabids in farmland ecosystems (Winder 2004). It is also an inexpensive way of marking insects and relatively easy compared with other more sophisticated methods.

Paint and dyes. These markers are commonly used, and date back to the 1920s when they were utilised to study insect population dynamics in the field (Geiger et al. 1919; Dudley and Searles 1923). The mark can be applied in several ways, but the most sensitive technique seems to be the use of hand atomisers or sprayguns, using water-soluble paints or dyes (Figure 7.2) (Porter and Jorgensen 1980; Schellhorn et al. 2004). This provides an easy and relatively inexpensive way of mass-marking natural enemies. However, using a brush to apply a single paint-mark is less likely to damage sensitive areas of the insect as eyes or antennae (Frampton et al. 1995; Mauremootoo et al. 1995).

There are different types of paints and dyes that can be used; however, resin-based dyes seem to be generally more stable than water-based ones (see Schellhorn et al. 2004). Although using paints or dyes as markers is simple, they can be used only on a single life-stage because insects lose the mark when they moult.

Protein marking. Natural enemies can be externally marked with specific vertebrate proteins which, following insect recapture, can be detected by a protein-specific sandwich enzyme-linked inmunoabsorbent assay (ELISA) (Hagler et al. 1992; Hagler and Jackson 1998; Hagler et al. 2002; Hagler and Miller 2002). Mass-reared biological control agents can be sprayed with specific immunoglobulin G (IgG) using a standard hand-sprayer, then dried. The use of vertebrate proteins for marking natural enemies is a promising alternative to many conventional marking procedures (Hagler and Jackson 2001). Absorbance is measured with a microplate reader and the positively-marked insects are identified (Hagler et al. 1992). Recent improvements reduce the labour required to conduct ELISA by soaking the insect instead of homogenising it in sample buffer (Hagler 2004). These and future developments should bring down the cost and labour involved in protein-marking studies.

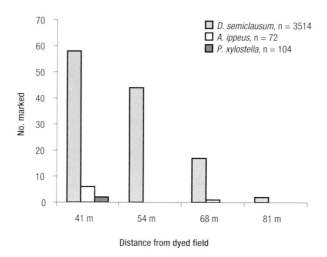

Figure 7.2: The number of *D. semiclausum, A. ippeus* and *P. xylostella,* marked with a fluorescent pigment, captured on yellow-sticky-bucket traps placed 20 cm above ground, at 4 distances at St Kilda, South Australia. The number of individuals captured at each distance from 41–81 m was 1542, 1129, 898 and 121 respectively.

Reproduced with permission from Schellhorn (SARDI) et al. *The use of dusts and dyes to mark populations of benefical insects in the field;* published by Taylor and Francis 2004.

Internal markers

Pollen. Through dissection of specimens, ingested pollen grains can be visualised. Often the use of a stain (such as saffranin, e.g. Wratten et al. 1995, Hagler and Jackson 2001) will help in the identification of the pollen grains. Pollen as an internal marker can be used when studying only those natural enemies that ingest it. It is currently unclear if parasitoids use pollen as a source of food or if pollen intake occurs only in association with nectar feeding (Jervis et al. 1992). However, many predatory insects such as hoverflies are pollen-feeders. Several studies have shown that gut examination for pollen grain types and quantity is feasible for hoverflies that have been caught in yellow-water traps (Wratten et al. 1995; Hickman and Wratten 1996; Lövei et al. 1998a; Irvin et al. 1999; Hickman et al. 2001; Wratten et al. 2003). For rapid assessments of the presence or absence of particular pollen types (plant families or species) in insect guts, the methods of Wratten et al. (2003) may be adequate. In that work, the insect abdomen was removed (on a microscope slide) and teased apart with mounted needles, then saffranin stain was added, followed by a coverslip. Preliminary scanning under a binocular microscope, using a range of magnifications, is ideally followed by examination using compound microscopy (Irvin et al. 1999; Wratten et al. 2003).

C3/C4. Carbon isotope ratios have been used to examine coccinellid (*Hippodamia convergens*) movement in a multicrop system involving C3 and C4 plants. The ratios were transferred from the herbivores feeding on the plants to their predators, and movement from one photosynthetic plant type to another could be detected, revealing the importance of refuge sources in insect pest management (Prasifka and Heinz 2004).

Rubidium. Rare-earth or trace elements such as rubidium occur naturally in soil, plants and insects, and techniques using these have been developed to mark both herbivorous and entomophagous insects in the field (Payne and Wood 1984; Hopper and Woolson 1991; Corbett et al. 1996; Fernandes et al. 1997; Long et al. 1998; Prasifka et al. 2001). Although trace elements for insect marking can include strontium, cesium, manganese, hafnium and iridium; rubidium is most frequently used in marking studies (see Hagler and Jackson 2001). Predators and

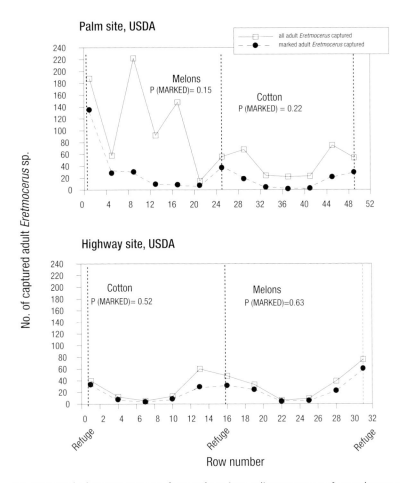

Figure 7.3: Dispersal of *Eretmocerus* sp. from refuge into adjacent crops of cantaloupe and cotton IVRS, Brawley, California. Shown are cumulative captures from April to June 1995. Insects were self-marked with RbCl.

Reproduced with permission from Picket et al. *A natural enemy refuge, marking with rubidium and movement of Bemisia parasitoids*; published by Taylor and Francis 2004.

parasitoids can be marked (Figure 7.3) or self-mark by feeding on floral nectar or pollen from rubidium-marked plants, or on prey/hosts that have fed on rubidium-marked plants. An aqueous solution of rubidium chloride (RbCl) is used to artificially raise the natural level of rubidium in insects (Berry et al. 1972). The solution can be applied to a specific habitat and, because it is absorbed by plants, can move through the phloem and be ingested with nectar or pollen or directly by herbivores, then transferred to predators or parasitoids (Graham et al. 1978; Corbett et al. 1996; Prasifka et al. 2001). Rubidium is located between potassium and cesium in the periodic table and can serve as a chemical analog of potassium. The last element is replaced by rubidium in the tissues of insects feeding on treated plants (Berry et al. 1972). At low concentrations, rubidium has little if any impact on survivorship or behaviour (Jackson et al. 1988; Hopper and Woolson 1991).

Rubidium content of insects is normally determined using flame emission spectroscopy (Jackson et al. 1988; Hopper and Woolson 1991; Corbett et al. 1996; Corbett and Rosenheim 1996). Chemical digestion can be achieved via a two-step wet oxidation procedure using concentrated nitric acid and hydrogen peroxide (see Corbett et al. 1996 for details), microwaving samples in nitric acid (Hopper and Woolson 1991) or by ashing under high temperatures then dissolving in diluted nitric acid (Jackson et al. 1988).

Sugar. The anthrone test, high-performance liquid chromatography, gas chromatography and thin-layer chromatography are all useful tools for identification of sugars within insect guts (Olson et al. 2000; Heimpel and Jervis 2004). Research on a range of groups of biting flies using these methods has been able to separate flies that had fed on honeydew, nectar or both (Wallbanks et al. 1991; Burgin and Hunter 1997; Burkett et al. 1999; Hunter and Ossowski 1999; Russell and Hunter 2002) and the technique has recently been applied to parasitoids (Heimpel et al. 2004). Elucidating the origin of sugars in insect guts can allow sugar-feeding habits in the field to be characterised. It can also help to determine the importance of different sugar resources available to natural enemies and determine the spatial scale needed for pest management.

Protein marking. Analysis of insect predators' guts can also be done with ELISA. Specific proteins can be fed to insects, which are released and later recaptured, and analysed by a protein-specific ELISA (Hagler 1997). Parasitoids can also be marked internally by feeding them honey containing a specific protein (Hagler and Jackson 1998). In the case of predators eating prey, releasing laboratory-reared aphids, for example, as an aggregation into a crop in which they do not occur naturally could enable the movement of aphid predators to be tracked. Pitfall trapping of carabids at distances from the artificial aphid aggregation would be an appropriate method. An example would be inserting laboratory-derived cultures of the pea aphid *Acyrthosiphon pisum* on broad bean (*Vicia faba*) into the ground in a cereal field. For more details and examples of the use of protein markers see Hagler et al. (1992), Hagler (1997), Hagler and Jackson (1998), Hagler and Jackson (2001), Hagler et al. (2002) and Hagler and Miller (2002).

Molecular markers. These markers have been used to study dispersal, mating, immigration source, population relationships, genetic structure and diversity and insecticide resistance at the individual, population or species level (Wagner and Selander 1974; Bartlett 1982; Hoy 2003). Although molecular markers have proven effective not only for studies of physical dispersal but also to examine gene flow, this requires extensive preparation, has a high cost and is time-consuming (Hagler and Jackson 2001). However, the cost of using allozymes and DNA sequencing has decreased greatly, compared with other molecular techniques (Caterino et al. 2000; Hoy 2003).

These markers can be visualised on gels after electrophoresis of samples, using various techniques such as allozymes, DNA fingerprinting by analysis of microsatellites, amplified fragment length polymorphisms (AFLPs), random amplified polymorphism (RAPDs), restriction fragment length polymorphism (RFLPs) of mitochondrial or nuclear DNA, allele-specific PCR, double-strand conformation polymorphism (DSCP), heteroduplex analysis (HDA), microarrays and DNA (nuclear and mitochondrial) sequencing. Detailed description of each technique for use in insect molecular ecology can be found in Hoy (2003).

Various molecular marker techniques have been used for following the movements of insects in space and time; these include allozymes, microsatellites, mtDNA sequences, RAPDs, AFLPs and introns (MacDonald and Loxdale 2004).

Allozyme analysis is one of the most cost-effective techniques available and is relatively easy to perform, compared with other molecular methods (Hunter et al. 1996; Atanassova et al. 1998; Kitthawee et al. 1999), but this method may not detect sufficient variation to answer some ecological questions. Also, because proteins are less stable than is DNA, they are more sensitive to handling and storage (Hoy 2003).

Microsatellite markers can be used to reveal differentiation among closely related populations in the field (Burke 1989; Wang et al. 1999). They can be a useful tool to study dispersal and establishment of specific/biotypes, including those with low levels of protein variation, such as hymenopteran parasitoids (Hoy 2003). Because microsatellite sequences vary between species, sequence data must be obtained for each species being studied, making it expensive to develop and time-consuming. However, there are many published microsatellite loci for a variety of organisms (Neve and Meglecz 2000).

The RAPD approach is often used for population studies, particularly for asexual organisms and those such as haplo-diploid hymenopteran species suspected of low genetic variation. In this way the structure and dynamics of the population of parasitoid *Diaeretiella rapae* in an environment with two hosts available for oviposition (the cabbage aphid *Brevicoryne brassicae* on crucifers, and the Russian wheat aphid *Diuraphis noxia*) was studied. This showed that *D. rapae* adult parasitoids, having located habitats with aphids, disperse only short distances (Vaughn and Antolin 1998).

Examples where primers have been designed for sequencing specific DNA regions include *Aphidius ervi* (Hufbauer et al. 2001), *Cotesia congregata* (Jensen et al. 2002), *Diaeretiella rapae* (MacDonald et al. 2004) and several predatory beetles (Brouat et al. 2002; Haddril et al. 2002; Keller and Largiader 2003). For more examples of the use of RFLP and AFLP for insect marking see MacDonald and Loxdale (2004).

Tracking techniques

When conducting long-term ecological investigations, or studies involving insects that move great distances, simple low-cost markers may not be the best option. The use of tracking devices such as electronic tags and portable harmonic radar to study insect movement may be required. Recent developments in 'entomological' radars have greatly improved the practicability of observing both migratory and non-migratory flights of insects, including beneficial species (Chapman et al. 2004). There are few examples of studies using radar tracking systems to study predators and parasitoids. Some work using harmonic direction-finders has been able to unravel behaviour of night-active carabids (Wallin and Ekbom 1988; van der Ent 1989; Wallin 1991; Kennedy 1994; Wallin and Ekbom 1994; Lövei et al. 1998b). Another study conducted with two beneficial coccinellid beetles (*H. convergens* and *Harmonia axyridis*) revealed that these beneficial insects could retain their ability to fly, being good candidates for studies using electronic tagging (Boiteau and Colpitts 2004).

Although there are few studies showing the use of radar technology in predator and parasitoid studies, there are many examples of this technique with other arthropod groups that can be easily adapted for predation studies.

Conclusions

Marking and tracking techniques have been greatly advanced in recent years, and continue to evolve. Each technique can be adapted to new situations, depending on the questions asked. Not every technique can be used in every system; some techniques have not yet been used in studying movement of insect predators and parasitoids, but the potential is clear.

This chapter has emphasised some of the available techniques that have been used or that seem of most potential for use in studies of predator and parasitoid movement, involving habitat manipulation. For a more detailed analysis, see Lavandero et al. (2004).

References

Atanassova, P., Brookes, C.P., Loxdale, H.D. and Powell, W. (1998). Electrophoretic study of five aphid parasitoid species of the genus Aphidius (Hymenoptera: Braconidae), including evidence for reproductively isolated sympatric populations and a cryptic species. *Bulletin of Entomological Research* 88: 3–13.

Barbosa, P. and Wratten, S.D. (1998). Influence of plants on invertebrate predators: implications to conservation biological control. In *Conservation Biological Control* (P. Barbosa, ed.), pp. 83–100. Academic Press, San Diego.

Bartlett, A.C. (1982). Genetic markers: discovery and use in insect population dynamics studies and control programs. In *Sterile Insect Technique and Radiation in Insect Control*, pp. 461–465. International Atomic Energy Agency, Vienna.

Berndt, L.A., Wratten, S.D. and Hassan, P.G. (2002). Effects of buckwheat flowers on leafroller (Lepidoptera: Tortricidae) parasitoids in a New Zealand vineyard. *Agricultural and Forest Entomology* 4: 39–45.

Berry, W., Stimmann, M.W. and Wolf, W. (1972). Marking of native phytophagous insects with rubidium: a proposed technique. *Annals of the Entomological Society of America* 65: 236–238.

Boiteau, G. and Colpitts, B. (2004). The potential of harmonic radar technology for tracking beneficial insects. In *Effective Marking and Tracking Techniques for Monitoring the Movement of Predator and Parasitoid Arthropods* (B.I. Lavandero, S.D. Wratten and J. Hagler, eds). Special issue: *International Journal of Pest Management*. In press.

Brickle, N.W., Harper, D.G.C., Aebischer, N.J. and Cockayne, S.H. (2000). Effects of agriculture intensification on the breeding success of corn buntings *Miliaria calandra*. *Journal of Applied Ecology* 37 (5): 742–755.

Brouat, C., Mondor-Genson, G. and Audiot, P. (2002). Isolation and characterization of microsatellite loci in the ground beetle *Carabus nemoralis* (Coleoptera: Carabidae). *Molecular Ecology Notes* 2: 119–120.

Bryant, V., Pendleton, M., Murray, R., Lingren, P. and Raulston, J. (1991). Techniques for studying pollen adhering to nectar-feeding corn earworm (Lepidoptera: Noctuidae) moths using scanning electron microscopy. *Journal of Economic Entomology* 84: 237–240.

Burgin, S.G. and Hunter, F.F. (1997). Nectar versus honeydew as sources of sugar for male and female black flies (Diptera: Simuliidae). *Journal of Medical Entomology* 34: 605–608.

Burke, T. (1989). DNA fingerprinting and other methods for the study of mating success. *Trends in Ecology and Evolution* 4: 139–144.

Burkett, D.A., Kline, D.L. and Carlson, D.A. (1999). Sugar meal composition of five North Central Florida mosquito species (Diptera: Culicidae) as determined by gas chromatography. *Journal of Medical Entomology* 36: 462–467.

Caterino, M.S., Cho, S. and Sperling, F.A.H. (2000). The current state of insect molecular systematics: a thriving tower of Babel. *Annual Review of Entomology* 45: 1–54.

Chapman, J.W., Reynolds, D.R. and Smith, A.D. (2004). Migratory and foraging movements in beneficial insects: a review of radar monitoring and tracking methods. In *Effective Marking and Tracking Techniques for Monitoring the Movement of Predator and Parasitoid Arthropods* (B.I. Lavandero, S.D. Wratten and J. Hagler, eds). Special issue: *International Journal of Pest Management*. In press.

Corbett, A. and Rosenheim, J.A. (1996). Impact of a natural enemy overwintering refuge and its interaction with the surrounding landscape. *Ecological Entomology* 21: 155–164.

Corbett, A., Murphy, B.C., Rosenheim, J. and Bruins, P. (1996). Labeling and egg parasitoid, *Anagrus epos* (Hymenoptera: Mymaridae), with rubidium within an overwintering refuge. *Environmental Entomology* 25 (1): 29–38.

DeGrandi-Hoffman, G., Thorp, R., Loper, G. and Eisikowitch, D. (1992). Identification and distribution of cross-pollinating honeybees on almonds. *Journal of Applied Ecology* 29: 238–246.

Dent, D. (2000). *Insect Pest Management*. CAB International, London.

Dudley, J. and Searles, E. (1923). Color marking of the striped cucumber beetle (*Diabrotica vittata* Fab.) and preliminary experiments to determine its flight. *Journal of Economic Entomology* 16: 363–368.

Faegri, K. and Iversen, J. (1975). *Textbook of Pollen Analysis*. Blackwell Scientific, Oxford.

Fernandes, O., Wright, R., Baumgarten, K. and Mayo, Z. (1997). Use of rubidium to label *Lysiphlebus testaceipes* (Hymenoptera: Braconidae), a parasitoid of greenbugs (Homoptera: Aphididae), for dispersal studies. *Environmental Entomology* 26: 1167–1172.

Frampton, G.K., Cilgi, T., Fry, G. and Wratten, S.D. (1995). Effects of grassy banks on the dispersal of some carabid beetles (Coleoptera: Carabidae) on farmland. *Biological Conservation* 71: 347–355.

Garcia, A.F., Griffiths, G.J.K. and Thomas, C.F.G. (2000). Density, distruibution and dispersal of the carabid beetle *Nebria brevicollis* in two adjacent cereal fields. *Annals of Applied Biology* 137 (2): 89–97.

Geiger, J., Purdy, W. and Tarbett, R. (1919). Effective malarial control in a rice field district with observations on experimental mosquito flights. *Journal of the American Medical Association* 72: 844–847.

Graham, H., Wolfenbarger, D. and Nosky, J. (1978). Labeling plants and their insects fauna with rubidium. *Environmental Entomology* 7: 379–383.

Griffiths, E. and Wratten, S.D. (1979). Intra- and inter-specific differences in cereal aphid low-temperature tolerance. *Entomologia Experimentalis et Applicata* 26: 161–167.

Gurr, G.M. and Wratten, S.D. (2000). *Biological Control: Measures of Success*. Kluwer Academic, Dordrecht.

Gurr, G.M., Wratten, S.D. and Barbosa, P. (2000). Success in conservation biological control of arthropods. In *Biological Control: Measures of Success* (G.M. Gurr and S.D. Wratten, eds), pp. 105–132. Kluwer Academic, Dordrecht.

Gurr, G.M., Wratten, S.D. and Luna, J.M. (2003). Multi-function agricultural biodiversity: pest management and other benefits. *Basic and Applied Ecology* 4: 107–116.

Gurr, G.M., Wratten, S.D., Tylianakis, J.K.J. and M. Keller (2004). Providing plant foods for insect natural enemies in farming systems: Balancing practicalities and theory. In *Plant-derived Food and Plant–carnivore Mutualism* (F.L. Wäckers, P. van Rijn and J. Bruin, eds). Cambridge University Press, Cambridge. In press.

Haddril, P., Majerus, M. and Mayes, S. (2002). Isolation and characterization of highly polymorphic microsatellite loci in the two-spot ladybird, *Adalia bipunctata*. *Molecular Ecology Notes* 2: 316–319.

Hagler, J. (1997). Protein marking insects for mark-release-recapture studies. *Trends in Entomology* 1: 105–115.

Hagler, J. (2004). Improvements on protein marking technology. In *Effective Marking and Tracking Techniques for Monitoring the Movement of Predator and Parasitoid Arthropods* (B.I. Lavandero, S.D. Wratten and J. Hagler, eds). Special issue: *International Journal of Pest Management*. In press.

Hagler, J. and Jackson, C.G. (1998). An immunomarking technique for labeling minute parasitoids. *Environmental Entomology* 27: 1010–1016.

Hagler, J.R. and Jackson, C.G. (2001). Methods for marking insects: current techniques and future prospects. *Annual Review of Entomology* 46: 511–543.

Hagler, J. and Miller, E. (2002). An alternative to conventional insect marking procedures: detection of a protein mark on pink bollworm by ELISA. *Entomologia Experimentalis et Applicata* 103: 1–9.

Hagler, J. and Naranjo, S. (2004). The use of multiple protein markers to label *Hippodamia convergens* for investigating their intercrop dispersal patterns. In *Effective Marking and Tracking Techniques for Monitoring the Movement of Predator and Parasitoid Arthropods* (B.I. Lavandero, S.D. Wratten and J. Hagler, eds). Special issue: *International Journal of Pest Management*. In press.

Hagler, J., Cohen, A.C., Bradley-Dunlop, D. and Enriquez, F. (1992). New approach to mark insects for feeding and dispersal studies. *Environmental Entomology* 21: 20–25.

Hagler, J., Jackson, C.G., Henneberry, T. and Gould, J. (2002). Parasitoid mark-release-recapture techniques: II. Development and application of a protein marking technique for *Eretmocerus* spp., parasitoids of *Bemisia argentifolii*. *Biocontrol Science and Technology* 12 (6): 661–675.

Hassell, M.P. (2000). Host–parasitoid population dynamics. *Journal of Animal Ecology* 69: 543–566.

Heimpel, G.H. and Jervis, M.A. (2004). An evaluation of the hypothesis that floral nectar improves biological control by parasitoids. In *Plant-provided Food and Plant–carnivore Mutualism* (F. Wäckers, P. van Rijn and J. Bruin, eds). Cambridge University Press, Cambridge. In press.

Heimpel, G.E., Lee, J., Wu, Z., Weiser, L., Wäckers, F. and Jervis, M. (2004). Tracking biting flies and parasitoids by identifying the sugars in their gut. In *Effective Marking and Tracking Techniques for Monitoring the Movement of Predator and Parasitoid Arthropods* (B.I. Lavandero, S.D. Wratten and J. Hagler, eds). Special issue: *International Journal of Pest Management*. In press.

Hickman, J.M. and Wratten, S.D. (1996). Use of *Phacelia tanacetifolia* flower strips to enhance biological control of aphids by hoverfly larvae in cereal fields. *Journal of Economic Entomology* 89: 832–840.

Hickman, J.M., Wratten, S.D., Jepson, P.C. and Frampton, C.M. (2001). Effect of hunger on yellow water trap catches of hoverfly (Diptera: Syrphidae) adults. *Agricultural and Forest Entomology* 3: 1–6.

Hopper, K.R. and Woolson, E.A. (1991). Labeling a parasitic wasp, *Microplitis croceipes* (Hymenoptera: Braconidae), with trace elements for mark-recapture studies. *Annals of the Entomological Society of America* 84 (3): 255–262.

Hossain, Z., Gurr, G.M. and Wratten, S.D. (2001). Habitat manipulation in lucerne (*Medicago sativa* L.): strip harvesting to enhance biological control of insect pests. *International Journal of Pest Management* 47 (2): 81–88.

Hossain, Z., Gurr, G.M., Wratten, S.D. and Raman, A. (2002). Habitat manipulation in lucerne *Medicago sativa*: arthropod population dynamics in harvested and 'refuge' crop strips. *Journal of Applied Ecology* 39 (3): 445–454.

Hoy, M.A. (2003). *Insect Molecular Genetics: An Introduction to Principles and Applications*. Academic Press, San Diego.

Hufbauer, R., Bogdanowicz, S., Perez, L. and Harrison, R. (2001). Isolation and characterization of microsatellites in *Aphidius ervi* (Hymenoptera: Braconidae) and their applicability to related species. *Molecular Ecology Notes* 1: 197–199.

Hunter, F.F. and Ossowski, A.M. (1999). Honeydew sugars in wild-caught female horse flies (Diptera: Tabanidae). *Journal of Medical Entomology* 36: 896–899.

Hunter, M.S., Antolin, M.F. and Rose, M. (1996). Courtship behaviour, reproductive relationships, and allozyme patterns of three North American populations of *Eretmocerus* nr. *californicus* (Hymenoptera: Aphelinidae) parasitizing the whitefly *Bemisia* sp, *tabaci* complex (Homoptera: Aleyrodidae). *Proceedings of the Entomological Society of Washington* 98: 126–137.

Irvin, N.A., Wratten, S.D., Frampton, C.M., Bowie, M.H., Evans, A.M. and Moar, N.T. (1999). The phenology and pollen feeding of three hover fly (Diptera: Syrphidae) species in Canterbury, New Zealand. *New Zealand Journal of Zoology* 26: 105–115.

Jackson, C.G., Cohen, A.C. and Verdugo, C.L. (1988). Labeling *Anaphes ovijentatus* (Hymenoptera: Mymaridae), and egg parasite of *Lygus* spp. (Hemiptera: Miridae), with rubidium. *Annals of the Entomological Society of America* 81 (6): 919–922.

Jensen, M., Kester, K., Kankare, M. and Brown, B. (2002). Characterization of microsatellite loci in the parasitoid, *Cotesia congregata* (Say) (Hymenoptera: Braconidae). *Molecular Ecology Notes* 2: 346–348.

Jervis, M.A., Kidd, N.A.C. and Walton, M. (1992). A review of methods for determining dietary range in adult parasitoids. *Entomophaga* 37 (4): 565–574.

Kean, J., Wratten, S.D., Tylianakis, J. and Barlow, N. (2003). The population consequences of natural enemy enhancement, and implications for conservation biological control. *Ecology Letters* 6: 1–9.

Kearns, C.A. and Inouye, D.W. (1993). *Techniques for Pollination Biologists*. University Press of Colorado, Colorado.

Keller, I. and Largiader, C. (2003). Five microsatellite DNA markers for the ground beetle *Abax parallelepipedus* (Coleoptera: Carabidae). *Molecular Ecology Notes* 3: 113–114.

Kennedy, P.J. (1994). The distribution and movement of ground beetles in relation to set-aside arable land. In *Carabid Beetles: Ecology and Evolution* (K. Desender, M. Dufrêne, M. Loreau, M.L. Luff and J.P. Maelfait, eds), pp. 439–444. Kluwer Academic, Dordrecht.

Kitthawee, S., Julsilikul, D., Sharpe, R.G. and Baimai, V. (1999). Protein polymorphism in natural populations of *Diachasmimorpha longicaudata* (Hymenoptera : Braconidae) in Thailand. *Genetica* 105: 125–131.

Krebs, C.J. (1985). *Ecology: The Experimental Analysis of Distribution and Abundance*. Harper & Row, New York.

Krebs, J., Wilson, J., Bradbury, R. and Siriwardena, G. (1999). The second silent spring? *Nature* 400: 611–612.

Landis, D.A., Wratten, S.D. and Gurr, G.M. (2000). Habitat management to conserve natural enemies of arthropod pests in agriculture. *Annual Review of Entomology* 45: 175–201.

Lavandero, B.I., Wratten, S.D., Hagler, J.R. and Jervis, M. (2004). The importance of effective marking and tracking techniques for monitoring the movement of predator and parasitoid arthropods. In *Effective Marking and Tracking Techniques for Monitoring the Movement of Predator and Parasitoid Arthropods* (B.I. Lavandero, S.D. Wratten and J. Hagler, eds). Special issue: *International Journal of Pest Management*. In press.

Lei, G., Camara, M.D. and Lei, G. (1999). Behaviour of a specialist parasitoid, *Cotesia melitaearum*: from individual behaviour to metapopulation processes. *Ecological Entomology* 24 (1): 59–72.

Leibold, M.A. (1996). A graphical model of keystone predators in food webs: Trophic regulation of abundance, incidence, and diversity patterns in communities. *American Naturalist* 147: 784–812.

Long, R., Corbett, A., Lamb, C., Reberg-Horton, C., Chandler, J. and Stimmann, M.W. (1998). Movement of beneficial insects from flowering plant to associated crops. *California Agriculture* 52: 23–36.

Lövei, G.L., McDougall, D., Bramley, G., Hodgson, D.J. and Wratten, S.D. (1992). Floral resources for natural enemies: the effect of *Phacelia tanacetifolia* (Hydrophyllaceae) on within-field distribution of hoverflies (Diptera: Syrphidae). *Proceedings of the 45th New Zealand Crop Protection Conference*, pp. 60–61. New Zealand Plant Protection Society, Rotoroa.

Lövei, G.L., MacLeod, A. and Hickman, J.M. (1998a). Dispersal and effects of barriers on the movement of the New Zealand hoverfly *Melanostoma fasciatum* (Diptera: Syrrphidae) on cultivated land. *Journal of Applied Entomology* 122: 115–120.

Lövei, G.L., Stringer, I.A.N., Devine, C.D. and Cartellieri, M. (1998b). Harmonic radar – a method using inexpensive tags to study invertebrate movement on land. *New Zealand Journal of Ecology* 21: 187–193.

MacDonald, C. and Loxdale, H. (2004). Molecular markers to study population structure and dynamics in beneficial insects. In *Effective Marking and Tracking Techniques for Monitoring the Movement of Predator and Parasitoid Arthropods* (B.I. Lavandero, S.D. Wratten and J. Hagler, eds). Special issue: *International Journal of Pest Management*. In press.

MacDonald, C., Brookes, C.P., Edwards, K., Baker, D., Lockton, S. and Loxdale, H.D. (2004). Microsatellite isolation and characterization in the beneficial parasitoid wasp *Diaeretiella rapae* (M'Intosh) (Hymenoptera: Braconidae). *Molecular Ecology Notes*. In press.

MacLeod, A., Wratten, S.D., Sotherton, N.W. and Thomas, M. (2004). 'Beetle banks' as refuges for beneficial arthropods in farmland: long-term changes in predator communities and habitat. *Agricultural and Forest Entomology*. In press.

Mauremootoo, J., Wratten, S.D., Worner, S. and Fry, G. (1995). Permeability of hedgerows to predatory carabid beetles. *Agriculture, Ecosystems and Environment* 52: 141–148.

McLachlan, A.R.G. and Wratten, S.D. (2003). Abundance and species richness of field-margin and pasture spiders (Araneae) in Canterbury, New Zealand. *New Zealand Journal of Zoology* 30: 57–67.

Neve, G. and Meglecz, E. (2000). Microsatellite sequences in different taxa. *Trends in Ecology and Evolution* 15: 376–377.

Oksanen, L. (1991). Trophic levels and trophic dynamics: a consensus emerging? *Trends in Ecology and Evolution* 6: 58–60.

Olson, D.M., Fadamiro, H., Lundgren, J.G. and Heimpel, G.E. (2000). Effects of sugar feeding on carbohydrate and lipid metabolism in parasitoid wasps. *Physiological Entomology* 25: 17–26.

Payne, J. and Wood, B. (1984). Rubidium as a marking agent for the hickory shuckworm, *Cydia caryana* (Lepidoptera: Tortricidae). *Environmental Entomology* 13: 1519–1521.

Pickett, C.H., Roltsch, W.J. and Corbett, A. (2004). Tracking parasitoid movement using rubidium chloride and the role of an overwintering refuge. In *Effective Marking and Tracking Techniques for Monitoring the Movement of Predator and Parasitoid Arthropods* (B.I. Lavandero, S.D. Wratten and J. Hagler, eds). Special issue: *International Journal of Pest Management*. In press.

Porter, S. and Jorgensen, C. (1980). Recapture studies of the harvester ant, *Pogonomyrmex owyheei* Cole, using a fluorescent marking technique. *Ecological Entomology* 5: 263–269.

Prasifka, J.R. and Heinz, K.M. (2004). The use of C3 and C4 plants to study natural enemy ecology and movement. In *Effective Marking and Tracking Techniques for Monitoring the Movement of Predator and Parasitoid Arthropods* (B.I. Lavandero, S.D. Wratten and J. Hagler, eds). Special issue: *International Journal of Pest Management*. In press.

Prasifka, J., Heinz, K. and Sansone, C. (2001). Field testing rubidium marking for quantifying intercrop movement of predatory arthropods. *Environmental Entomology* 30: 711–719.

Root, R.B. (1973). Organization of a plant–arthropod association in simple and diverse habitats : the fauna of collards (*Brassica oleracea*). *Ecological Monographs* 43: 95–124.

Russell, C.B. and Hunter, F.F. (2002). Analyses of nectar and honeydew feeding in *Aedes* and *Ochlerotatus* mosquitoes. *Journal of the American Mosquito Control Association* 18: 86–90.

Schellhorn, N.A. and Silberbauer, L.X. (2002). The role of surrounding vegetation and refuges: increasing the effectiveness of predators and parasitoids in cotton and broccoli systems. *Proceedings for the International Symposium on the Biological Control of Arthropods, Honolulu*, pp. 235–243. US Department of Agriculture, Forest Service, Morgantown, WV, FHTET–2003–05.

Schellhorn, N., Harmon, J.P. and Andow, D.A. (2000). Using cultural practices to enhance insect pest control by natural enemies. In *Insect Pest Management: Techniques for Environmental Protection* (J. Recheigl and N. Recheigl, eds). CRC Press, Boca Raton.

Schellhorn, N.A., Siekmann, G., Paull, C. and Furness, G. (2004). Using dyes and dust to mark natural populations of insects in the field. In *Effective Marking and Tracking Techniques for Monitoring the Movement of Predator and Parasitoid Arthropods* (B.I. Lavandero, S.D. Wratten and J. Hagler, eds). Special issue: *International Journal of Pest Management*. In press.

Silberbauer, L., Gregg, P., Wratten, S.D., Del Socorro, A., Bowie, M. and Yee, M. (2004). Pollen as a marker to track movement of predatory insects in agroecosystems. In *Effective Marking and Tracking Techniques for Monitoring the Movement of Predator and Parasitoid Arthropods* (B.I. Lavandero, S.D. Wratten and J. Hagler, eds). Special issue: *International Journal of Pest Management*. In press.

Southwood, T.R.E. (1978). *Ecological Methods: With Particular Reference to the Study of Insect Populations*. Chapman & Hall, London.

Southwood, T.R.E. and Henderson, P.A. (2000). *Ecological Methods*. Blackwell Science, Oxford.

Thies, C. and Tscharntke, T. (1999). Landscape structure and biological control in agroecosystems. *Science* 285: 893–895.

Thomas, C.D. (2001). Scale, dispersal and population structure. In *Insect Movement Mechanisms and Consequences* (I.P. Woiwod, D.R. Reynolds and C.D. Thomas, eds), pp. 321–336. CAB International, London.

Thomas, C.F.G. (1995). A rapid method for handling and marking carabids in the field. In *Arthropod Natural Enemies in Arable Land I. Density, Spatial Heterogeneity and Dispersal* (S. Toft and W. Riedel, eds). *Acta Jutlandica* 70: 57–59.

Thomas, C.F.G., Green, F. and Marshall, E.J.P. (1997). Distribution, dispersal and population size of ground beetles, *Pterostichus melanarius* (Illiger) and *Harpalus rufipes* (Degeer)(Coleoptera, Carabidae), in field margin habitats. *Biological Agriculture and Horticulture* 15 (1–4): 337–352.

Thomas, C.R., Noordhuis, R., Holland, J.M. and Goulson, D. (2002). Botanical biodiversity of beetle banks: effects of age and comparison with conventional arable field margins in southern UK. *Agriculture, Ecosystems and Environment* 93 (1–3): 403–412.

Thomas, M., Wratten, S.D. and Sotherton, N.W. (1992). Creation of 'island' habitats in farmland to manipulate populations of beneficial arthropods: predator densities and species composition. *Journal of Applied Ecology* 29: 524–531.

Thomas, M.B., Wratten, S.D. and Sotherton, N.W. (1991). Creation of 'island' habitats in farmland to manipulate populations of beneficial arthropods: predator densities and emigration. *Journal of Applied Ecology* 28: 906–917.

Tuda, M. and Shima, K. (2002). Relative importance of weather and density dependence on the dispersal and on-plant activity of the predator *Orius minutus*. *Population Ecology* 44 (3): 251–257.

Turnock, W., Chong, J. and Luit, B. (1978). Scanning electron microscopy: a direct method of identifying pollen grains on moths (Noctuidae: Lepidoptera). *Canadian Journal of Zoology* 56: 2050–2054.

Tylianakis, J.M., Didham, R.K. and Wratten, S.D. (2004). Improved fitness of aphid parasitoids receiving resource subsidies. *Ecology*. 85: 658–666.

van der Ent, L.J. (1989). Individual walking-behaviour of the carabid *Carabus problematicus* Herbst in a Dutch forest has been recorded by a radar-detection-system. *Netherlands Journal of Zoology* 39 (1–2): 104.

Van Driesche, R.G. and Bellows, T.S.J. (1996). *Biological Control*. Chapman & Hall, New York.

Vaughn, T.T. and Antolin, M.F. (1998). Population genetics of an opportunistic parasitoid in an agricultural landscape. *Heredity* 80: 152–162.

Wagner, R.P. and Selander, R.K. (1974). Isozymes in insects and their significance. *Annual Review of Entomology* 19: 117–138.

Wallbanks, K.R., Moore, J.S., Bennett, L.R., Soren, R., Molyneux, D.H., Carlin, J.M. and Perez, J.E. (1991). Aphid derived sugars in the neotropical sandfly – *Lutzomyia peruensis*. *Tropical Medicine and Parasitology* 42: 60–62.

Wallin, H. (1991). Movement patterns and foraging tactics of a caterpillar hunter inhabiting alfalfa fields. *Functional Ecology* 5: 740–749.

Wallin, H. and Ekbom, B.S. (1988). Movements of carabid beetles (Coleoptera: Carabidae) inhabiting cereal fields: a field tracing study. *Oecologia* 77: 39–43.

Wallin, H. and Ekbom, B.S. (1994). Influence of hunger level and prey densities on movement patterns in three species of Pterostichus beetles (Coleoptera: Carabidae). *Environmental Entomology* 23: 1171–1181.

Wang, R., Kafatos, F.C. and Zheng, L. (1999). Microsatellite markers and genotyping procedures for *Anopheles gambiae*. *Parasitology Today* 15: 33–37.

Weisser, W.W. (2000). Metapopulation dynamics in an aphid-parasitoid system. *Entomologia Experimentalis et Applicata* 97 (1): 83–92.

Winder, L. (2004). Marking and recapturing carabid beetles. In *Effective Marking and Tracking Techniques for Monitoring the Movement of Predator and Parasitoid Arthropods* (B.I. Lavandero, S.D. Wratten and J. Hagler, eds). Special issue: *International Journal of Pest Management*. In press.

Wratten, S.D., Watt, A.D., Carter, N. and Entwistle, J.C. (1990). Economic consequences of pesticide use for grain aphid control in winter wheat in 1984 in England. *Crop Protection* 9: 73–78.

Wratten, S.D., White, A.J., Bowie, M.H., Berry, N.A. and Weigmann, U. (1995). Phenology and ecology of hoverflies (Diptera: Syrphidae) in New Zealand. *Environmental Entomology* 24 (3): 595–600.

Wratten, S.D., Bowie, M.H., Hickman, J.M., Evans, A.M., Sedcole, J.R. and Tylianakis, J.M. (2003). Field boundaries as barriers to movement of hover flies (Diptera: Syrphidae) in cultivated land. *Oecologia* 134: 605–611.

Wratten, S.D., Hochuli, D.F., Gurr, G.M., Tylianakis, J.M. and Scarratt, S. (2004). Conservation biology, biodiversity and IPM. In *Perspectives in Ecological Theory and Integrated Pest Management* (M. Kogan and P. Jepson, eds). Cambridge University Press, Cambridge. In press.

Chapter 8

Precision agriculture approaches in support of ecological engineering for pest management

Moshe Coll

Introduction

Applied entomologists have long recognised the role of homogeneous vegetation in pest outbreaks. Such outbreaks can also be brought about by the uniform application of insecticides over large areas. Pesticide applications simplify arthropod communities, impoverish them of many beneficial species and reduce their clustered distribution. Until recently, mechanised large-scale crop production systems did not allow for a judicious deployment of insecticides and herbicides at a scale smaller than a field. In the last decade or so, however, new technologies have become available to allow the fine-tuning of inputs in the field, through the implementation of precision agriculture.

Precision agriculture, also known as precision or site-specific farming, is a set of management practices that vary inputs spatially within a field. For the purpose of this chapter, precision agriculture is defined as the management of cropping systems at the appropriate spatial and temporal scales based on predicted economic and ecological outcomes. Thus, this definition includes ecological as well as economic criteria, and unlike most definitions also includes the temporal scale so important for insect population management. The major conceptual novelty of this cultural practice is the recognition that there is a great deal of variability within agricultural fields and that this variability results in large yield differences across the field. Accordingly, the general goal of precision agriculture is to vary management inputs in different parts of the field so that crop response (i.e. yield) is optimised.

This chapter explores how the efficacy of pest control measures could be improved and adverse environmental impact be minimised by considering the spatial variability in pest occurrence and their habitat. This approach to ecological engineering makes use of new technologies that allow farmers to spatially vary inputs in the field and that are being adopted in precision agriculture. The chapter first considers the ecological effects of plant diversity and pesticide application on the density and population dynamics of insect herbivores, then describes some of the new technologies used in precision agriculture. These technologies include remote sensing (RS), yield monitors (YM), variable rate technology (VRT), global positioning systems (GPS) and geographic information systems (GIS). Finally, the applicability of these technologies, which are now used primarily for crop production, is examined in relation to the management of insect pests and weeds. The results of recent research are used to illustrate how these new tools could be used to optimally arrange crop plants in the landscape in order to reduce pest-inflicted damage.

Effects of plant diversity

Vegetation heterogeneity occurs at three levels in agroecosystems: genetic heterogeneity within a monospecific plant stand (e.g. a field), taxonomic heterogeneity (species composition) in and near fields, and landscape heterogeneity which includes vegetational differences among fields and non-crop land. Because of the commercial production of selected genotypes (seeds) and the large-scale mechanised cultivation practised in modern agriculture, vegetation in agricultural systems is often genetically uniform on a large scale, i.e. a single genotype of a given crop may cover hundreds or even thousands of hectares. In contrast to vegetation diversity (see below), relatively little is known about the influence of landscape vegetation patterns on insect populations (Power and Kareiva 1990; Landis and Marino 1999; Stevenson 2002; Schmidt et al., ch. 4 this volume). Nonetheless, it is clear that the spatial arrangement of crops is an important factor influencing the population ecology of pest species.

Much attention has been devoted to the homogeneity of agricultural systems brought about by the use of mono-genotypic crops and monocultures. Because genetic and vegetation uniformity often underlies pest problems, applied entomologists have promoted vegetation diversification through, for example, the use of multi-line cultivars, mixtures of varieties and intercropping (Dempster and Coaker 1974; Cromartie 1981; Pickett and Bugg 1998 and references therein). Various approaches and many examples of vegetation manipulation are included in this volume. These approaches address the adverse effects of vegetation uniformity on pest populations and suggest how manipulation of vegetation in and around fields may reduce crop losses. Precision agriculture in ecological engineering for pest management offers scope to rationalise the intensity of weed management, restoring a degree of within-crop vegetational diversity.

Effects of pesticide use

Habitat homogeneity can be brought about by the uniform application of insecticides over large areas. Accordingly, it is important to consider how the efficacy of pest-control measures could be improved and adverse environmental impact be minimised by responding to the spatial variability in pest occurrence and habitats. Uniform applications of insecticides across the field and on an area-wide basis drastically simplify arthropod communities and impoverish them of many beneficial species (Johnson and Tabashnik 1999). When present, these species often reduce the density of pest populations through predation, parasitism and competitive interactions. Likewise, herbicide applications that eradicate all non-crop plants in the field and its vicinity may make it easier for pests to locate crop plants and deprives beneficial predators and parasitoids of important food sources (Shelton and Edwards 1983). Fine-tuning of pesticidal inputs, so that applications are made only at 'hot-spots' where pest and weed densities reach their respective action thresholds, would create a mosaic of communities that differ in species composition and interspecific interactions, and thus are more likely to retard pest population outbreaks. This ecological engineering approach would also greatly reduce the release of toxic chemicals into the environment (Weisz et al. 1996; Brenner et al. 1998).

The direct lethal and sublethal effects of insecticide applications on insect pests are well documented. Insecticide applications can also lead to the resurgence of pest populations. In many cases, this effect can be attributed to the removal of natural enemies of the target pest through the deployment of broad-spectrum insecticides (Hardin et al. 1995). Minimising the exposure of insect predators and parasitoids to insecticides through site-specific applications, for example, would favour biological control and subsequently reduce pest damage. Another common effect of insecticide use is the development of resistance in pest populations. Site-specific insecticide application would create spatial refuges of susceptible pests unexposed to the toxins and conserve natural enemies that slow the rate of selection of resistant pest populations. Thus, site-specific pest control could reduce the appearance of resistant pest populations (e.g. Midgarden et al. 1997; Fleischer et al. 1999).

Insecticide applications may also affect the stability of pest populations and facilitate outbreaks simply by changing the distribution of insects in the field. Distribution of insect pests in the environment is usually uneven; herbivores tend to concentrate in patches of high-quality resources provided by their host plants. Moreover, random events, such as the landing site of a single colonising aphid female, also lead to clumped distribution of pests in the field. Thus, in a given field, each insect species occurs as a set of local sub-populations linked together through dispersal to form a meta-population. Current thinking holds that the buffering effect of inter-patch dispersal would make pest population outbreaks less likely because it links sink and source populations. Field-wide application of insecticide may act to homogenise pest density across the field (Trumble 1985) and therefore facilitate pest outbreaks. Furthermore, alteration of pest distribution patterns in the field may have considerable influence on sampling procedures. Far less sampling effort is necessary to estimate pest density when it approaches random or even distribution.

Insecticide applications also affect natural enemies of the target pests. This effect involves not only the direct insecticide-induced reduction of pest and enemy densities but also acts through changes in pest distribution, as insecticide applications often make pest distribution more regular and less clumped (Trumble 1985). A change in pest distribution is likely to influence enemy efficiency by altering the ability of parasites and predators to optimally forage and exploit surviving hosts and prey. For example, in the wake of insecticide applications small patches containing a few surviving hosts may not be detectable by enemies, which will then leave the field (Waage 1989). Theoretically, the homogenising effect of insecticides on pest distribution is also important in destabilising predator–prey interactions (Hassell and May 1974), though Murdoch et al. (1985) have argued that a stable equilibrium of predator and prey populations is not necessary for satisfactory biological control.

Finally, area-wide application of toxic insecticides could also change the age structure of pest and enemy populations. Insecticides typically kill susceptible life-stages. For pests, this is often the exposed, sessile larval stage, as adults are capable of migrating to untreated areas. Similarly, pesticides kill mostly exposed adult parasitoids and immature predators. This differential action of insecticides desynchronises pest and natural enemy population cycles and may result in pest outbreaks (Godfray and Hassell 1987). The effect is expected to be more pronounced where overlapping generations of pest populations occur (Waage 1989 and references therein).

Technologies deployed in precision agriculture

The wide range of technologies that have recently become available provide the tools necessary to make precision agriculture a realistic farming practice. The main tools used in precision agriculture today include:

- yield monitors (YM), sensors used to monitor yield during the harvest in order to quantify yields across the field;
- variable rate technologies (VRT) that are mounted on application equipment such as fertiliser applicators and sprayers to control their delivery rates in different parts of the field;
- global positioning system (GPS), a satellite-based locating system which identifies an earth-based position using longitude, latitude and, in some cases, elevation;

- geographic information system (GIS), a data-management system specifically designed to store spatial data and create variable-intensity maps.

These allow the delivery of variable levels of a specific management practice to various parts of the field and in a single field operation. For precision pest control, these tools should be combined with a decision-support system.

A few examples will clarify the manner in which these different components are linked in the practice of precision agriculture. To take into account variation in soil texture and history of land use across the field, data on soil nutrient status are collected through grid-based soil testing. The location of each soil sample is determined using the GPS, and the data are stored in a GIS. The GIS-generated management map is then connected to a GPS-linked VRT, so that variable rates of fertiliser are applied to different parts of the field, according to actual levels of nutrients in the soil. Similarly, lime can be applied at differential rates across the field, according to soil and yield profiles (using GPS-linked YM). The GIS also allows the overlaying of several maps, such as a soil nutrient map and crop yield map, to test for any relationships between the two data sets. Similarly, farmers could use the GIS to overlay yield maps collected from the same field over several years in order to identify problematic areas of consistently low yields.

These examples represent one approach to the implementation of precision agriculture, namely the use of map-based methods (Figure 8.1). In this approach, data obtained by grid sampling are used to generate a site-specific map, which is then coupled with a variable-rate applicator in the field. Similar site-specific maps can be generated from aerial or satellite images (remote sensing, RS). By using various filters and imaging techniques (e.g. infrared photography), variations in the health and stand of crop plants can be detected, making it possible to correct nutrient deficiencies in specific parts of the field.

A second approach to precision agriculture is founded upon sensor-based methods. Real-time sensors are used to measure the spatial variation in a variable, and this information is immediately utilised to control a VR applicator. For example, a sensor mounted in front of a

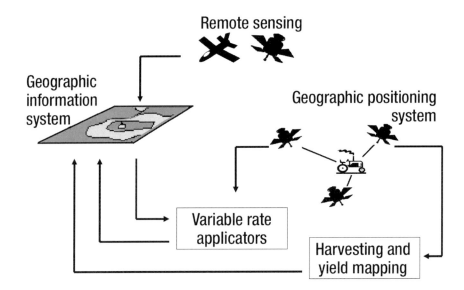

Figure 8.1: Components of map-based precision agricultural systems.

tractor may be used to detect weed stands. The sensor is connected to the sprayer behind the tractor, switching it on when weed density exceeds a predetermined threshold.

To date, a growing number of farm operators have adopted precision agriculture methodologies, primarily for crop management purposes; information on soil and plant characteristics are used to differentially apply fertiliser, gypsum, lime and water to various parts of the field. These site-specific agricultural inputs are clearly economical in large-scale operations and greatly reduce the release of contaminants into the environment.

Further detail on these methodologies is given by Mulla (1997) and the journal *Precision Agriculture* covers newly developed techniques and applications. The following section explores how precision agriculture methods could be used to conserve and enhance the activity of natural enemies and to manage pest populations.

Applicability of precision agriculture for pest management

In the past, pests were hand-picked – no doubt the ultimate in precision pest control. More recently, pesticide application to infested hot-spots has been practised. While this is still done in some systems, the move to large-scale pest-control operations have made it impractical to intensively monitor pest populations across the field. Moreover, the use of delivery equipment that covers large areas make it impossible to manipulate control measures on a small scale.

For the last 30 years or so, integrated pest management (IPM) has been the approach of choice for pest control. Several principles of IPM are highly compatible with the ideology behind precision agriculture. These are:

- take corrective measures based on economic and ecological criteria, i.e. the use of action thresholds;
- maximise sustainability by the use of resistant varieties, and biological and cultural control measures;
- minimise off-the-farm inputs, such as pesticide applications.

Like IPM, precision agriculture acts to minimise economic and environmental damage through the optimisation of, for example, fertiliser and pesticide inputs. Yet IPM lacks the spatial component so central to precision agriculture.

In IPM, decisions are based on routine monitoring of pest populations. Collecting reliable data on pest density across the field that would allow the creation of management maps and control VR applicators is probably the most challenging step toward the use of precision agricultural technologies for pest control (Fleischer et al. 1999). Thus, precision agriculture, because it has been designed for use in crop production, must be adapted for use in pest management. This section proposes several approaches for the use of precision agricultural tools for insect pest management and discusses the control of weeds because these plants greatly affect insect pests both directly and indirectly, through their influence on natural enemies. Various authors have discussed the application of remote sensing to plant pathology (Jackson 1986; Nilsson 1995), insect pest monitoring (Pathak and Dhaliwal 1985), weed detection (Zwiggelaar 1998) and plant protection in general (Hatfield and Pinter 1993; Ellsbury et al. 2000). The use of remote-sensing techniques for pest monitoring is not, therefore, considered.

The application of precision agriculture to pest management has been a slow process because of technological constraints stemming from the dynamic and cryptic nature of disease agents and insect pests. In spite of this, the clustered distribution of crop pests (pathogens, insects and weeds) does make them suitable for this type of management. In light of the severe adverse effects of pesticides on the environment, and the potential for drastically reducing the release of toxic chemicals, a careful examination of ways by which precision-farming methods could be adapted for pest control is certainly justified.

In some systems, the use of RS to create pest-management maps may be advantageous. RS can detect infested hot-spots by monitoring either the health of the plants or visible by-products of infestations, such as defoliation and sooty mould-contaminated honeydew that is secreted by homopteran pests (Hart and Myers 1968; Chaing et al. 1976). Target-oriented control measures, such as the use of biological control agents and selective insecticides, could then be applied to the detected hot-spots. However, these visual clues are likely to be detectable only after it is too late to take corrective measures. Thus, the RS-based approach to precision agriculture will be most suited to relatively stable systems such as forests and orchards, characterised by a high tolerance to pest infestation, particularly when the targets are indirect pests – i.e. pests that do not feed on the harvested parts of the plants. Finally, technical difficulties must also be overcome. For instance, when aerial photos were used to detect sooty mould on aphid-infested corn plants, black areas such as shadows were mistaken for mould and the actual level of mould cover or its relation to pest density could not be established (Jackson et al. 1974; Wallen et al. 1976). Similar difficulties were encountered when infrared aerial photography was used to monitor brown soft scale populations in citrus (Hart and Myers 1968).

RS may prove useful for weed control. Distribution of weeds in the field is relatively stable, and since most crops form a regular pattern when planted in rows, weed patches could be detected through RS (Zwiggelaar 1998). In addition, some delay in the application of control measures may be tolerated. Because herbicides represent the biggest single variable cost in crop protection, site-specific application could result in significant savings (Brightman 1998). Late in the season, however, when the crop canopy closes, weeds may no longer be detectable. Also, RS does not always allow the identification of weeds (e.g. gramineous vs broad-leaved species). This approach would therefore not necessarily contribute to the selection of target-oriented control measures, such as selective herbicides.

RS may nevertheless prove useful for controlling pests in many systems. Detection of consistent patterns of pest infestation over several years will enable the preparation of multi-year management maps that could be used to apply pre-emptive control measures (see discussion of this approach in Blackmore 2000). With this approach, data collected through RS are transferred to a GIS and used to create, for example, maps of defoliation attributable to a pest such as the gypsy moth. After several years, the existence of a consistent distribution pattern of defoliated trees could be established. If indeed particular areas of the forest are seen to be defoliated repeatedly over several years, control measures could be applied to these susceptible areas before pest populations reach damaging (and detectable) levels (e.g. Liebhold et al. 1998).

An alternative approach involves creating management maps for other variables that are found to be correlated with pest density or pest-induced damage. These variables may include field topography (when pest and weed infestations follow contours in the landscape), soil nitrogen level (when high levels result in high pest density) and soil type. A case in point is the infestation of nearly three times more tubers in loess than sandy soils by the potato tuber moth, *Phthorimaea operculella* (Coll et al. 2000). This difference has been attributed to the greater tendency of loess soil to crack upon drying, allowing the pest larvae easy access to tubers. Similarly, soil physical properties may be correlated with spatial distribution of northern corn rootworms (Ellsbury et al. 2000). Because soil type does not change with time, it may provide a convenient tool for predicting areas likely to suffer from high levels of some pest-induced damage.

In some cases, monitoring of pests and/or yields across the field may also be used to detect areas likely to harbour dense pest populations, particularly when their presence does not correlate with variations in other field parameters. Every year, for example, many pest species re-invade fields planted to annual crops. This early-season migration into the field often results in greater infestations at the margins than at the centre of the fields (Coll et al. 2000 and references therein). If this pattern can be shown to be consistent, pre-emptive measures can be applied to susceptible parts of the field. GIS is a highly suitable tool for detecting such consistent infestation patterns.

Finally, GIS may prove useful for determining the optimal arrangement of crops in area-wide pest management programs. This was the objective of a study conducted in a primary field-crop production region in Israel (Sarid 1998). Over three growing seasons, three major polyphagous pest species (the African bollworm *Helicoverpa armigera*, the Egyptian cotton-worm *Spodoptera littoralis* and the tobacco whitefly *Bemisia tabaci*) were monitored in cotton, sunflower, chickpea, maize and watermelon crops on 6000 ha in the Judean foothills. Some 40 insect-monitoring stations consisting of a pheromone trap for each of the moth species and a yellow sticky trap for the whitefly were distributed throughout the study area and visited twice weekly. The pests were also monitored on crop plants. Data concerning the spatial arrangement of crops and uncultivated areas, soil characteristics, pest-trapping data, pesticide applications and yield records were compiled using a GIS (Arc Info and Arc View, ESRI, USA), by digitising aerial photos and soil maps, and by using a GPS to locate the samples. The following *H. armigera* data are presented to illustrate the effect of crop association and rotation on pest density, pesticide use and yield in cotton.

Significantly more moths were trapped in cotton fields adjacent to sunflowers than in those near other cotton fields (27.2±4.3 and 12.5±3.7 moths per trap per night, respectively). Similar results were recorded for cotton fields near chickpea. The three crops are known host plants of *H. armigera,* and the pest appears to move into cotton after it builds up a population on the earlier sunflower and chickpea crops. This effect of neighbouring crops on *H. armigera* density in cotton was positively correlated with the size of neighboring chickpea and sunflower fields. In turn, significantly more insecticide applications were deployed against the pest in cotton plots that neighboured sunflower or chickpea plots than in those near other cotton fields (6.1±1.2, 4.3±1.9 and 3.5±1.4 applications per season, respectively). The effect of neighbouring crops on *H. armigera,* and perhaps on other shared pests such as *B. tabaci,* resulted in higher cotton yields in cotton adjacent to other cotton fields than in those near sunflower and chickpea (588±57, 562±32 and 547±35 kg per 1000 m², respectively). Interestingly, the yield in cotton fields near sunflower or chickpea was lower even though these fields received more insecticide applications. Finally, significantly more insecticide applications were recorded when crop rotation was not practised – i.e. when cotton was planted in the same field over two consecutive growing seasons – than when a different crop was grown in the field a year earlier (2.8±0.7 and 1.7±0.5 applications per season, respectively). Results therefore show that, from the pest management perspective, an optimal arrangement of these crops in a particular region – ecological engineering at a landscape level – would minimise the number of cotton fields that are adjacent to sunflower and chickpea plots, and would maximise crop rotation.

GIS and GPS-linked data collection could be used in this context to optimally arrange crops in an area. In addition to crop association and rotation, soil characteristics should also be taken into account to match crop requirements. Thus, once it is determined what proportion of the cultivated land is to be allocated to each crop (based on economic and cultural criteria), their assignment to the fields could be optimised based on soil type, for example, the avoidance of certain crop associations and the maximisation of crop rotation. The system could be also used to allocate areas for conventional crops as refuges near Bt-expressing transgenic crops. This would act to retard the evolution of resistance in insect pests. The complexity of the procedure would increase as further issues, such as the avoidance of planting crops that are treated with insecticides from the air near human inhabitants or waterways, are considered. For further discussion and more examples of the operational use of precision agricultural methods for pest management see Willers et al. (1999), Dupont et al. (2000) and Ellsbury et al. (2000).

Conclusions

The goal of modern agriculture in general, and pest control in particular, is to increase food production while reducing the release of contaminants into the environment. Precision agriculture uses new technologies to provide us with a unique opportunity to add a spatial dimension to our pest control measures. However, attaining greater precision in plant protection programs requires a more intimate knowledge of the system at hand. We need to understand how biotic and abiotic factors affect pest populations, yield and the fate of pesticides in the environment. The small size, cryptic nature, rapid change in spatial occurrence and dynamic infestation pattern characteristic of insect pests present exceptional challenges.

In light of these obstacles, have we no choice but to continue controlling pests in a uniform way across fields? Not necessarily. Herbicide sprayers equipped with sensors to detect weed patches and control delivery rate are already on the market. A soil sampler that will characterise not only nutrient content but also potato cyst nematode levels in the soil is currently in development (Legg and Stafford 1998). Although the cost of these technologies is currently prohibitive, they may become standard farm equipment in the future. The relative importance that society attaches to economic and environmental forces will determine how rapidly new technologies will be developed, or existing ones improved and modified for precision control of insect pests. New tools are needed for the monitoring of pests on a small spatial scale (Riley 1989; Fleischer et al. 1999), for the delivery of variable rates of pest-control measures other than pesticides, such as mass release of biological control agents, and for geostatistical analyses to detect spatial relations between variables in the environment (Liebhold et al. 1993).

This chapter has discussed several approaches to pest control that would deploy new technologies used in precision agriculture. At the present, technological constraints retard the adoption of RS and tractor-mounted sensors, as they do not allow the timely detection of pests or damage. However, site-specific control of insect pests may be possible based on pest biology (e.g. typical within-field distribution patterns) and parameters correlated with pest infestation (e.g. soil/plant nitrogen levels, climatic factors). Finally, precision agriculture methodologies, such as the GPS, RS and GIS, may be used for area-wide management of pests through farm landscaping. Peculiarities of each system will require the deployment of various approaches. In some cases, manipulation of host plant quality through careful adjustments of fertiliser delivery to various parts of the field may be important. In other systems, optimisation of crop association in neighbouring fields may be the preferred line of action. Thus, biological characteristics of pest infestation in a particular system, together with economic criteria, will point to the most effective approach.

Acknowledgements

I thank A. Nevo and A. Ben-Nun for their assistance with GIS, R. Yonah for help with manuscript preparation, and two anonymous reviewers for helpful comments. The GIS work was supported in part by a research grant from the Israel Cotton Production and Marketing Board.

This chapter is based on a talk 'Pest management in the era of precision agriculture' presented at the symposium on 'Plant Protection and the Environment', Lecture Series in Agricultural Sciences, held on January 19 2000 at the Faculty of Agricultural, Food and Environmental Quality Sciences, Hebrew University of Jerusalem, Rehovot, Israel (Coll 2000).

References

Altieri, M.A. and Letourneau, D.K. (1982). Vegetation management and biological control in agroecosystems. *Crop Protection* 1: 405–430.

Andow, D.A. (1991). Vegetational diversity and arthropod population response. *Annual Review of Entomology* 36: 561–586.

Bach, C.E. (1980). Effects of plant density and diversity on the population dynamics of a specialist herbivore, the striped cucumber beetle, *Acalymma vittata*. *Ecology* 61: 1515–1530.

Bach, C.E. (1984). Plant spatial pattern and herbivore population dynamics: plant factors affecting the movement patterns of a tropical cucurbit specialist *Acalymma vittata*. *Ecology* 65: 175–190.

Blackmore, B.S. (2000). The interpretation of trends from multiple yield maps. *Computers and Electronics in Agriculture* 26: 37–51.

Brenner, R.J., Focks, D.A., Arbogas, R.T., Weaver, D.K. and Shuman, D. (1998). Practical use of spatial analysis in precision targeting for integrated pest management. *American Entomologist* 44: 79–101.

Brightman, D.K. (1998). Precision in practice – will it be cost effective? In *Proceedings of the International Brighton Crop Protection Conference: Pests & Diseases,* vol. 3, pp. 1151–1158. 16–19 November 1998, Brighton, UK.

Chaing, H.C., Meyer, M.P. and Jensen, M.S. (1976). Armyworm defoliation in corn as seen on IR aerial photographs. *Entomologia Experimentalis et Applicata* 20: 301–303.

Coll, M. (1991). Effects of vegetation texture on the mexican bean beetle and its parasitoid, *Pediobius foveolatus*. PhD thesis, University of Maryland at College Park. 145 pages.

Coll, M. (2000). Plant protection in the era of precision agriculture. *Phytoparasitica* 28: 194.

Coll, M. and Bottrell, D.G. (1994). Effects of nonhost plants on an insect herbivore in diverse habitats. *Ecology* 75: 723–731.

Coll, M., Gavish, S. and Dori, I. (2000). Population biology of the potato tuber moth, *Phthorimaea operculella* (Lepidoptera: Gelechiidae), in two potato cropping systems in Israel. *Bulletin of Entomological Research* 90: 309–315.

Cromartie, W.J. (1981). The environmental control of insects using crop diversity. In *CRC Handbook of Pest Management in Agriculture* (D. Pimentel, ed.), vol. 1, pp. 223–250. CRC Press, Boca Raton, Florida.

Dempster, J.P. and Coaker, T.H. (1974). Diversification of crop ecosystems as a means of controlling pests. In *Biology in Pest and Disease Control* (J.D. Price and M.E. Solomon, eds), pp. 106–114. Blackwell Scientific, Oxford.

Dupont, J.K., Willers, J.L., Campanella, R., Seal, M.R. and Hood, K.B. (2000). Spatially variable insecticide applications through remote sensing. *Proceedings of the Beltwide Cotton Conference*.

Ellsbury, M.M., Clay, S.A., Fleischer, S.J., Chandler, L.D. and Schneider, S.M. (2000). Use of GIS/GPS systems in IPM: progress and reality. In *Emerging Technologies for Integrated Pest Management: Concepts, Research and Implementation* (G.C. Kennedy and T. Sutton, eds), pp. 419–438. American Phytopathological Society, St Paul, Minnesota.

Fleischer, S.J., Blom, P.E. and Weisz, R. (1999). Sampling in precision IPM: When the objective is a map. *Phytopathology* 89: 1112–1118.

Godfray, H.C.J. and Hassell, M.P. (1987). Natural enemies may be a cause of discrete generations in tropical insects. *Nature* 327: 144–147.

Hardin, M.R., Benrey, B., Coll, M., Lamp, W.O., Roderick, G.K. and Barbosa, P. (1995). Arthropod pest resurgence: an overview of potential mechanisms. *Crop Protection* 14: 3–18.

Hart, W.G. and Myers, V.I. (1968). Infrared aerial color photography for detection of populations of brown soft scale in citrus groves. *Journal of Economic Entomology* 61: 617–624.

Hassell, M.P. and May, R.M. (1974). Aggregation in predators and insect parasites and its effects on stability. *Journal of Animal Ecology* 43: 567–594.

Hatfield, J.L. and Pinter, P.J. (1993). Remote sensing for crop protection. *Crop Protection* 12: 403–413.

Jackson, H.R., Wallen, V.R., Galway, D. and MacDiarmid, S.W. (1974). Corn aphid infestation computer analyzed from aerial color-IR. *Photogrammetric Engineering* 40: 943–952.

Jackson, R.D. (1986). Remote sensing of biotic and abiotic plant stress. *Annual Review of Phytopathology* 24: 265–287.

Johnson, M.W. and Tabashnik, B.E. (1999). Enhanced biological control through pesticide selectivity. In *Handbook of Biological Control: Principles and Applications of Biological Control* (T.S. Bellows and T.W. Fisher, eds), pp. 297–317. Academic Press, San Diego.

Kareiva, P. (1983). Influence of vegetation texture on herbivore populations: resource concentration and herbivore movement. In *Variable Plants and Herbivores in Natural and Managed Systems* (R.F. Denno and M.S. McClure, eds), pp. 259–289. Academic Press, New York.

Landis, D.A. and Marino, P.C. (1999). Landscape structure and extrafield processes: impact on management of pests and beneficials. In *Handbook of Pest Management* (J. Ruberson, ed.), pp. 74–104. Marcel Dekker, New York.

Legg, B.J. and Stafford, J.V. (1998). Precision agriculture – new technologies. In *Proceedings of the International Brighton Crop Protection Conference: Pests & Diseases*, vol. 3, pp. 1143–1150. 16–19 November 1998, Brighton, UK.

Liebhold, A.M., Rossi, R.E. and Kemp, W.P. (1993). Geostatistics and geographic information systems in applied insect ecology. *Annual Review of Entomology* 38: 303–327.

Liebhold, A., Luzaader, E., Reardon, R., Roberts, A., Ravlin, F.W., Sharov, A. and Zhou, G. (1998). Forecasting gypsy moth (Lepidoptera: Lymantriidae) defoliation with a geographical information system. *Journal of Economic Entomology* 91: 464–472.

Midgarden, D., Fleischer, S.J., Weisz, R. and Smilowitz, Z. (1997). Site-specific integrated pest management impact on development of esfenvalerate resistance in Colorado potato beetle (Coleoptera: Chrysomelidae) and on densities of natural enemies. *Journal of Economic Entomology* 90: 855–867.

Mulla, D.J. (1997). Geostatistics, remote sensing and precision farming. In *Precision Agriculture: Spatial and Temporal Variability of Environmental Quality* (J.V. Lake, G.R. Bock and J.A. Goode, eds), pp. 100–119. Ciba Foundation Symposium 210, Wiley, New York.

Murdoch, W.W., Chesson, J. and Chesson, P.L. (1985). Biological control in theory and practice. *American Naturalist* 125: 344–366.

Nilsson, H.E. (1995). Remote sensing and image analysis in plant pathology. *Annual Review of Phytopathology* 15: 489–527.

Pathak, M.D. and Dhaliwal, G.S. (1985). Remote sensing and monitoring of insect pest problems in rice. In *Application of Remote Sensing for Rice Production* (A. Deepak and K.R. Pao, eds), pp. 77–87. A. Deepak Publishing, New Delhi.

Pickett, C.H. and Bugg, R.L. (eds) (1998). *Enhancing Biological Control: Habitat Management to Promote Natural Enemies of Agricultural Pests*. University of California Press, Berkeley. 422 pages.

Power, A.G. and Kareiva, P. (1990). Herbivorous insects in agroecosystems. In *Agroecology* (C.R. Carroll, J.H. Vandermeer and P. Rosset, eds), pp. 301–327. McGraw-Hill, New York.

Riley, J.R. (1989). Remote sensing in entomology. *Annual Review of Entomology* 34: 247–271.

Risch, S. (1981). Insect herbivore abundance in tropical monocultures and polycultures: an experimental test of two hypotheses. *Ecology* 62: 1325–1340.

Root, R.B. (1973). Organization of a plant-arthropod association in simple and diverse habitats: the fauna of collards (*Brassica oleraceae*). *Ecological Monographs* 43: 95–120.

Sarid, Y. (1998). Area wide dynamics of field crop pests. M.Sc. thesis, Hebrew University of Jerusalem (in Hebrew with English abstract). 57 pages.

Shelton, M.D. and Edwards, C.R. (1983). Effects of weeds on the diversity and abundance of insects in soybeans. *Environmental Entomology* 12: 296–298.

Stevenson, F.C. (2002). Landscape patterns and pest problems. In *Encyclopedia of Pest Management* (D. Pimentel, ed.), pp. 437–438. Marcel Dekker, New York.

Trumble, J.T. (1985). Implications of changes in arthropod distribution following chemical application. *Researches on Population Biology* 27: 277–285.

Waage, J. (1989). The population ecology of pest-pesticide-natural enemy interactions. In *Pesticides and Non-target Invertebrates* (P.C. Jepson, ed.), pp. 81–93. Intercept, Wimborne, UK.

Wallen, V.R., Jackson, H.R. and MacDiarmid, S.W. (1976). Remote sensing of corn aphid infestation, 1974 (Hemiptera: Aphididae). *Canadian Entomologist* 108: 751–754.

Weisz, R., Fleischer, S. and Smilowitz, Z. (1996). Site-specific integrated pest management for high-value crops: impact on potato pest management. *Journal of Economic Entomology* 89: 501–509.

Willers, J.L., Seal, M.R. and Luttrell, R.G. (1999). Remote sensing, line-intercept sampling for tarnished plant bugs (Heteroptera: Miridae) in mid-south cotton. *Journal of Cotton Science* 3: 160–170.

Zwiggelaar, R. (1998). A review of spectral properties of plants and their potential use for crop/weed discrimination in row-crops. *Crop Protection* 17: 189–206.

Chapter 9

Effects of agroforestry systems on the ecology and management of insect pest populations

Miguel A. Altieri and Clara I. Nicholls

Introduction

Agroforestry is an intensive land-management system that combines trees and/or shrubs with crops and/or livestock on a landscape level to achieve optimum benefits from biological interactions. Agroforestry also balances ecosystem demands to sustain diversity and productivity, while meeting multiple-use and sustained-yield needs of agriculture (Nair 1993). The systems are often applied successfully by indigenous farmers in the developing world who usually understand land-use interactions in their local ecosystems. Many of the benefits of agroforestry systems are derived from the increased diversity of these systems compared with corresponding monocultures of crops or trees. Despite the fact that little research has been conducted on pest interactions within agroforestry systems, agroforestry has been assumed to reduce pest outbreaks usually associated with monocultures. Although the effects of various agroforestry designs on pest populations can be of a varied nature (microclimatic, nutritional, natural enemies etc.), regulating factors do not act in isolation from each other.

The few reviews on pest management in agroforestry (Rao et al. 2000; Schroth et al. 2000) stipulate that the high plant diversity associated with agroforestry systems provide some level of protection from pest and disease outbreaks. To explain such regulation, these authors use the same theories advanced by agroecologists to explain lower pest levels in annual polycultural agroecosystems (Andow 1991; Altieri 1994, Nicholls and Altieri, ch. 3 this volume). Some authors caution that the use of high plant diversity as a strategy to reduce pest and disease risks in agroforestry systems meets considerable technical difficulties, as the design and management of complex systems is cumbersome. Like orchard situations, agroforestry systems can be considered semipermanent, relatively undisturbed systems, with no fallow, and crop rotation does not apply in the short term, so particular biological situations affecting insects occur in these systems. Insect populations are more stable in complex agroforestry systems because a diverse and more permanent habitat can maintain an adequate population of the pest and its enemies at critical times (van den Bosch and Telford 1964). For most entomologists, the relative permanency of agroforestry systems affords the opportunity of manipulating the components of the habitat to the benefits of ecologically sound pest-management practices (Prokopy 1994). Such practices include the manipulation of groundcover vegetation and/or of shade trees to enhance biological control of arthropod pests.

This chapter focuses on the effects of vegetationally diverse agroforestry systems on the ecology of insect pests, concentrating more specifically on the actual/potential mechanisms underlying pest reduction in agroforestry systems. Although most of the entomological research on these has been conducted in tropical areas, the results from such research are also applicable to temperate regions, obviously taking into consideration the differences in biological diversity, climate, soils and so on.

Biodiversity, biotic interactions and ideas for pest management

Two distinct components of biodiversity can be recognised in agroforestry systems (Vandermeer and Perfecto 1995). The first component, *planned biodiversity,* includes the crops and livestock purposely included in the agroforestry system by the farmer, which will vary depending on the management inputs and crop spatial/temporal arrangements. The second component, *associated biodiversity*, includes all soil flora and fauna, herbivores, carnivores, decomposers etc. that colonise the agroecosystem from surrounding environments and that may thrive in the agroecosystem depending on its management and structure. The relationship of both types of biodiversity components is illustrated in Figure 9.1. Planned biodiversity has a direct function, as illustrated by the bold arrow connecting the planned biodiversity box with the ecosystem function box. Associated biodiversity also has a function, but it is influenced strongly by the planned biodiversity. Thus, planned biodiversity also has an indirect function, illustrated by the dotted arrow in the figure, which is realised through its influence on the associated biodiversity. For example, the trees in an agroforestry system create shade, which makes it possible to grow only sun-intolerant crops. So the direct function of this second species (the trees) is to create shade. Yet along with the trees might come wasps that seek out the nectar in the tree's flowers. These wasps may in turn be the natural parasitoids of pests that normally attack crops. The wasps are part of the associated biodiversity. The trees then create shade (direct function) and attract wasps (indirect function) (Vandermeer and Perfecto 1995).

Complementary interactions between the various biodiversity components can also be of a multiple nature. Some of these interactions can be used to induce positive and direct effects on

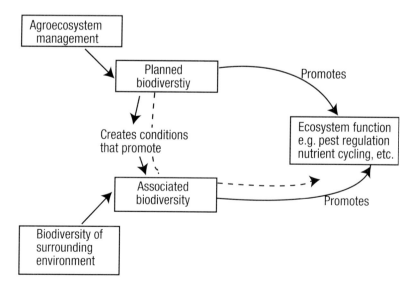

Figure 9.1: Types of biodiversity and their role in pest regulation in agroforestry systems.

the biological control of specific crop pests, soil fertility regeneration and/or enhancement and soil conservation. The exploitation of these interactions in real situations involves agroecosystem design and management and requires an understanding of the numerous relationships between soils, microorganisms, plants, insect herbivores and natural enemies.

According to agroecological theory (Altieri 1995), the optimal behaviour of agroecosystems depends on the level of interactions between the various biotic and abiotic components. By assembling a functional biodiversity, it is possible to initiate synergies that subsidise agroecosystem processes by providing ecological services such as the activation of soil biology, the cycling of nutrients, the enhancement of beneficial arthropods and antagonists, and so on (Gliessman 1999), all important in determining the sustainability of agroecosystems.

In modern agroecosystems, experimental evidence suggests that biodiversity can be used for improved pest management (Andow 1991; Altieri 1994). Several studies have shown that it is possible to stabilise the insect communities of agroecosystems by designing and constructing vegetational architectures that support populations of natural enemies or have direct deterrent effects on pest herbivores. The key is to identify the type of biodiversity that it is desirable to maintain and/or enhance in order to carry out ecological services such as biological pest control, nutrient cycling, water and soil conservation, and then to determine the best practices that will encourage the desired biodiversity components (Figure 9.2). There are many agricultural

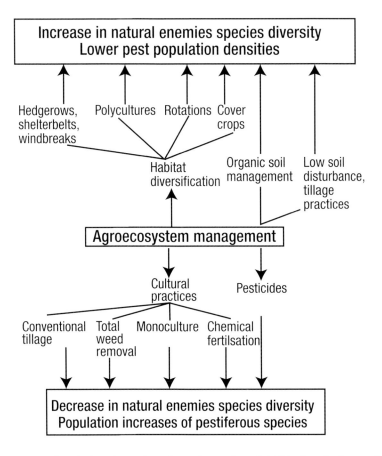

Figure 9.2: Assortment of agricultural practices that enhance beneficial biodiversity in agroforestry systems.

practices and designs that have the potential to enhance functional biodiversity, and others that negatively affect it. Although many of these strategies apply to agricultural systems, the strategy is to apply the best management practices in order to enhance or regenerate the kind of biodiversity that can subsidise the sustainability of agroforestry systems by providing ecological services. The role of agroecologists should be to encourage those agricultural practices that increase the abundance and diversity of above- and below-ground organisms, which in turn provide key ecological services to agroforestry systems.

Thus, a key strategy of agroecology is to exploit the complementarity and synergy that result from the various combinations of crops, trees and animals in agroecosystems that feature spatial and temporal arrangements such as polycultures, agroforestry systems and crop–livestock mixtures. In real situations, the exploitation of these interactions involves agroecosystem design and management and requires an understanding of the numerous relationships among soils, microorganisms, plants, insect herbivores and natural enemies.

Effects of trees in agroforestry on insect pests and associated natural enemies

The deliberate association of trees with agronomic crops can result in insect management benefits due to the structural complexity and permanence of trees and to their modification of microclimates and plant apparency within the production area. Individual plants in annual cropping systems are usually highly synchronised in their phenology and short-lived. In such systems, the lack of temporal continuity is a problem for natural enemies because prey availability is limited to short periods of time. Refugia and other resources, such as pollen, nectar and neutral insects, are not consistently available. The addition of trees of variable phenologies or diverse age structure through staggered planting can provide refuge and a more constant nutritional supply to natural enemies because resource availability through time is increased. Trees can also provide alternate hosts to natural enemies, as in the case of the planting of prune trees adjacent to grape vineyards to support overwintering populations of the parasitoid *Anagrus epos* (Murphy et al. 1996).

Shade effects

Shade from trees may markedly reduce pest density in understorey intercrops. Hedgerows or windbreaks of trees have a dramatic influence on microclimate; almost all microclimate variables (heat input, wind speed, soil desiccation and temperature) are modified downwind of a hedgerow. Tall intercrops or thick groundcovers can also alter the reflectivity, temperature and evapotranspiration of shaded plants or at the soil surface, which in turn could affect insects that colonise according to 'background' colour or that are adapted to specific microclimatological ranges (Cromartie 1991). Both immature and adult insect growth rates, feeding rates and survival can be dramatically affected by changes in moisture and temperature (Perrin 1977).

The effect of shade on pests and diseases in agroforestry has been studied quite intensively in cocoa and coffee systems undergoing transformation from traditionally shaded crop species to management in unshaded conditions. In cocoa plantations, insufficient overhead shade favours the development of numerous herbivorous insect species, including thrips (*Selenothrips rubrocinctus*) and mirids (*Sahlbergella, Distantiella* etc.). Even in shaded plantations, these insects concentrate at spots where the shade trees have been destroyed, for example by wind (Beer et al. 1997). Bigger (1981) found an increase in the numbers of Lepidoptera, Homoptera, Orthoptera and the mirid *Sahlbergella singularis* and a decrease in the number of Diptera and parasitic Hymenoptera from the shaded towards the unshaded part of a cocoa plantation in Ghana.

In coffee, the effect of shade on insect pests is less clear than in cocoa, as the leafminer (*Leucoptera meyricki*) is reduced by shade, whereas the coffee berry borer (*Hypothenemus hampei*) may increase under shade. Similarly, unshaded tea suffers more from attack by thrips and mites, such as the red spider mite (*Oligonychus coffeae*) and the pink mite (*Acaphylla theae*), whereas heavily shaded and moist plantations are more severely damaged by mirids (*Helopeltis* spp.) (Guharay et al. 2000).

Although in Central America coffee berry borer appears to perform equally well in open sun and managed shade, naturally occurring *Beauveria bassiana* (an entomopathogenic fungus) multiplies and spreads more quickly with greater humidity, therefore fungus applications should coincide with peaks in rainfall (Guharay et al. 2000). After a study of how the microclimate created by multistrata shade management affected herbivores, diseases, weeds and yields in Central American coffee plantations, Staver et al. (2001) defined the conditions for minimum expression of the pest complex. For a low-elevation dry coffee zone, shade should be managed between 35% and 65%, as shade promotes leaf retention in the dry season and reduces *Cercospora coffeicola*, weeds and *Planococcus citri*. Obviously, the optimum shade conditions for pest suppression differ with climate, altitude and soils. The selection of tree species and associations, density and spatial arrangements as well as shade-management regimes are critical considerations for shade strata design.

The complete elimination of shade trees can have an enormous impact on the diversity and density of arthropods, especially ants. Studying the ant community in a gradient of coffee plantations going from systems with high density of shade to shadeless plantations, Perfecto and Vandermeer (1996) reported a significant decrease in ant diversity. Although a relationship between ant diversity and pest control is not always apparent, research suggests that a diverse ant community can offer more safeguards against pest outbreaks than a community dominated by just a few species. In Colombia, preliminary reports point to lower levels of coffee borer, the main coffee pest in the region, in shaded coffee plantations. There is indication that a non-dominant small ant species is responsible for the control. Apparently, this species does not live in unshaded plantations. In cocoa, ant species that flourish under shaded conditions have been very successful in controlling various pests. One of the most obvious consequences of pruning or shade elimination, with regard to the ant community, is the change in microclimatic conditions. In particular, microclimate becomes more variable with more extreme levels of humidity and temperature, which in turn promotes changes in the composition of the ant community (Perfecto and Vandermeer 1996).

Crop attractiveness

Chemical cues used by herbivores to locate host plants may be altered in an agroforestry system. Trees may exhibit a dramatically different chemical profile than annual herbaceous plants intercropped in the system, masking or lessening the impact of the chemical profile produced by the annual crop. Several studies have demonstrated olfactory deterrence as a factor in decreasing arthropod abundance (Risch 1981). The attractiveness of a plant species for the pests of another species can be usefully employed in agroforestry associations in the form of trap crops which concentrate the pests or disease vectors into a place where they cause less damage or can be more easily neutralised (e.g. by spraying or collecting) (see also Mensah and Sequeira, ch. 12, and Khan and Pickett, ch. 10 this volume). Such trap crops are an interesting option when they attract pests from the primary crop within the field (local attraction), but not when they attract pests from areas outside the field (regional attraction). Nascimento et al. (1986) demonstrated the strong attraction of the citrus pest *Cratosomus flavofasciatus* by the small tree *Cordia verbe-*

nacea in Bahia, Brazil, and recommended the inclusion of this tree at distances of 100–150 m in citrus orchards. They speculated that pests of several other fruit crops could similarly be trapped by this tree species.

Nutritional and cover effects

In certain agroforestry systems, such as alley cropping, which usually include leguminous shade trees, relatively large quantities of nitrogen-rich biomass may be applied to crops. Luxury consumption of nitrogen may result in reduced pest resistance of the crops. The reproduction and abundance of several insect pests, especially Homoptera, are stimulated by high concentration of free nitrogen in the crop's foliage (Luna 1988).

The manipulation of groundcover vegetation in tropical plantations can significantly affect tree growth by altering nutrient availability, soil physics and moisture, and the prevalence of weeds, plant pathogens, insect pests and associated natural enemies (Haynes 1980). A number of entomological studies conducted in these systems indicate that plantations with rich floral undergrowth exhibit a significantly lower incidence of insect pests than clean cultivated orchards, mainly because of an increased abundance and efficiency of predators and parasitoids, or other effects related to habitat changes. In the Solomon Islands, O'Connor (1950) recommended the use of a cover crop in coconut groves to improve the biological control of coreid pests by the ant *Oecophylla smaragdina subnitida*. In Ghana, coconut gave light shade to cocoa and, without apparent crop loss, supported high populations of *Oecophylla longinoda*, keeping the cocoa crop free from cocoa capsids (Leston 1973).

Wood (1971) reported that in Malaysian oil palm (*Elaeis guineensis*) plantations, heavy groundcover, irrespective of type, reduced damage to young trees caused by rhinoceros beetle (*Oryctes rhinoceros*). The mode of action is not certain, but it appears that the groundcover impedes flight of the adult beetles or restricts their movement on the ground. Economic control of this pest was possible by simply encouraging the growth of weeds between the trees.

Plant diversity and natural enemies

In Kenyan studies assessing the effects of nine hedgerow species on the abundance of major insect pests of beans and maize (also known as corn), and associated predatory/parasitic arthropods, Girma et al. (2000) found that beanfly (*Ophiomyia* spp.) infestation was significantly higher in the presence of hedgerows (35%) than in their absence (25%). Hedgerows did not influence aphid (*Aphis fabae*) infestation of beans. In contrast, maize associated with hedgerows experienced significantly lower stalk borer (*Busseola fusca* and *Chilo* spp.) and aphid (*Rhopalosiphum maidis*) infestations than did pure maize, the margin of difference being 13% and 11% respectively for the two pests. Ladybird beetles closely followed their prey, aphids, with significantly higher catches in sole-cropped plants than in hedgerow-plots and away from hedgerows. Activity of wasps was significantly greater close to hedgerows than away from them. Spider catches during maize season were 77% greater in the presence of hedgerows than in their absence, but catches during other seasons were similar between the two cropping systems.

In one of the few studies of the influence of temperate agroforestry practices on beneficial arthropods, Peng et al. (1993) confirmed the increase in insect diversity and improved natural enemy abundance in an alley-cropping system over that of a monoculture-crop system. Their study examined arthropod diversity in control plots sown to peas (*Pisum sativum* va. Sotara) versus peas intercropped with four tree species (walnut, ash, sycamore and cherry) and hazel bushes. They found greater arthropod abundance in the alley-cropped plots compared to the control plots, and natural enemies were more abundant in the tree lines and alleys than in the controls. The authors attributed the increase in natural enemies to the greater availability of overwintering sites and shelter in the agroforestry system. In fact, Stamps and Linit (1997) argued that agroforestry holds promise for increasing insect diversity and reducing pest problems because the combination of trees and crops provides greater niche diversity and complexity in both time and space than does polyculture of annual crops.

Designing natural successional analog agroforestry systems

At the heart of the agroecology strategy is the idea that an agroecosystem should mimic the diversity and functioning of local ecosystems, thus exhibiting tight nutrient cycling, complex structure and enhanced biodiversity. The expectation is that such agricultural mimics, like their natural models, can be productive, pest-resistant and conservative of nutrients and biodiversity. Ewel (1986) argued that natural plant communities have several traits (including pest suppression) that would be desirable to incorporate into agroecosystems. To test this idea, he and others (Ewel et al. 1986) studied productivity, growth, resilience and resource-utilisation characteristics of tropical successional plant communities. The researchers contended that desirable ecological patterns of natural communities should be incorporated into agriculture by designing cropping systems that mimic the structural and functional aspects of secondary succession. Thus the prevalent coevolved, natural, secondary plant associations of an area should provide the model for the design of multi-species crop mixtures.

This succession analog method requires a detailed description of a natural ecosystem in a specific environment and the botanical characterisation of all potential crop components. When this information is available, the first step is to find crop plants that are structurally and functionally similar to the plants of the natural ecosystem. The spatial and chronological arrangement of the plants in the natural ecosystem are then used to design an analogous crop system (Hart 1980). In Costa Rica, researchers conducted spatial and temporal replacements of wild species by botanically/structurally/ecologically similar cultivars. Thus, successional members of the natural system such as *Heliconia* spp., cucurbitaceous vines, *Ipomoea* spp., legume vines, shrubs, grasses and small trees were replaced by plantain, squash varieties and yams. By years two and three, fast-growing tree crops (Brazil nuts, peach, palm, rosewood) may form an additional stratum, thus maintaining continuous crop cover, avoiding site degradation and nutrient leaching, and providing crop yields throughout the year (Ewel 1986).

Under a scheme of managed succession, natural successional stages are mimicked by intentionally introducing plants, animals, practices and inputs that promote the development of interactions and connections between component parts of the agroecosystem. Plant species (both crop and non-crop) are planted that capture and retain nutrients in the system and promote good soil development. These plants include legumes, with their nitrogen-fixing bacteria, and plants with phosphorus-trapping mycorrhizae. As the system develops, increasing diversity, food-web complexity and level of mutualistic interactions all lead to more effective feedback mechanisms for pest and disease management. The emphasis during the development process is on building a complex and integrated agroecosystem with less dependence on external inputs.

There are many ways that a farmer, beginning with a recently cultivated field of bare soil, can allow successional development to proceed beyond the early stages. One general model is to begin with an annual monoculture and progress to a perennial tree crop system, following the steps outlined below.

- The farmer begins by planting a single annual crop that grows rapidly, captures soil nutrients, gives an early yield and acts as a pioneer species in the developmental process.
- As a next step (or instead of the previous one), the farmer can plant a polyculture of annuals that represent different components of the pioneer stage. The species would differ in their nutrient needs, attract different insects, have different rooting depths and

return a different proportion of their biomass to the soil. One might be a nitrogen-fixing legume. All these early species would contribute to the initiation of the recovery process, and would modify the environment so that non-crop plants and animals, especially the macro and microorganisms necessary for developing the soil ecosystem, can also begin to colonise.

- Following the initial stage of development, short-lived perennial crops can be introduced. Taking advantage of the soil cover created by the pioneer crops, these species can diversify the agroecosystem in important ecological aspects. Deeper root systems, more organic matter stored in standing biomass, and greater habitat and microclimate diversity all combine to advance the successional development of the agroecosystem.
- Once soil conditions are stabilised, the ground is prepared for planting longer-lived perennials, especially orchard or tree crops, with annual and short-lived perennial crops maintained in the areas between them. While the trees are in their early growth, they have limited impact on the environment around them. At the same time, they benefit from having annual crops around them, because in the early stages of growth they are often more susceptible to interference from aggressive weedy species that would otherwise occupy the area.
- As the tree crops develop, the space between them can continue to be managed with annuals and short-lived perennials.
- Once the trees reach full development, the end point in the developmental process is achieved. This last stage is dominated by woody plants, which are key to the site-restoring process mediated by vegetation with deep and permanent root systems.

Once a successionally developed agroecosystem has been created, the problem becomes how to manage it. The farmer has three basic options.

- Return the entire system to the initial stages of succession by introducing a major disturbance, such as clear-cutting the trees in the perennial system. Many of the ecological advantages that have been achieved will be lost and the process must begin anew.
- Maintain the system as a tree-crop-based agroecosystem.
- Reintroduce disturbance into the agroecosystem in a controlled and localised manner, taking advantage of the dynamics that such patchiness introduces into an ecosystem. Small areas in the system can be cleared, returning those areas to earlier stages in succession, and allowing a return to the planting of annual or short-lived crops. If care is taken in the disturbance process, the below-ground ecosystem can be kept at a later stage of development, whereas the above-ground system can be made up of highly productive species that are available for harvest removal.

According to Ewel (1999), the only region where it pays to imitate natural ecosystems rather than struggle to impose simplicity through high inputs in ecosystems that are inherently complex, is the humid tropical lowlands. This area epitomises environments of low abiotic stress but overwhelming biotic intricacy. The keys to agricultural success in this region are to (a) channel productivity into outputs of nutritional and economic importance, (b) maintain adequate vegetational diversity to compensate for losses in a system simple enough to be agronomically manageable, (c) manage plants and herbivores to facilitate associational resistance and (d) use perennial plants to maintain soil fertility, guard against erosion and make full use of resources.

To many, the ecosystem-analog approach is the basis for the promotion of agroforestry systems, especially the construction of forest-like agroecosystems that imitate successional vegetation, which exhibit low requirements for fertiliser, high use of available nutrients and high protection from pests (Sanchez 1995).

Ecological principles for design

As traditional farmers have done, natural successional communities can be used as models for agroecosystem design because they offer several traits of potential value to agriculture: high resistance to pest invasion and attack, high retention of soil nutrients, enhanced agrobiodiversity and reasonable productivity (Ewel 1999).

As stated by Gliessman (1999), a major challenge in the tropics is to design agroecosystems that, on the one hand, take advantage of some of the beneficial attributes of the early stages of succession yet, on the other hand, incorporate some of the advantages gained by allowing the system to reach the later stages. Only one desirable ecological characteristic of agroecosystems – high net primary productivity – occurs in the early stages of development, an important reason to create more permanent agroecosystems through the inclusion of perennials. The application of the following principles can lead to the design of more mature, complex and pest-stable agroforestry systems.

1 Increase species diversity, as this promotes fuller use of resources (nutrients, radiation, water etc.), protection from pests and compensatory growth. Many researchers have highlighted the importance of various spatial and temporal plant combinations to facilitate complementary resource use or to provide intercrop advantage, such as in the case of legumes facilitating the growth of cereals by supplying extra nitrogen. Compensatory growth is another desirable trait; if one species succumbs to pests, weather or harvest, another species fills the void, maintaining full use of available resources. Crop mixtures also minimise risks, especially by creating the sort of vegetative texture that suppresses specialist pests.

2 Enhance longevity through the addition of perennials that contain a thick canopy, thus providing continual cover that protects the soil. Constant leaf-fall builds organic matter and allows uninterrupted nutrient circulation. Dense and deep root systems of long-lived woody plants is an effective mechanism for nutrient capture, offsetting the negative losses through leaching.

3 Impose a fallow to restore soil fertility through biomass accumulation and biological activation, and to reduce agricultural pest populations as lifecycles are interrupted with a rotation of fallow vegetation and crops.

4 Enhance addition of organic matter by including legumes, biomass-producing plants and incorporating animals. Accumulation of organic matter is key to activating soil biology, improving soil structure and macroporosity and elevating the nutrient status of soils.

5 Increase landscape diversity by having a mosaic of agroecosystems representative of various stages of succession. Risk of complete failure is spread among, as well as within, the various farming systems. Improved pest control is also linked to spatial heterogeneity at the landscape level.

Conclusions and further research

The effects of agroforestry designs and technologies on pests and diseases can be divided into biological (species-related) and physical effects of components (e.g. microclimate). The former are highly specific for certain plant–pest or plant–disease combinations and have to be studied on a case-by-case basis. The latter are easier to generalise, but even they depend on the regional climatic conditions. Based on results from intercropping studies, agroforesters expect that agroforestry systems may provide opportunities to noticeably increase arthropod diversity and lower pest populations compared with the polyculture of annual crops or trees by themselves (Schroth et al. 2000). But more work is needed in specific areas of research such as studies of the differences in arthropod populations between agroforestry and traditional agronomic systems, research into the specific mechanisms behind enhancement of pest management with agroforestry practices, and basic research into the life-histories of target pests and potential natural enemies. An understanding of what aspects of trees modify pest populations – shelter, food or host resources for natural enemies, temporal continuity, microclimate alteration or apparency – should help in determining future agroforestry design practices (Rao et al. 2000).

Well-designed agroforestry techniques can reduce crop stress by providing the right amount of shade, reducing temperature extremes, sheltering from strong winds and improving soil fertility, thereby improving the tolerance of crops against pest and disease damage, while at the same time influencing the developmental conditions for pest and disease organisms and their natural enemies. Poorly designed systems, on the other hand, may increase crop susceptibility to pests.

It is important to realise that the majority (75%) of agroforests are located in developing countries managed by traditional farmers who cultivate a few hectares of land. They rely on low-energy, labour-intensive production methods and few agrochemicals (Altieri 1995). These resource-poor farmers have practised agroforestry for centuries: they used trees for fences and pest control, as well as for food, fodder, construction materials and fuel (Altieri and Farrell 1984; Greathead 1988). These small farmers cannot afford high-input technologies or expensive agroforestry designs. The key challenge is to maintain a highly diverse farm with woodlands, forests and herbaceous edges, which allow traditional agroforesters to regulate pest populations by providing food and habitat for birds, spiders, parasites and other natural enemies of pests.

Although small farmers may lack the research tools used by scientists in industrial countries, traditional agroforesters do have valuable knowledge to contribute towards the design of sustainable agroforestry systems. They have developed practical systems for identifying damaging stages of pests, understand their biologies and apply management techniques to suppress their populations. This knowledge can be tapped through participatory research schemes whereby farmers and researchers engage in a true collaborative partnership.

References

Altieri, M.A. (1994). *Biodiversity and Pest Management in Agroecosystems*. Haworth Press, New York.

Altieri, M.A. (1995). *Agroecology: The Science of Sustainable Agriculture*. Westview Press, Boulder.

Altieri, M.A. and Farrell, J. (1984). Traditional farming systems of south-central Chile, with special emphasis on agroforestry. *Agroforestry Systems* 2: 3–18.

Andow, D.A. (1991). Vegetational diversity and arthropod population response. *Annual Review of Entomology* 36: 561–586.

Beer, J., Muschler, R., Kass, D. and Somarriba, E. (1997). Shade management in coffee and cacao plantations. In *Forestry Sciences: Directions in Tropical Agroforestry Research* (P.K.R. Nair and C.R. Latt, eds), pp. 139–164. Kluwer Academic, Dordrecht.

Bigger, M. (1981). Observations on the insect fauna of shaded and unshaded Amelonado cocoa. *Bulletin of Entomological Research* 71: 107–119.

Cromartie, W.J. (1991). The environmental control of insects using crop diversity. In *CRC Handbook of Pest Management in Agriculture* (D. Pimentel, ed.), vol. 1, pp. 223–251. CRC Press, Boca Raton.

Ewel, J.J. (1986). Designing agricultural ecosystems for the humid tropics. *Annual Review Ecology and Systematics* 17: 245–271.

Ewel, J.J. (1999). Natural systems as models for the design of sustainable systems of land use. *Agroforestry Systems* 45: 1–21.

Ewel, J., Benedict, F., Berish, C. and Brown, B. (1986). Leaf area, light transmission, roots and leaf damage in nine tropical plant communities. *Agro-ecosystems* 7: 305–326.

Girma, H., Rao, M.R. and Sithanantham, S. (2000). Insect pests and beneficial arthropod populations under different hedgerow intercropping systems in semiarid Kenya. *Agroforestry Systems* 50 (3): 279–292.

Gliessman, S.R. (1999). *Agroecology: Ecological Processes in Agriculture*. Ann Arbor Press, Michigan.

Greathead, D.J. (1988). Crop protection without chemicals: pest control in the third world. *Aspects of Applied Biology.* Wellesbourne, Warwick: Association of Applied Biologists. 17: 19–28.

Guharay, F., Monterrey, J., Monterroso, D. and Staver, C. (2000). *Manejo integrado de plagas en el cultivo de café*. CATIE. Managua, Nicaragua.

Hart, R.D. (1980). *Agroecosistemas*. CATIE, Turrialba, Costa Rica.

Haynes, R.J. (1980). Influence of soil management practice on the orchard agroecosystem. *Agro-ecosystems* 6: 3–32.

Leston, D. (1973). The ant mosaic-tropical tree crops and the limiting of pests and diseases. *PANS* 19: 311.

Luna, J. (1988). Influence of soil fertility practices on agricultural pests. In *Global Perspectives in Agroecology and Sustainable Agricultural Systems* (P. Allen and D. van Dusen, eds), pp. 589–600. Proceedings of the VI IFOAM International Scientific Conference. University of California, Santa Cruz.

Murphy, B.C., Rosenheim, J.A. and Granett, J. (1996). Habitat diversification for improving biological control: abundance of *Anagrus epos (Hymenoptera: Mymaridae)* in grape vineyards. *Environmental Entomology* 25 (2): 495–504.

Nair, P.K. (1993). *An Introduction to Agroforestry*. Kluwer Academic, Dordrecht.

Nascimento, A.S., Mesquita, A.L.M. and Caldas, R.C. (1986). Population fluctuation of the citrus borer, *Cratosomus flavofasciatus* Guerin (Coleoptera: Curculionidae), on the trap plant, *Cordia verbenacea* (Boraginaceae). *Anais da Sociedade Entomologica do Brasil* 15: 125–134.

O'Connor, B.A. (1950). Premature nutfall of coconuts in the British Solomon Islands Protectorate. *Agriculture Journal*, Fiji Department of Agriculture 21: 1–22.

Peng, R.K., Incoll, L.D., Sutton, S., Wright, L.C. and Chadwick, A. (1993). Diversity of airborne arthropods in a silvoarable agroforestry system. *Journal of Applied Ecology* 30 (3): 551–562.

Perfecto, I. (1995). Biodiversity and the transformation of a tropical agroecosystem: ants in coffee plantations. *Ecological Applications* 5: 1084–1097.

Perfecto, I. and Vandermeer, J.H. (1996). Microclimatic changes and the indirect loss of ant diversity in a tropical agroecosystem. *Oecologia* 108: 577–582.

Perrin, R.M. (1977). Pest management in multiple cropping systems. *Agro-ecosystems* 3: 93–118.

Prokopy, R.C. (1994). Integration in orchard pest and habitat management: a review. *Agriculture, Ecosystems and Environment* 50: 1–10.

Rao, M.R., Singh, M.P. and Day, R. (2000). Insect pest problems in tropical agroforestry systems: contributory factors and strategies for management. *Agroforestry Systems* 50 (3): 243–277.

Risch, S.J. (1981). Insect herbivore abundance in tropical monocultures and polycultures: an experimental test of two hypotheses. *Ecology* 62: 1325–1340.

Sanchez, P.A. (1995). Science in agroforestry. *Agroforestry Systems* 30: 5–55.

Schroth, G., Krauss, U., Gasparotto, L., Aguilar, J., Duarte, A. and Vohland, K. (2000). Pests and diseases in agroforestry systems of the humid tropics. *Agroforestry Systems* 50: 199–241.

Stamps, W.T. and Linit, M.J. (1997). Plant diversity and arthropod communities: implications for temperate agroforestry. *Agroforestry Systems* 39 (1): 73–89.

Staver, C., Guharay, F., Monteroso, D. and Muschler, R.G. (2001). Designing pest-suppressive multistrata perennial crop systems: shade-grown coffee in Central America. *Agroforestry Systems* 53: 151–170.

van den Bosch, R. and Telford, A.D. (1964). Environmental modification and biological control. In *Biological Control of Insect Pests and Weeds* (P. DeBach, ed.), pp. 459–488. Chapman & Hall, London.

Vandermeer, J. and Perfecto, I. (1995). *Breakfast of Biodiversity*. Food First Books, Oakland, California.

Wood, B.J. (1971). Development of integrated control programs for pests of tropical perennial crops in Malaysia. In *Biological Control* (C.B. Huffaker, ed.), pp. 422–457. Plenum Press, New York.

Chapter 10

The 'push–pull' strategy for stemborer management: a case study in exploiting biodiversity and chemical ecology

Zeyaur R. Khan and John A. Pickett

Introduction

Lepidopteran stemborers are the most important biotic constraints to maize (also known as corn) production throughout Africa. They seriously limit potentially attainable maize yields by infesting the crop throughout its growth, from seedling stage to maturity. Seventeen species in two families (Pyralidae and Noctuidae) have been found to attack maize in various parts of Africa. However, in eastern Africa, *Chilo partellus* (Pyralidae) and *Busseola fusca* (Noctuidae) are of significant importance. Maize yield losses, caused by stemborers, vary widely in different regions and comprise 20–40% of potential output, depending on agroecological conditions, crop cultivar, agronomic practices and intensity of infestation (Ampofo 1986; Seshu Reddy and Sum 1992; Khan et al. 1997b, 2000). Reducing losses caused by stemborers through improved management strategies would significantly increase maize production, and result in better nutrition and purchasing power of many maize farmers. In this chapter, we review approaches to providing a stemborer-management system involving attractant trap crops and repellent intercrops. In presenting as complete an account as possible, we rely heavily on our previous and in-press publications to provide further details for the successes described, with all statements corroborated therein, if not here.

Use of chemical pesticides is the main method of stemborer control recommended by Ministries of Agriculture to farmers in eastern and southern Africa. However, chemical control of stemborers is uneconomical and impractical to many resource-poor, small-scale farmers. Stemborers are difficult to control by chemicals, largely because of the cryptic and nocturnal habits of adult moths and the protection provided by the stem of the host crop in the developing stages. This is aggravated by the stemborers' ability to develop resistance to the pesticides. The adverse effects of these chemicals on human and animal health, the environment and animal and plant biodiversity are well-known.

In eastern and southern Africa, cultural control methods such as burning stemborers' wild hosts and dried maize stalks are recommended. However, these methods destroy not only plant biodiversity but also beneficial arthropods. Although for many years there have also been major efforts in eastern Africa to breed maize cultivars resistant or tolerant to stemborers, little progress has been made in this direction (Ampofo 1986; Seshu Reddy and Sum 1992).

Management strategies that exploit an understanding of the insect–plant interactions have been under investigation for many years (Khan et al. 1997a, b). Based on information gathered about the interactions between stemborers and their host and non-host plants, a novel 'push–pull', or stimulo-deterrent, diversionary strategy has been developed to manage cereal stemborers in maize-based farming systems in eastern and southern Africa. Push–pull strategies involve attempts to modify the environment so that pest population densities are lowered and natural enemies are increased (Khan et al. 1997a). The stemborer push–pull strategy involves the combined use of a trap and a repellent fodder plant of economic importance, whereby adult stemborers are simultaneously attracted to a trap crop and repelled from the maize crop.

Several plant species which could be used as trap or repellent plants in a push–pull strategy have been identified. Napier grass (*Pennisetum purpureum*) and Sudan grass (*Sorghum vulgare sudanense*) have shown value as trap plants, whereas molasses grass (*Melinis minutiflora*) and silverleaf desmodium (*Desmodium uncinatum*) repel ovipositing stemborers (Khan et al 1997b, 2000). Molasses grass intercropped with maize reduces infestation of the maize by stemborers and increases stemborer parasitism by a natural enemy, *Cotesia sesamiae* (Khan et al. 1997a). All four plants are of economic importance to farmers in eastern Africa as livestock fodder.

Development of a stemborer push–pull strategy

Cereal stemborers are generally polyphagous and their host plant range includes several members of the Poaceae, including wild hosts as well as more than one cultivated crop (Ingram 1958; Seshu Reddy 1983; Khan et al. 1991; Polaszek and Khan 1998). In tropical and subtropical parts of Africa, the stemborers, under original conditions, attacked only wild grasses and other plants (Mally 1920). The wild hosts often harbour food sources for many pest insect species, and may encourage insect invasion and outbreaks in neighbouring agroecosystems (van Emden 1990). During the non-growing season, stemborers remain present on wild host plants and the diapausing populations in crop residues. On the other hand, there are reports that several wild host plants buffer the crop against attacks by some stemborer species (Schulthess et al. 1997). Although some information on wild habitats as hosts of cereal stemborers in Africa is available (Ingram 1958; Bowden 1976; Seshu Reddy 1983), from the applied perspective it was important to study these interactions for development of sustainable integrated pest-management strategies. Our main aim was to discover grasses that were more preferred (than maize) for stemborer oviposition and that do not support pest development.

Host range of stemborers

To expand knowledge on the range of wild hosts of cereal stemborers, destructive sampling of more than 500 wild plant species belonging to families Poaceae, Cyperaceae and Typhaceae was conducted. The sampling was undertaken in three ecological zones in Kenya: hot and humid (Kenyan coast, up to 300 m altitude), warm and semiarid (Lake Victoria shores, about 1200 m altitude), and high-altitude wet and cool area (Trans-Nzoia district, more than 1800 m altitude). The sampling was done from eight different locations in each ecological zone, at three-month intervals during 1994–95. Stemborer larvae collected from each host were brought to the laboratory for rearing on the same host for three generations, to allow host confirmation. Observations were recorded on host suitability for insect development and oviposition. Thirty-three species of wild host plants belonging to three families were recorded as hosts to one or more species of stemborers, in different agroecologies in Kenya (Khan et al. 1997b; Polaszek and Khan 1998).

Wild grasses as reservoirs of natural enemies

Wild hosts adjacent to cultivated crops can provide extremely important refugia for natural enemies as well as sources of nectar, pollen and host/alternate prey (Altieri and Whitcomb 1979;

Altieri 1981; Altieri and Letourneau 1984; Landis et al. 2000; Nicholls and Altieri, ch. 3 this volume). To find out if wild hosts adjacent to cultivated crops provided alternative hosts for parasitoids, larvae of borers associated only with wild host plants were collected and reared on their respective hosts for three generations. Adults and larvae of these borers were sent to National Museums in Kenya and to the British Museum for identification. One hundred field-collected larvae of each 'wild borer' species were incubated in separate vials for 14 days and any parasitoids emerging from them were recorded, as described in Khan et al. (2001). To confirm whether these 'wild borers' were also hosts to parasitoids of crop borers, 100 third instar, laboratory-reared larvae of each species were exposed separately to 100 adults of two species of larval parasitoids, *Cotesia flavipes* and *Cotesia sesamiae*. The 'wild borer' larvae were incubated for 14 days and emergence of parasitoids was recorded. The experiment was replicated three times (Khan et al. 1997b).

Three species of stemborers, *Bactra stragnicolana*, *Phragmataecia boisduvalii* and *Poeonama* sp. were recorded to be associated with several species of gramineous and other wild plants in various parts of Kenya. *Bactra stranicolana* is a pyralid stemborer associated with *Cyperus immensus* and *C. distans*. This borer is an alternate host to *C. flavipes*, a parasitoid of cereal stemborers. *Phragmataecia boisduvalii*, a noctuid associated with *Echinochloa pyramidalis* and *Phragmites* sp., is an alternate host for *C. sesamiae*. *Poeonoma* sp., is also a noctuid associated with *Pennisetum purpureum*, and *P. macrourum* is an alternate host for *C. sesamiae* (Khan et al. 1997b).

Selection of trap and repellent plants

Triplicate small (4 m × 4 m) plots of the 33 plant species selected from the survey and several non-host plants were grown at the field station of the International Center of Insect Physiology and Ecology (ICIPE) at Mbita Point, on the banks of Lake Victoria, Suba district, Kenya. After establishment, samples of the different plants (30 plant tillers) were examined randomly for stemborer oviposition and infestation by splitting open the stems and counting the different stages of larvae present. Stemborer colonisation, which results from an initial choice by ovipositing adults, was assessed and the most attractive (trap crops) and least attractive (repellent inter-crops) plant species were selected.

Two species attracted considerably more oviposition than did maize. These were both forage crops: Napier grass, which is related to pearl millet, and Sudan grass, which is related to sorghum. Although Napier grass attracted more oviposition than did maize, most of the larvae did not survive (Khan et al. 2000). However, Sudan grass allowed development of stemborers and also had a very high parasitisation rate, with 70–80% of the larvae being killed (Khan et al. 1997b). Notably, the molasses grass, *Melinis minutiflora*, an indigenous gramineous plant which has been developed as a source of cattle forage throughout the world, particularly in South America, attracted no oviposition at all (Khan et al. 1997a; Khan 2002). *M. minutiflora* possesses some anti-tick properties, causing ticks feeding on cattle to detach when the cattle are in contact with the grass (Prates et al. 1993; Mwangi et al. 1995).

Legumes are not attacked by cereal stemborers. Therefore, assessment of stemborer activity was made in association with maize as the host plant. Out of these studies, silverleaf *Desmodium uncinatum* and greenleaf *Desmodium intortum* were shown to repel ovipositing stemborers (Khan et al. 2001).

Field evaluation of 'push–pull' against stemborers

With candidate 'push' and 'pull' plants having been selected in 1996, on-station field trials began in 1996 at the Mbita Point field station in the Suba district and in the Trans-Nzoia district of Kenya, in collaboration with staff at the Kenyan Agricultural Research Institute (KARI) and

Ministry of Agriculture, Kenya (Khan et al. 1997b). Initially, 40 m × 40 m plots of maize were grown either as monocultures or surrounded by a 1 m border of Napier or Sudan grass to act as a trap crop. Again, assessment was made by splitting the stems and counting the larvae present inside stems of maize and trap plants. In plots with the border of Napier grass, there was considerably more oviposition and early larval development in the trap crops compared with maize. Nevertheless, only 20% of the larvae survived on the Napier grass whereas 80% survived through to adults on the maize. This additional control effect was caused by production of sticky sap by Napier grass tissues in response to penetration by first- and second-instar stemborer larvae (Figure 10.1). Most of the larvae were trapped in the sticky fluid and killed.

The Sudan grass border trap crop contained eight times more larvae than did maize, and stemborer numbers within the maize were reduced to one-third compared with the maize monoculture (measured at the 12-week plant growth stage) (Khan et al. 1997b). In 1997, scientist-managed trials on small-scale farmers' fields in Trans-Nzoia region showed that the Napier grass trap crop system reliably gave significant yield improvements of approximately 15–18%. In 1998–99, farmer-managed trials with Napier grass and Sudan grass on poorly yielding farms in the Trans-Nzoia and Suba regions showed similar successes, with yield increases in the order of 20–25%.

Separate studies in both regions, initially at experimental sites, moving into scientist-managed and then farmer-managed trials, demonstrated the effectiveness of intercropping *M. minutiflora*. For example, numbers of *C. partellus* larvae found in maize stems were reduced from 80 in the monoculture plots to five in plots having a row of *M. minutiflora* planted between each row of maize (six replicate samples) (Khan et al. 1997b). The traditional intercrop, cowpea (*Vigna sinensis*), only reduced numbers of larvae to 45. Similar reductions were found for the other stemborer species. Although not as effective as *M. minutiflora*, *D. uncinatum* as intercrops also reduced stemborer damage, and the results were always significantly better than the maize monoculture or maize intercropped with other legumes, including cowpea. The planting rate for *M. minutiflora* with maize was subsequently reduced from a 1:1 intercrop:crop ratio to 1:10 with excellent results (Khan et al. 2000, 2001). Parasitisation by *Cotesia* sp. of stemborers in

Figure 10.1: Production of sticky sap by Napier grass tissues in response to penetration by first- and second-instar stemborer larvae.

maize fields was significantly higher in the intercrops than in maize monocultures (Khan et al. 1997a).

The strategy of combining crop, trap crop and intercrop to control stemborer infestation has been highly successful, from the initial experimental trials through to scientist-managed farm trials and farmer-managed trials. For example, the full push–pull system with maize, *D. uncinatum* and Napier grass was implemented in on-farm field trials on a relatively poorly yielding farm in the Suba district during the long rainy season in 1999. Results showed that control plots produced 1.8 tonnes/ha compared with about 4 tonnes/ha for the push–pull plots, a statistically significant increase (Khan et al. 2001). Similar results were obtained in 1999 in the Trans-Nzoia region, where more productive farms also saw significant yield benefits, i.e. 6–7 tonnes/ha for the treated plots compared with 4–5 tonnes/ha for control plots. In 2003, significant increases in maize yields were recorded from 10 districts in Kenya, where now more than 1500 farmers are practising the push–pull approach (Figure 10.2). The associated specific effects on stemborer populations, percent damage and level of parasitisation is presented in greater detail in Khan et al. (2001).

It has been a general principle that plants used in push–pull pest-management strategies must themselves have value for the communities involved. In the work described here, the trap crops and intercrops can all be used as forage for livestock. Indeed, the luxuriant stands of Napier grass and Sudan grass allowed farmers to improve their cattle husbandry and many have increased the number of cows in zero grazing systems (Figure 10.3). In the regions where zero grazing is the usual method for cattle husbandry, such forage is extremely important.

Cost–benefit analysis, performed by socioeconomists, has shown returns by a factor of over 2.2, with the maize monoculture returning less than 0.8 and even pesticide intervention systems less than 1.8 (Khan et al. 2001). Returns from maize and Napier grass alone were comparable with those fields where insecticide was applied to control stemborers.

Role of semiochemicals for stemborer control

The next step was to investigate the volatiles produced by the plants. Samples of host-plant volatiles have been investigated by gas chromatography coupled-electroantennography (GC-EAG) on the antennae of stemborers (Khan et al. 2000). GC peaks consistently associated with

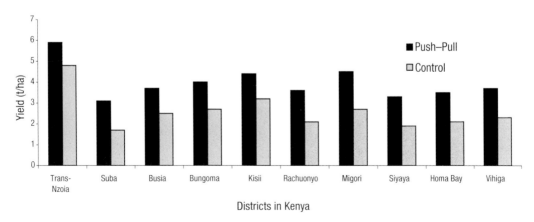

Figure 10.2: Average increases in maize yields in 'push–pull' fields in ten districts of Kenya in 2003. Differences in yields were significant at $P < 0.05$ for Trans-Nzoia and Kissi districts, and at $P < 0.01$ for the other regions.

EAG activity were tentatively identified by GC coupled-mass spectrometry (GC-MS) and identity was confirmed using authentic samples. Six active compounds were identified: octanal, nonanal, naphthalene, 4-allylanisole, eugenol and linalool. Behavioural tests, employing oviposition onto an artificial substrate treated with individual compounds, demonstrated positive activity for all these compounds.

Coupled GC-EAG with volatiles from the intercrop plant *M. minutiflora* showed a wide range of peaks associated with EAG activity. A general hypothesis developed during our work on insect pests is that non-hosts are recognised by colonising insects through release of repellent or masking semiochemicals, although it is almost inevitable that compounds also produced by hosts will be present. In this case, the host cereal plants and the non-host *M. minutiflora* would be expected to have a number of volatiles in common, as they are all members of the Poaceae. For *M. minutiflora*, five new peaks with EAG activity were identified, in addition to the attractant compounds and others normally produced by members of the Poaceae (Kimani et al. 2000; Khan et al. 2000). These comprised: (*E*)-ß-ocimene, α-terpinolene, ß-caryophyllene, humulene and (*E*)-4,8-dimethyl-1,3,7-nonatriene. The ocimene and nonatriene had already been encountered as semiochemicals produced during damage to plants by herbivorous insects (Turlings et al. 1990, 1995). It is likely that these compounds, being associated with a high level of stemborer colonisation and, in some circumstances, acting as foraging cues for parasitoids, would be repellent to ovipositing stemborers, as was subsequently demonstrated in behavioural tests. Investigating the legume volatiles, it was shown that *D. uncinatum* also produced the ocimene and nonatriene, together with large amounts of other sesquiterpenes, including α-cedrene (Khan et al. 2000).

Exploiting natural enemies in the push–pull system

Planting grass around maize fields significantly increased parasitisation of *C. partellus* and *B. fusca* (Khan et al. 1997b). Therefore, exploitation of natural enemies through trap and repellent plants represents a clear benefit of the push–pull strategy. Compared with the maize monocultures, where only 4.8% *C. partellus* and 0.5% *B. fusca* larvae were parasitised, 18.9 % *C. partellus* and 6.17% *B. fusca* larvae were parasitised in the maize field surrounded by Sudan grass (Khan et al. 1997b).

It was also noted in field trials that intercropping with *M. minutiflora* not only reduced populations of stemborers in maize, but there were more stemborers parasitised in these plots than in maize monocultures (parasitised in maize monocrop, 5.4%; intercropped with *M. minutiflora*, 20.7%) (Khan et al. 1997a). The chemical compounds identified as reducing stemborer attack, such as the nonatriene, were also responsible for the increased parasitoid foraging. Indeed, when this compound was presented to the parasitoid *C. sesamiae* in a Y-tube olfactometer at a level similar to that found in the volatiles from *M. minutiflora*, it accounted for most of the attractiveness of the natural sample. This suggests that intact plants with an inherent ability to release such stimuli could be used in new crop protection strategies.

Although the nonatriene and ocimene are also released by *D. uncinatum*, and responsible for its repellency to stemborers, there was no detectable increase in parasitism in intercropped plots. It may be that other components produced by *D. uncinatum*, including large amounts of α-cedrene, interfere with this effect. To test this hypothesis, a new collection of *Desmodium* species has been established at the Mbita Point field station.

Potential benefits of the push–pull strategy

The push–pull tactics conducted in Kenya have helped more than 1500 participating farmers to increase their maize yields by an average of 17% in Trans-Nzoia district (a high-potential-yield area) and by 25% in Suba district (a low-potential-yield area). Increased maize yields accompanied by the following additional features have contributed to high farmer adoption rates.

Dairy and livestock production. The push–pull habitat manipulation tactics have increased livestock production (milk and meat) by providing more fodder and different crop residues, especially on small farms where competition for land is high. For example, the Suba district of Kenya, a milk-deficit region on the shores of Lake Victoria, produces only 7 million litres of milk, far short of the estimated annual demand of 13 million litres, and has mostly indigenous livestock (zebu). In this district, a major constraint to keeping improved dairy cattle for milk production is the unstable availability and seasonality of feed, often of low quality. The habitat manipulation strategy, adopted by 250 farmers in this district, is facilitating efforts by agricultural authorities to promote livestock production and improve milk supply. Through the combined effort, the number of improved dairy cattle in the district increased from four in 1997 to 350 in September 2002 (Figure 10.3). With this rate of growth in improved dairy cattle, it is expected that Suba will be self-sufficient in milk production by 2005.

Soil conservation and fertility. Soil erosion and low fertility are very common problems in eastern Africa. The habitat manipulation strategy has placed some of the existing practices addressing these problems into a multifunctional context. For example, the cultivation of Napier grass for livestock fodder and soil conservation now assumes an additional rationale as a trap plant for stemborer management. Similarly, *Desmodium*, a nitrogen-fixing legume, grown for improving soil fertility and for quality fodder, is also an effective striga weed suppressant (Khan et al. 2002).

Enhancing biodiversity. The push–pull strategy contributes to the promotion and conservation of biodiversity. A recent study with the University of Haifa, Israel, has demonstrated that the numbers of beneficial soil arthropods in maize–*Desmodium* fields were significantly higher than the numbers in maize monocrops. The destruction of biodiversity is linked to the expansion of crop monoculture at the expense of diverse vegetation (Khan et al. unpublished).

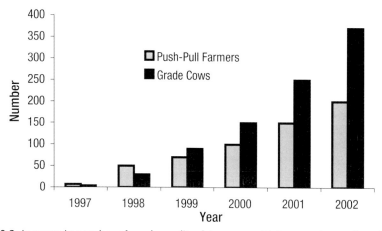

Figure 10.3: Increase in number of grade quality dairy cows with increase in number of farmers using push–pull habitat manipulation in Suba district.

Protecting fragile environments. Existing evidence indicates that higher crop yields and improved livestock production resulting from habitat manipulation strategies have the potential to support many rural households under existing socioeconomic and agroecological conditions. Thus, there will be less pressure for human migration to environments designated for protection. Moreover, farmers using such strategies have less motivation to use pesticides.

Income generation and gender empowerment. The habitat manipulation strategies described here have shown promise of not only significantly enhancing farm incomes, but also of gender empowerment through the sale of farm grain surpluses, fodder and *Desmodium* seed. In addition to targeting women farmer-groups, improved rural productivity and quality of life is expected to affect youth groups in the rural areas and help to stem migration from rural to urban areas.

Conclusion and future of the push–pull strategy

So far, there has been a limited research effort to integrate intercropping in pest-management programs in general, and biological control programs in particular. Although information on the effect of habitat diversity on natural enemies and herbivores has increased greatly in the past 20 years, most studies do not attempt to explain why population densities differ between monoculture and polycultures. Thus, these studies do not contribute to our understanding of the mechanisms involved. The work reviewed in this chapter has expanded understandings of insect–plant and pest–natural enemy interactions.

Push–pull strategies for controlling insect pests were first described by Pyke et al. (1987) and Miller and Cowles (1990). However, neither of those studies exploited natural enemies. Moreover, they used an added chemical deterrent or toxin to repel or kill the pest. The present stemborer push–pull strategy does not use any added chemical deterrents or toxins, and attempts exploitation of natural enemies through trap and repellent plants.

Using the push–pull strategy, field trials are being initiated in South Africa on the development of resistance-management strategies for transgenic maize. This strategy will use fodder grasses, such as Napier grass, as trap plants, which are highly attractive to ovipositing stemborers but do not allow stemborer development. On the other hand, the repellent plants, such as molasses grass, which not only repel ovipositing stemborers (push) from the main crop but are highly attractive to natural enemies, are used. We believe that by combining the push–pull strategy with genetically engineered Bt maize (see Altieri et al., ch. 2 this volume), the selection pressure on genetically engineered maize by stemborer population could be reduced. The increased population of natural enemies in Bt maize fields could also help significantly reduce the number of Bt-resistant stemborer larvae, which would have developed into Bt-resistant moths.

The push–pull approach described here is now expanding into Ethiopia, Uganda, Malawi and Tanzania. A pilot program has been initiated in southern Africa, addressing stemborer control in the arid and semi-arid areas of the Northern Province of South Africa. Each region has, in addition to varying climatic conditions and use of alternative cultivars, some differences in crops that must be taken into account, and considerable experience has been gained in this aspect by the pilot studies in various countries. However, wherever these approaches are developed for the specific needs of local farming practices and communities, it is essential that the scientific basis of the modified systems be completely elucidated, otherwise there will be a drift from effectiveness and justifiable dissatisfaction among the practising farmers. It is also essential that the approach, as used on-farm, is continually monitored in order to detect, as early as possible, any deleterious effects that might arise from the possible enhancement of other pest problems, so that remedial measures can be devised.

Acknowledgements

The push–pull strategy for stemborer control was developed by ICIPE in close collaboration with Rothamsted Research, UK, the Kenya Agricultural Research Institute and the Ministry of Agriculture, Kenya, with funding from the Gatsby Charitable Foundation, UK. Rothamsted Research receives grant-aided support from the Biotechnology and Biological Sciences Research Council of the UK.

References

Altieri, M.A. (1981). Weeds may augment biological control of insects. *California Agriculture* 35: 22–24.

Altieri, M.A. and Letourneau, D.K. (1984). Vegetation diversity and outbreaks of insect pests. *CRC Critical Reviews in Plant Sciences* 2: 131–169.

Altieri, M.A. and Whitcomb, W.H. (1979). The potential use of weeds in the manipulation of beneficial insects. *HortScience* 14: 12–18.

Ampofo, J.K.O. (1986). Maize stalk borer (Lepidoptera: Pyralidae) damage and plant resistance. *Environmental Entomology* 15: 1124–1129.

Bowden, J. (1976). Stem borer ecology and strategy for control. *Annals of Applied Biology* 84: 107–111.

Ingram, W.R. (1958). The lepidopterous stalk borers associated with Graminae in Uganda. *Bulletin of Entomological Research* 49: 367–383.

Khan, Z.R. (2002). Cover crops. In *Encyclopedia of Pest Management* (D. Pimentel, ed.), pp. 155–158. Marcel Dekker, New York.

Khan, Z.R., Litsinger, J.A., Barrion, A.T., Villanueva, F.F.D., Fernandez, N.J. and Taylor, L.D. (1991). *World Bibliography of Rice Stem Borers, 1794–1990*. International Rice Research Institute, Los Banos, Philippines/International Centerof Insect Physiology and Ecology, Nairobi, Kenya. 415 pages.

Khan, Z.R., Ampong-Nyarko, K., Chiliswa, P., Hassanali, A., Kimani, S., Lwande, W., Overholt, W.A., Pickett, J.A., Smart, L.E., Wadhams, L.J. and Woodcock, C.M. (1997a). Intercropping increases parasitism of pests. *Nature* 388: 631–632.

Khan, Z.R., Chiliswa, P., Ampong-Nyarko, K., Smart, L.E., Polaszek, A., Wandera, J. and Mulaa, M.A. (1997b). Utilisation of wild gramineous plants for management of cereal stemborers in Africa. *Insect Science and its Application* 17: 143–150.

Khan, Z.R., Pickett, J.A., van den Berg, J., Wadhams, L.J. and Woodcock, C.M. (2000). Exploiting chemical ecology and species diversity: stem borer and striga control in Africa. *Pest Management Science* 56: 957–962.

Khan, Z.R., Pickett, J.A., Wadhams, L.J. and Muyekho, F. (2001). Habitat management strategies for the control of cereal stemborers and striga in Kenya in Maize. *Insect Science and its Application* 21: 375–380.

Khan, Z.R., Hassanali, A., Overholt, W.A., Khamis, T.M., Hooper, A.M., Pickett, J.A., Wadhams, L.J. and Woodcock, C.M. (2002). Control of witchweed *Striga hermonthica* by intercropping with *Desmodium* spp., and the mechanism defined as allelopathic. *Journal of Chemical Ecology* 28: 1871–1885.

Kimani, S.M., Chabbra, S.C., Lwande, W., Khan, Z.R., Hassanali, A. and Pickett, J.A. (2000). Airborne volatiles from *Melinis minutiflora* P. Beauv., a non-host plant of spotted stem borer. *Journal of Essential Oil Research* 12: 221–224.

Landis, D.A., Wratten, S.D. and Gurr, G.M. (2000). Habitat management to conserve natural enemies of arthropod pests in agriculture. *Annual Review of Entomology* 45: 175–201.

Mally, C.W. (1920). The maize stalk borer: *Busseola fusca* Fuller. Department of Agriculture, Union of South Africa, Bulletin no. 3. Cape Times, Government Printers. 111 pages.

Miller, J.R. and Cowles, R.S. (1990). Stimulo-deterrent diversion: a concept and its possible implication to onion maggot control. *Journal of Chemical Ecology* 16: 519–525.

Mwangi, E.N., Essuman, S., Kaaya, G.P., Nyandat, E., Munyinyi, D. and Kimondo, M.G. (1995). Repellency of the tick *Rhipicephalus appendiculatus* by the grass *Melinis minutiflora*. *Tropical Animal Health and Production* 27: 211–216.

Polaszek, A. and Khan, Z.R. (1998). Host plants. In *African Cereal Stem Borers: Economic Importance, Taxonomy, Natural Enemies and Control* (A. Polaszek, ed.), pp. 3–10. CAB International, Wallingford, UK.

Prates, H.T., Oliviera, A.B., Leite, R.C. and Craveiro, A.A. (1993). Anti-tick activity and chemical composition of *Melinis minutiflora* essential oil. *Presquisa Agropecuaria Brasileira* 28: 621–625.

Pyke, B., Rice, M., Sabine, B. and Zaluki, M. (1987). The push-pull strategy – behavioural control of Heliothis. *Australian Cotton Grower* May–July 1987: 7–9.

Schulthess, F., Bosque-Perez, N.A., Chabi-Olaye, A., Gounou, S., Ndemah, R. and Georgen, G. (1997). Exchange of natural enemies of lepidopteran cereal stemborers between African regions. *Insect Science and its Application* 17: 97–108.

Seshu Reddy, K.V. (1983). Studies on the stem borer complex of sorghum in Kenya. *Insect Science and its Application* 4: 3–10.

Seshu Reddy, K.V. and Sum, K.O.S. (1992). Yield-infestation relationship and determination of economic injury level of the stem borer, *Chilo partellus* (Swinhoe) in three varieties of maize, *Zea mays* L. *Maydica* 37: 371–376.

Turlings, T.C.J., Tumlinson, J.H. and Lewis, W.J. (1990). Exploitation of herbivore-induced plant odors by host-seeking parasitic wasps. *Science* 250: 1251–1253.

Turlings, T.C.J., Loughrin, J.H., McCall, P.J., Röse, U.S.R., Lewis, W.J. and Tumlinson, J.H. (1995). How caterpillar-damaged plants protect themselves by attracting parasitic wasps. *Proceedings of the National Academy of Sciences USA* 92: 4174.

van Emden, H.F. (1990). Plant diversity and natural enemy efficiency in agroecosystems. In *Critical Issues in Biological Control* (M. Mackauer, L.E. Ehler and J. Roland, eds), pp. 63–80. Intercept, Andover.

Chapter 11

Use of sown wildflower strips to enhance natural enemies of agricultural pests

Lukas Pfiffner and Eric Wyss

Introduction

Intensive agriculture and excessive use of agrochemicals have resulted in an impoverished wildlife in agricultural landscapes, especially in arable landscapes but also in perennial high-input crops. The elimination of semi-natural habitats, simplification of crop rotations as well as high input of fertilisers and pesticides is considered to be responsible for the severe decline of biological diversity that has been observed (e.g. Aebischer 1991). These practices can reduce habitat quality and remove the necessary habitat structure that is important to many natural enemies.

Moreover, agricultural landscapes are increasingly being simplified; with natural and semi-natural areas fragmented and replaced altogether by large monocultural fields. As a consequence, most of the natural enemies depending on such semi-natural habitats for overwintering (Sotherton 1985; Lys and Nentwig 1994; Pfiffner and Luka 2000) need to disperse further to reach summer feeding habitats like agricultural crops. Fragmentation and loss of suitable habitats has caused natural enemies to decline in species diversity and abundance, and has even resulted in extinctions (Fahrig 1997) and loss of biological control functions (Kruess and Tscharntke 1994). Such landscape-scale aspects in biological control are explored in more detail in chapter 4.

Nowadays, a desirable goal in agricultural landscapes is the enhancement of biotic diversity through the use of sustainable farming methods and the conservation and reestablishment of non-crop habitats. Agri-environmental programs have been established in several European countries (e.g. rural development and set-aside programs). Since 1993, the Swiss government has subsidised low-input and sustainable-farming methods (e.g. low-input integrated crop management, organic farming) and the establishment of non-crop habitats. Farmers are encouraged to increase the amount of these non-crop habitats including low-input habitats in order to reverse the observed decline of farmland fauna and to conserve or improve the functions and services of the agroecosystems.

It has been demonstrated that habitat manipulation of the environment can enhance the survival of natural enemies and thereby improve their efficiency as pest-control agents (Gurr et al. 1998; Landis et al. 2000, Nicholls and Altieri, ch. 3 this volume). Field margins are an important type of habitat that provides refuge and resources for many arthropods. Thus field margins play a key role in maintaining biological diversity on farmland (Fry 1994). In addition, it may be useful to combine these semi-natural habitats with low-input agriculture to enhance effects on fauna diversity and natural pest control (Pfiffner 2000; Pfiffner and Luka 2003).

Some habitat manipulation options are known to improve pest control in adjacent production systems. These include grassy beetle banks (Thomas and Marshall 1999; Collins et al. 2002); weedy strips; set-aside strips and field margins (see overview by Marshall and Moonen 2002). This chapter focuses on the use of sown wildflower strips (a synonymous term for weed strips) which may be located in field margins. These are used to augment natural enemies and to increase the diversity of flora and fauna in annual and perennial cropping systems.

One of the first initiatives to implement strip farming with sown wildflower strips in practice was taken by Nentwig (1989). Since 1993 this type of field margin has been encouraged and subsidised in the Swiss agri-environmental programs. Nowadays, more than 3000 ha of wildflower strips (in general with a width of 3–10 m) exist on farms all over Switzerland (Anonymous 2003). This chapter will detail how these strips are composed and how they can be established and managed in practice. Finally, their effects on natural enemies and pest control in annual and perennial systems are discussed.

Figure 11.1: Sown wildflower strip in Swiss farmland; senescent stems of teazle (*Dipsacus fullonum*) conspicuous (Photograph: G.M. Gurr).

Habitat manipulation using wildflower strips

To stop the decline of agrobiodiversity and the loss of biological control functions in intensively used landscapes, we consider that a minimum 10% of agricultural land should be allocated to 'ecological compensation areas'; Broggi and Schlegel (1989) suggested 12% for intensively agricultural regions in the Swiss lowland. Ecological compensation areas include non-crop habitats (such as hedges, set-asides, woodlots and weedy strips) and low-input agricultural habitats, for example low-input meadows such as a *Mesobrometum*, a dry, unfertilised meadow. However, most of the intensively used agricultural landscapes have less than 2–3% of area devoted to such ecological compensation areas, therefore additional non-crop habitats are needed. One option is sowing wildflower strips (Figure 11.1), which may also fulfil many goals of agroecology and nature conservation. Though artificial, these strips are a good supplement to natural and semi-natural habitats including other field margins.

Using wildflower strips sown with a mixture of different annual, biennial and perennial plant species, we intend to enhance biological pest control by providing various environmental requisites for natural enemies: supplementary foods (alternate host or prey, or in some cases pollen); complementary foods (nectar, pollen, honeydew); modified microclimate; and shelter, for example after agricultural practices, and wintering or nesting habitat. The strips must be integrated into landscape in a spatially and temporally favourable way for natural enemies, and must be practical for farmers to implement. As explained below, there are certain requirements for size, location/spacing, vegetation composition and management techniques if they are to be successful.

Location and composition

In general, wildflower strips are 3–10 m in width and can be sown on nutrient-rich and loamy soils with a poor seed bank. Those naturally regenerated on nutrient-rich soils often have a plant community dominated by grasses, with low species richness. On light soils, natural regeneration may be more appropriate, if autochthonous flora and no problematic perennial weeds exist. Sown strips can be situated at the edge of arable fields or used to divide large fields, and ideally connect natural and semi-natural habitats to form a network of ecological compensation areas. The importance of landscape connectivity to wildlife is stressed by Kinross et al. (ch. 13 this volume). The botanical composition of the weed mixture results from a long selection process in which specific characters of approximately 100 plant species have been evaluated according to aspects of agroecology (with a focus on natural pest control) and nature conservation (Nentwig 1989, 1998; Nentwig et al. 1998). The different properties of these plant species have been tested at the species and mixture level: effects on insect diversity and abundance of natural enemies, length of flowering period, survival capacity in agricultural soils, longevity in a diverse plant mixture as well as the tendency to invade adjacent crops (Frei and Manhart 1992; Heitzmann 1995; Wäckers and Stepphuhn 2003). In the early 1990s two different basic mixtures were used, one for annual and the other for perennial crops (Wyss 1994). The mixtures have been further developed, based on field experience on hundreds of farms in Switzerland. Field tests with similar seed mixtures were also performed in Germany (Vogt et al. 1998; Denys and Tscharntke 2002) and Austria (Kienegger and Kromp 2001). A careful choice of plant species is important, to avoid problems with pests and other noxious organisms. For example, slugs can considerably affect the development of wildflower strips (Frank 2003). Therefore, it is recommended that the seed quantity for certain plant species (e.g. *Centaurea cyanus, Papaver rhoeas*) be increased because they are highly susceptible to slug herbivory (Frank 2003).

Presently, the seed mixture consists of 24 (basic) up to 37 plant species (full mixture). This mixture includes annual, biennial and perennial wildflowers and cultivated plant species (Table 11.1) and is adapted to the Swiss plateau. Several commercial seed companies in Switzerland sell

Table 11.1: Composition of the seed mixture for wildflower strips (status 2003), adapted to the Swiss plateau and officially recommended by the Swiss Federal Research Station for Agroecology and Agriculture.

Species	Botanical type[2]	g/ha[3]
Annual species		
Agrostemma githago[1]	segetal	600
Anchusa arvensis	segetal	70
Buglossoides arvense	segetal	60
Camelina sativa	segetal	30
Centaurea cyanus	segetal	500
Consolida regalis[1]	segetal	30
Fagopyrum esculentum	culture	7245
Legousia speculum-veneris[1]	segetal	30
Misopates orontium[1]	segetal	30
Nigella arvensis[1]	segetal	30
Papaver dubium	segetal	20
Papaver rhoeas	segetal	150
Silene noctiflora[1]	segetal	30
Stachys annua[1]	segetal	60
Vaccaria hispanica[1]	segetal	70
Valerianella rimosa[1]	segetal	30
Biannual species		
Cichorium intybus	ruderal	120
Daucus carota	meadow	150
Dipsacus fullonum	ruderal	2
Echium vulgare	ruderal	200
Malva sylvestris	ruderal	60
Melilotus albus	ruderal	20
Pastinaca sativa	ruderal	80
Reseda lutea	ruderal	40
Silene alba	ruderal	100
Tragogpogon orientalis	meadow	100
Verbascum densiflorum	ruderal	50
Verbascum lychnitis	ruderal	30
Perennial species		
Achillea millefolium	meadow	20
Anthemis tinctoria	ruderal	20
Centaurea jacea	meadow	200
Leucanthemum vulgare	meadow	80
Hypericum perforatum	meadow	60
Malva moschata	ruderal	20
Onobrychis viciifolia	meadow	600
Origanum vulgare	meadow	60
Tanacetum vulgare	ruderal	3
Total		**11 kg/ha**

Notes
1 Endangered plant species   2 Segetal species: typical wild flora of arable land   3 Seed quantity for the full mixture with 37 species
Underlined species: Basic mixture containing 24 species

the mixture, that is officially recommended by the Swiss Federal Research Station for Agroecology and Agriculture. Regionally produced species and species from native populations are used, to avoid detrimental effects on the indigenous flora. The importance of that has been underlined by Keller (1999). He compared nine wildflower species from different origins throughout Europe with the local adapted species, and found differences in establishment success and phenological patterns. The missing local adaptations were partly responsible for the reduced establishment rates of the alien species.

Establishment and management of wildflower strips

Wildflower strips must be established on sites free of problematic weeds (e.g. *Rumex obtusifolius, Agropyron repens, Cirsium arvense* and *Convolvulus arvensis*) and that offer appropriate soil conditions (no soil compaction or waterlogging). If the preceding vegetation was grassland, it is difficult to establish strips rich in species. After careful soil preparation, similar to that used for sowing grass-clover, wildflower strips are usually sown in spring on mineral soils. In contrast, in organic soils, strips are sown in autumn to avoid spring-germinating weeds (e.g. *Galinsoga* sp., *Setaria* sp.). Marshall et al. (1994) found autumn sowing more effective under British weather conditions. Because of the different seed sizes, sowing is by hand or by a pneumatic sowing machine.

Wildflower strips tend to be rich in species during the initial phase. With succession, however, a decrease in the number of plant species has often been recorded (Figure 11.2) although biennials and perennials then appear. The dominating process determining species composition is succession. However, the speed and direction that succession takes is strongly influenced by site conditions and the potential species pool (Ullrich 2001). The high number of plant species in the seed mixture reduces the risk of poor establishment. Marshall et al. (1994) compared different seed mixtures and found that seed mixes with many species give higher initial diversity than simpler mixtures. By the introduction of a wildflower strip, a high level of floristic diversity consisting of species sown and recruited spontaneously can be achieved (Ullrich 2001).

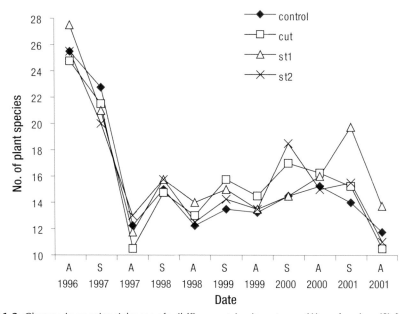

Figure 11.2: Change in species richness of wildflower strips in autumn (A) and spring (S) from 1st to 6th year under different management: cut, one cut; st1, one cut and a soil tillage with a harrow; st2 one cut and a soil tillage with a cultivator control, no cut or tillage (modified from Pfiffner and Schaffner 2000b).

Wildflower strips are subjected to fewer disturbances than are arable crops and this might make them beneficial to natural enemies. Ideally, mechanical disturbance and chemical inputs should be minimal, though they are affected by adjacent farming operations. Depending on the quality (floral diversity, dominant perennial species), wildflower strips may require maintenance for conserving structural and plant diversity in their succession. Since strips are not disturbed for many years, weed control is important if invasion of adjacent crops is to be avoided. Perennial rhizomatous weeds such as *Agropyron repens*, *Cirsium arvense* and *Convolvulus arvensis* pose a high risk for agricultural areas and require control. Where weed control becomes necessary, spot treatment can be used to target problem plants. To keep a diverse wildflower strip, management such as mowing or soil tillage is necessary to ensure the required plant community persists and to minimise weed problems. Experiences during the 1990s on farms (Günter 2000) as well as in a six-year randomised-plot trial (Pfiffner and Schaffner 2000) have shown that mowing, or mowing combined with soil tillage in autumn or spring, may positively influence plant diversity and its spatial structure. This management may be necessary every two to three years, depending on plant succession.

In general, a less-frequent autumn or spring cutting regime, avoiding a summer cut and leaving the hay in situ, is preferred. Patchy rotational management is a useful way to offer uncut areas for invertebrates, including natural enemies. Haughton et al. (1999) showed that cutting per se had negative side-effects on the abundance of spiders (Araneae), bugs (Heteroptera) and Auchenorrhyncha. Annual applications of glyphosate also decreased abundance of these arthropods.

Specific recommendations on seed mixtures, sowing date and appropriate management over the long term (mid-term) have to be carefully adjusted to site conditions (soil, weed flora and adjacent habitats) and to the specific goals.

Sown wildflower strips in annual cropping systems

Many of the difficulties for natural enemies in annual crops can be traced to the frequent and intense disturbances that characterise these agroecosystems. In these simplified systems, many ecological services associated with the maintenance or enhancement of biodiversity, such as biological control, pollination, erosion prevention, nutrient cycling etc., are compromised. The concept of restoring these functions by manipulating the landscape infrastructure may alleviate some of the problems and contribute to a more sustainable annual cropping system.

Effects on generalist and specialist predators

Generalist predators such as carabids and spiders are the predominant groups of epigaeic arthropod fauna in arable cropping systems and belong to the most abundant surface-dwelling predators in temperate agroecosystems (Thiele 1997; Wise 1994). Although generalist predators lack prey specificity, it is widely accepted that they can nevertheless exert a substantial predation impact on prey populations (Symondson et al. 2002). Several carabid species feed on key agricultural pests (reviewed by Sunderland 2002) such as aphids (Kielty et al. 1999) and slugs (Bohan et al. 2000), while others may alter weed densities through seed consumption (Lund and Turpin 1977). Interactions between specialist and non-specialist natural enemies usually result in complementary effects, enhancing pest control (Sunderland et al. 1997; Sunderland and Samu 2000). For example, foraging by specialists on the crop may displace pests off the plant to be killed by epigaeic carabids and spiders. A few studies show that an increased generalist predator density can result in improved pest control (Sunderland 1999; Lang 2003).

Diversity and abundance of generalist predators

There is evidence that wildflower strips enhance diversity and abundance of generalist predators in annual cropping systems (Lys and Nentwig 1992; Jmhasly and Nentwig 1995; Hausammann 1996a; Frank 1997; Pfiffner and Luka 2000; Pfiffner et al. 2000a). Overall, the species diversity of spiders and carabids in the strips was always higher than in the adjacent fields.

Pfiffner et al. (2000b) compared wildflower strips with different field margins and adjacent annual crops in a landscape (Grosses Moos, Switzerland) characterised by intensive land use of arable and vegetable crops. Studies using pitfall funnel traps in sown wildflower strips and low-input grass strips, as well as the adjacent cereal and vegetable crops, revealed that carabid beetles and spiders were the most abundant of the epigaeic beneficial arthropods in terms of species and individuals. Field margins had higher numbers of carabid beetle and spider species than the adjacent crops; at the crop level, vegetables harboured fewer species and individuals than cereal crops (Figure 11.3a, b). Abundance of carabids was highest in cereal crops during the growing season, dominated by two arable species (*Poecilus cupreus* and *Pterostichus melanarius*) which are known to feed on many arthropod pest species (Sunderland 2002). In addition, other agro-ecologically important large carabids such as species of the genus *Carabus* were abundant in wildflower strips and grass strips, whereas within the crop, significantly fewer large *Carabus* were found (Figure 11.4). Field margins could positively influence the abundance of endangered species and those requiring special microclimatic conditions, for example xero-thermophlic carabid beetles (Figure 11.5). Various investigations found different spatial distribution patterns of carabids (Pfiffner and Luka 2000, 2003) and spiders (Frank and Nentwig 1995) in relation to

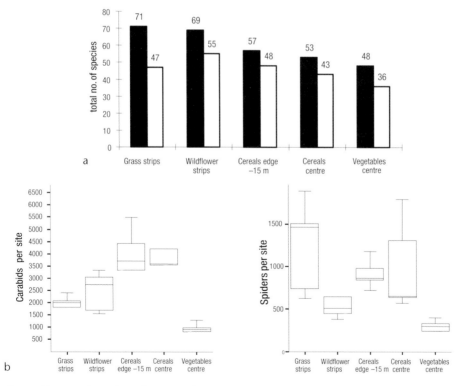

Figure 11.3: Total number of species of carabids (black) and spiders (unshaded) (a), and mean activity-density (b) of carabids (left) and spiders (right) found in two different field margins and adjacent arable crops of an intensively managed agricultural landscape. n = 5 per type of habitat. From Pfiffner et al. (2000b).

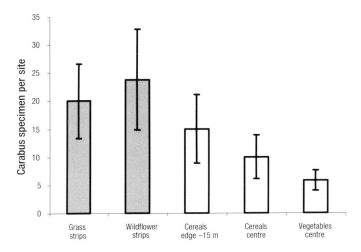

Figure 11.4: Enhancing effects of field margins on *Carabus* spp. predators. n = 5 per type of habitat. Pfiffner et al. (2000b).

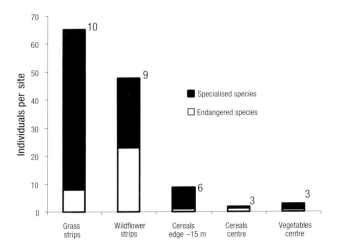

Figure 11.5: Occurrence of microclimatic specialists and endangered carabid species (mean activity density per site) in two different field margins and adjacent arable crops. Number of species are indicated beside the bars. From Pfiffner et al. (2000b).

the occurrence within the strips and the crop. Some species were nearly exclusive or much more abundant within these field margins, for example xero-thermophilous or stenoceous carabids such as *Brachinus* sp., *Harpalus* sp. and *Carabus* sp.

Specialist predators

Distinct effects of wildflower strips on specialised predators such as aphidophagous hoverflies (syrphids) and chrysopids have been reported; high densities of flowers constitute good foraging habitats for the adults (Weiss and Stettmer 1991; Salveter 1998; Frank 1999). Among the sown species of the strips, *Centaurea jacea* and *Pastinaca sativa* showed the highest syrphid larva densities (Salveter 1998). Hausammann (1996a) observed fewer aphids per aphidophagous syrphid larva near sown strips than in the middle of wheat fields. Hickmann and Wratten (1996)

found similar effects in winter wheat, with strips of *Phacelia tanacetifolia* harbouring significantly more eggs of aphidophagous syrphids and fewer aphids than did control plots. In
contrast, Salveter (1998) found no direct effects of wildflower strips on oviposition by syrphids
within fields in relation to the distance of the wildflower strip; furthermore, the strips were
neither significant for an early development of the first generation nor for an additional generation after wheat harvest. Effects may have been masked in that small-scale (range up to 1 km)
study as a result of the high mobility of syrphid adults.

Pest-control effects

The lack of intensive agricultural disturbance and the presence of a structurally complex,
species-rich vegetation in the strips provides favourable conditions for many groups of natural
enemies during the crop-growing season and winter period. There is some evidence that
wildflower strips can substantially improve the food supply and condition of natural enemies
(Zangger et al. 1994) and therefore support higher densities of predators and parasitoids within
fields. Better feeding conditions may augment and prolong the reproduction and may increase
searching activity of natural enemies (Morales-Ramos et al. 1996; Stapel et al. 1997; Bommarco
1998).

   The dispersal activity of natural enemies from non-crop habitat to crop habitat greatly alters
the natural pest-control effects. Lys et al. (1994) found higher activity in several carabid species
due to prolongation of reproductive period in the strip-managed area compared with the
control area. This effect is also attributed to better nutrition of natural enemies in the strip-
managed area. The number of spiders and web cover tended to decline with increasing distance
from the wildflower strip (Jmhasly and Nentwig 1995). For an abundant group of spiders in
arable land (*Erigone atra, E. dentipalpis, Oedothorax apicatus, Pardosa agrestis* and *P. palustris*)
dispersal from strip into adjacent fields was found. In spring they were more abundant in the
weed strip but until mid-year they moved into adjacent crops (Frank and Nentwig 1995; Lemke
and Poehling 2002). In addition, there is evidence that strips are providing source populations
that contribute to field invasion by these common spider species early in the season (Lemke and
Poehling 2002).

   A few investigations have shown positive effects on pest regulation. Hausammann (1996a)
found fewer cereal leaf beetles (*Oulema* sp.) in strip-managed fields than in fields without strips,
and a better predator–prey relationship between syrphids larvae and aphids near the strips.
Similarly, Collins et al. (2002) showed that grassy beetle banks have significant suppressing
effects on cereal aphids (*Sitobion avenae*) up to a distance of 83 m. Jmhasly and Nentwig (1995)
found higher predation rates of mainly Diptera and Aphidina by web spiders (*Araneidae,
Tetragnathidae* and *Linyphiidae*) in close vicinity to wildflower strips but with relatively small
turnover rates.

   Field margins are a valuable tool to increase the abundance of predators and parasitoids in
intensive vegetable productions. Pfiffner et al. (2003) investigated the parasitism rates of
lepidopteran cabbage pests in relation to presence or absence of species-rich wildflower strips.
Parasitism rates of caterpillars varied from less than 10% to 64%. Significantly higher rates of
parasitised *Pieris rapae* and *Mamestra brassicae* were found near the wildflower strip than the
control field without a strip. In addition, the diversity and abundance of parasitic Hymenoptera
in sweep-net catches were much higher in wildflower strips (up to three times more specimens
than in cauliflower; 14–16 families) than in the cabbage crops (7–11 families).

   More information is required on the dispersal of natural enemies to determine the optimal
spacing of strips. At the landscape scale, there are still many questions, for example which
features of infrastructure in relation to the cropping systems help reduce pest populations to
below the economic threshold.

Pests, herbivores and other arthropods in wildflower strips

Improved provision of food resources such as nectar may provide benefits to herbivores, parasitoids and predators. Therefore, careful selection of plants is important to avoid any risk of enhancing pest populations or offering an alternate host for a plant pathogen and other noxious organisms. A selective diversification with plants which are botanically unrelated to the crop, is needed. Gurr et al. (1998) proposed a checklist approach that allows a semi-quantitative assessment of hazards as well as economic and biological factors.

Various investigations into the effects on herbivores showed distinct effects of the strips. Aphids, phytophagous beetles and other herbivores (*Cicadellidae* and *Tenthredinidae*) can be enhanced by wildflower strips (Lethmayer 1995; Lethmayer et al. 1997). Lethmayer (1995) found 86 aphid species in strip-managed fields. Some of them were aerially transferred from other habitats (hedges, forest hedge), while about 50% of the species were monophagous (or oligophagous) and thus live only on a single host plant, mostly a non-crop plant. Therefore, these aphid species will not affect crops directly but can serve as alternative prey for antagonists. Furthermore, polyphagous, harmful aphid species such as *Aphis* sp., *Macrosiphum euphorbiae* and *Metopolophium dirhodum* were detected in relatively high numbers (about 42% of specimens). All these aphids are prey for predators and parasitoids and will probably attract their antagonists. Indeed, some studies have shown that aphid infestation of crops near wildflower strips or weedy fields was lower and the predator–prey relationship between specialist predators and aphids was better than in the control area (Hausammann 1996a). Among phytophagous beetles, the majority found in wildflower strips were harmless. Although many harmful beetles also feed on wild plants related to their host crop, no pest species occurred in higher numbers in the strips than in the fields (Frei and Manhart 1992; Hausammann 1996b). The only exception was *Meligethes aeneus*, which was more abundant in the strips after rape had finished flowering (Lethmayer et al. 1997). In addition, overwintering studies revealed that wildflower strips offer suitable habitats for a few pests such as weevils (e.g. *Sitona* sp., *Ceutorhynchus* sp.), but these were less abundant than in the natural field boundaries (Bürki and Hausammann 1993).

Another important consideration is the effect of strips on other pests such as molluscs, voles and moles. Slugs in wildflower strips are a potential risk for adjacent crops. The strips are favourable habitat because they provide a good food source, an optimal microclimate and encourage reproduction. This leads to high densities of certain slug species (e.g. *Arion lusitanicus*) in the strips and to a dispersal of adult slugs from the strips into the adjacent crops (Frank 1998). Crop damage caused by slugs was observed in different crops and complete crop loss within 2 m was found in rape, caused mostly by *A. lusitanicus*. No economic loss was detected in wheat adjacent to strips (Frank 1997; Frank 1998). However, an application of a molluscicide (metaldehyde) in a 50 cm wide band has been shown to protect rape and other crops from slug damage near the strip (Friedli and Frank 1998).

Baumann (1996) studied small mammals in strips and found strips to be suitable as temporary habitats for voles (*Microtus arvalis, Arvicola terrestris terrestris*) and moles (*Talpa europea, Crocidura russula*). However, low densities of overwintering voles and moles were recorded and no increase in crop damage was found. Briner (2002) indicated that the strips are a high-quality habitat for *M. arvalis,* and therefore can sustain a high population density of voles without increasing the risk of dispersal into adjacent fields.

There is evidence that wildflower strips can numerically enhance populations of some herbivores and invertebrate pests, which provide important food resources for the build-up of natural enemy populations. This is useful in periods when only few prey (pests, herbivores and neutral organisms) for natural enemies occur in field crops, for example at the beginning of the growing period or after harvest. This can be important for pest control, especially if generalist natural enemies are more abundant at an earlier stage of the growing period. Despite many beneficial effects, in some cases economic loss in susceptible crops is possible. Observed damage has been concentrated in crop edges close to the strip, but this has not been seen to lead to a higher abundance of pests in the rest of the field (e.g. Frank 1997, 1998; Briner 2002).

Possible mechanisms enhancing important agroecosystem functions

Strips as additional overwintering sites for beneficial arthropods
Many beneficial arthropods use field margins and semi-natural habitats as overwintering sites (Sotherton 1985; Andersen 1997). This has been shown in wildflower strips for beneficial arthropods (Pfiffner and Luka 2000) and for generalist predators (Lys and Nentwig 1994; Lemke and Poehling 2002). For many natural enemy species inhabiting arable crops, the presence of less-disturbed sites adjacent to or near the crops may be crucial to increase populations and from which to recolonise the crops in spring. For stenoceous species (e.g. xero-thermophilic), these movements are possible only if habitat quality meets their specific requirements (Pfiffner and Luka 2003). Some investigations have shown that high mean densities up to 1300 arthropods per m^2 were found in wildflower strips (Lys and Nentwig 1994), and field margins contained up to six times more arthropod species (13–37 species) than did crop habitats with three to six species (Pfiffner and Luka 2000). Due to their high botanical and structural diversity and a rich litter layer, wildflower strips are highly attractive to many beneficial arthropods. One of the most important factors inducing high densities of generalist predators seems to be the density or biomass of the vegetation layer. Bürki and Hausammann (1993) observed lowest densities of arthropods under plant species with little vegetation cover such as *Agrostemma githago* or *Chenopodium polyspermum*, and very high densities of predators (more than 2100 per m^2) under highly attractive, larger plants such as *Achillea millefolium* and *Arctium minus*. The permanent, undisturbed vegetation of wildflower strip habitats can reduce diurnal temperature fluctuations and diminish the mortality rates of arthropods in the uppermost soil of wildflower margins, as Bürki and Hausammann (1993) have shown. Therefore, the temperature-buffering properties of the semi-natural habitats could be a key abiotic factor in overwintering site selection.

In conclusion, wildflower strips are important overwintering sites for arthropods and significantly enhance the number of species and the abundance of beneficial arthropods on agricultural land. This fact is due to their structural complexity, botanical diversity and the permanent and relatively undisturbed vegetation layer.

Strips improve food resources
Wildflower strips provide food resources (nectar, pollen, alternative prey, honeydew-producing insects). Given sufficient alternative prey, populations of generalist predators may establish within a crop before the arrival and seasonal increase of pests (van Emden 1990).

Zangger et al. (1994) revealed morphological and physiological effects by monitoring a carabid species (*Poecilus cupreus*) in a cereal field with a wildflower strip and in a cereal monoculture. Various effects on the carabid were found in the cereal field adjacent to the wildflower strip. Females had longer elytra, were heavier and more satiated and egg production started earlier and lasted longer than in the carabids in the cereal monoculture. Other studies showed similar effects of habitat complexity and non-crop vegetation on body size, feeding condition and fecundity of two carabid species (Bommarco 1998). Barone and Frank (2003) investigated possible effects on a predator species of succession process in strips. Comparing one-year-old with four-year-old wildflower strips, they found that beetle reproductive potential and nutritional conditions increased with the succession age of strips. Thus, the importance of wildflower strips as a reservoir for generalist epigaeic predator may increase with habitat age.

Furthermore, at the landscape scale, a higher agroecosystem complexity and low input agriculture (organic farming) can positively influence the fecundity and feeding of beneficial arthropods (Bommarco 1998; Schmidt et al., ch. 4 this volume). Bommarco found significantly larger ground beetles (*Poecilus cupreus*) and three times higher fecundity in a more complex and organically managed landscape than in a less diverse, conventionally farmed landscape in Sweden.

Parasitoids also benefit from a diversified agroecosystem. Adult parasitoids must not only find hosts for reproductive purposes but also locate food to meet their short-term nutritional needs. Nectar, honeydew and pollen appear to be the most commonly exploited food sources under field conditions (Jervis et al. 1996; Jervis et al., ch. 5 this volume). In many parasitoids, sugar sources increase longevity (Dyer and Landis 1996; Gurr and Nicol 2000), fecundity (Morales-Ramos et al. 1996) and searching activity (Stapel et al. 1997). Several sown plant species of wildflower strip are attractive to hymenopteran parasitoids (Jervis et al. 1993) and offer valuable sugar sources. Wäckers and Stepphuhn (2003) have screened some wildflower plant species, characterising nutritional state and food source use of braconid parasitoids. Using an HPLC sugar analysis of whole insect bodies, they found that habitat diversification with wildflower strips enhanced the nutritional state of *Cotesia glomerata*, a larval parasitoid of *P. rapae*.

Farming intensity alters effects of non-crop habitats

Large disturbances, by both physical and chemical means, can reduce the abundance and diversity of beneficial species, and ecosystem functions can be altered (Giller et al. 1997). Declines in the distribution and abundance of beneficial arthropods because of intensive farming methods have been documented in corn fields (Dritschilo and Wanner 1980), potatoes (Kromp 1989) and wheat (e.g. Pfiffner and Niggli 1996). Numerous comparative investigations have studied effects of conventional and organic farming on carabids (e.g. Dritschilo and Wanner 1980; Kromp 1989; Pfiffner and Niggli 1996), spiders (Glück and Ingrisch 1990) and other beneficial arthropods (Moreby et al. 1994). In most cases, higher diversity and abundance in arable crops, mostly in cereal fields, was found under organic farming. A few studies performed in perennial crops such as orchards (Paoletti et al. 1995) or vegetable crops (Hokkanen and Holopainen 1986; Clark 1999) found similar differences.

In a three-year study, Pfiffner and Luka (2003) analysed the effects of two low-input farming systems on the communities of carabids and spiders. It was performed in six different landscapes and also considered the nearby non-crop and semi-natural habitats. The whole set of practices on organic farms appeared to be more favourable to the carabid fauna and (partly) to the spider community (Lycosids) even compared to the low-input integrated crop management. Farming intensities altered the spatial occurrence of generalist predators. Several generalist predator species of semi-natural habitats (wildflower strips, low-input meadows), which are typical for these habitats, were significantly more abundant in organic fields than under low-input, integrated crop management. Predators such as the carabids *Agonum muelleri* and *Poecilus cupreus* and some Lycosids (*Pardosa agrestis*, *P. palustris* and *Trochosa ruricola*) were significantly more abundant in organic fields. This may indicate a more close interaction between organic farming and non-crop/semi-natural habitats. Most of these species were also more abundant in the organically managed plots than in the conventional/integrated plots during a long-term trial (Pfiffner and Niggli 1996). Thus, low-input systems such as organic farming benefit from higher densities of generalist predators and, if combined with non-crop habitats, synergies may be possible. A better pest control in organic fields may result.

Sown wildflower strips in perennial crops

Contribution of wildflower strips to pest control in fruit crops

Conservation of natural enemies involves manipulating the cropping system to enhance the persistence and activity of predators and parasitoids. This strategy takes various forms, based on the requirements of the individual natural enemies (Hajek 2002). By maintaining or improving the quality of habitat adjacent to or within the crop, more favourable conditions for natural enemies can be provided. Intercropping or planting cover crops, hedgerows and set-aside strips are possibilities for fruit growers.

Sown and naturally regenerated wildflower strips were developed first for annual crops such as cereals. An extension of this approach to perennial crops such as apple seemed to be tractable, as perennial systems are less disturbed by agricultural operations and so have favourable characteristics for successful biological control (Hajek 2002).

A lot of work has been done to enhance the effects of predators and parasitoids in orchards by managing the existing groundcover (Altieri and Schmidt 1985; Flexner et al. 1991; Fye 1983; Alston 1994; Coli et al. 1994; Liang and Huang 1994, Brown and Schmitt 1996). Despite some encouraging results, some researchers have found enhancement of orchard pests like the mullein bug, *Campylomma verbasci* (Thistlewood et al. 1990). In a review, Prokopy (1994) concluded that groundcovers within an orchard can lead to counterproductive effects such as additional insect or vertebrate pests.

Some work has been done to enhance beneficials in orchards or vineyards by sowing flowering plants (Altieri and Whitcomb 1979; Remund et al. 1992; Wyss 1995; Boller et al. 1997; Solomon 1999; Vogt and Weigel 1999). Most of the published data showed significant positive effects on the number and diversity of predators and, consequently, suppressive effects on the number of pests, but few showed no or opposite results. This trend should be interpreted with caution, however, as negative results may have remained unpublished. The following sections discuss factors for an efficient enhancement of beneficial insects within apple orchards.

Requirements of sown wildflower strips in apple orchards

To ensure that the ecological requirements of natural enemies are met in the orchard ecosystem three main factors have to be optimised:

- alternative prey or hosts have to be offered in sufficient number within or near the crop when pests become temporarily scarce;
- alternative food sources such as nectar and pollen have to be offered for adult predators and parasitoids;
- shelter or undisturbed habitats must be created to ensure refuges and overwintering sites.

With sown wildflower strips, all requirements are fulfilled. They offer alternative prey and food sources with the mixture of different annual, biannual and perennial plant species during the whole period of vegetation and provide shelter or overwintering sites.

Early experiments in apple orchards in Switzerland by Wyss (1994) showed that the number of plant species within the mixture of annual, biannual and perennial species must be large enough to ensure a satisfactory density of the sown species. Based on experiments, Wyss (1994) developed a seed mixture adapted to perennial crops. However, the special conditions in or near apple orchards, with their perennial groundcover, limit life of the flowering strips because grass and other perennial weeds invade and may cause a loss of flowering species. Therefore, strips must be wide enough to ensure a diverse and long-lasting blossom over several years. In addition, mowing and/or a soil treatment at the end of the second year can be used to encourage botanical diversity. Unmanaged strips have the potential to give shelter to vertebrate pests, such as rodents, which without control could become a severe problem in the orchard.

To establish flowering strips in orchards only a few conditions have to be fulfilled:

- strips should be planned with a minimum width of 2–3 m;
- the preparation of the seedbed must be done carefully;
- sites with high occurrence of problematic weeds should be avoided;
- in regions where rodents occur, the strips should be planned at the edge of the orchards and should be regularly controlled (Wyss 1994).

General effects of sown wildflower strips on abundance and diversity of insects within apple orchards

The guilds of natural enemies associated with apple aphids and other apple pests have been extensively recorded (Leius 1967; Carroll and Hoyt 1984; Hagley and Allen 1990; Tourneur et al. 1991). They generally consist of a minimum of 50 species of arthropods belonging to several families. This diversity of beneficial insects encourages the view that biological control is feasible. Thus the exploitation of the entire natural enemy fauna is an attractive strategy.

In general, a more diverse plant canopy will enhance abundance and diversity of natural enemies by providing alternative food sources, shelter and overwintering sites. Besides natural enemies, a huge number of neutral insects is also attracted to the orchard by the diverse plant canopy of a wildflower strip. In an experimental apple orchard, with wildflower strips between tree rows and along the border in one part of the orchard (the other part served as control), Wyss (1996) found no significant difference in species diversity between the strip-managed part and the control. The strip-managed part was, however, colonised by predators of aphids and alternative prey (herbivores, but no additional apple pests) in higher numbers than in the control. Such build-up of predators of aphids and alternative prey resulting from sown flowering plants in orchards has also been reported by Leius (1967), Altieri and Schmidt (1985), Bugg and Dutcher (1989) and Bugg et al. (1990). However, recent work in grasslands by Koricheva et al. (2000) showed that specialised parasitoids are not affected by plant diversity directly, but indirectly via changes in plant biomass and cover. In addition, plant species composition affected the number of most invertebrate groups and is a more important determinant of invertebrate abundance in grasslands than is plant species richness. The mechanisms linking plant diversity and the abundance of these insects is still to be fully elucidated.

The diversity of the arthropod community can, however, be enhanced by a diverse plant community within an apple orchard. To show this effect, orchards at different sites and with a different plant composition have to be compared to avoid dispersal of arthropods between treatments. Several authors have shown an enhanced diversity of the arthropod community in unmanaged or extensively managed orchards compared to intensively managed orchards, due to a more diverse plant canopy (Altieri and Schmidt 1985; Brown and Schmitt 1996; Kienzle et al. 1997; Solomon et al. 1999).

It can be concluded that diversifying an apple orchard by introducing flowering plants often leads to an increased abundance of pest predators and alternative prey and to a more diverse arthropod community. The extent of pest suppression is explored below.

Effects of sown wildflower strips on fruit pests

Modern apple orchards typically are simplified communities. The soil below the trees is often kept bare and grass between the rows of trees is mown regularly. Therefore, it is not surprising that attempts have been made to control pests by increasing plant diversity in European and

North American orchards (Brown and Welker 1992; Wyss 1995; Brown and Schmitt 1996; Kienzle et al. 1997; Solomon et al. 1999; Vogt and Weigel 1999). In a three-year study, Wyss (1995) monitored the rosy apple aphid (*Dysaphis plantaginea*) infestation and aphidophagous predators in two parts of an organic orchard in Switzerland. In the first year, aphid infestation and predator abundance were identical in both parts. In the second year, a mixture of indige-nous flowering plants was sown in six 1 m wide strips located in one of the parts. The sown wild plants flowered successively from early spring to late autumn. Some also hosted aphids and other alternative prey when *D. plantaginea* was scarce on apple. Therefore, pollen, nectar and aphids were available to aphidophagous predators throughout the year. Later in the second and in the third year, aphidophagous predators were more abundant and aphids were less abundant on the trees in the part with the wildflower strips than in the control part (Figure 11.6). The impact on aphids by the aphidophagous predators was important during the flowering time of the sown strips in spring and summer. In autumn, web-building spiders were the dominant predators of aphids in the trees (Wyss et al. 1995). With a higher density of webs, spiders signif-icantly reduced the number of winged aphids returning from their summer host plants. Consequently, in spring fewer aphids were present on trees in the part with the wildflower strips. However, the potential for increasing aphid predators by sowing wildflower strips must be further investigated: the economic threshold was exceeded in years of high aphid occurrence and additional measures such as applications of insecticides would have been necessary.

When Vogt and Weigel (1999) repeated Wyss' experiment in a much smaller orchard in Germany, they recorded more *D. plantaginea* on the trees in the area with the wildflower strips than in the control area. However, they found much better synchronisation between aphid predators and the green apple aphid (*Aphis pomi*), which occurs on trees in early summer. This fact led to a significant suppressive effect for the green apple aphid. Vogt et al. (1998) mentioned

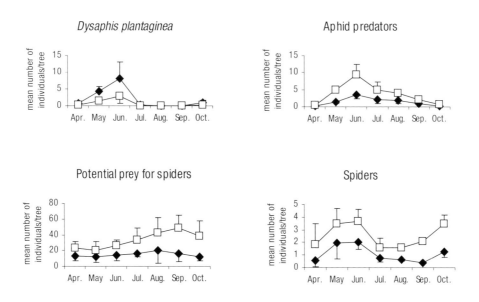

Figure 11.6: Mean number of rosy apple aphid *Dysaphis plantaginea*, potential prey for spiders, aphid predators and spiders on trees found in the apple orchard area with wildflower strips (white symbols) and in the control area (black symbols). In year 3 of the experiment significant differences were noted between the strip-managed part and the control part for all four arthropod categories (p <0.001, Mann-Whitney U test). After Wyss (1995, 1996) and Wyss et al. (1995).

several reasons for the different success of the wildflower strips on the rosy apple aphid in the two studies:

- different management of the wildflower strips (mowing vs no management) made overwintering impossible for beneficials in the German orchard;
- unmown strips in the Swiss orchard started flowering a month earlier than the mown strips in the German orchard;
- the smaller size of the German orchard;
- flower strips in the German orchard were very close to the tree rows and could therefore have influenced tree nutrition and consequently aphid development.

The two studies show that a greater abundance of natural enemies achieved by manipulating plant diversity does not automatically translate into aphid control. Many other factors may influence the efficacy of flowering strips. This also forces reconsideration of the link between diversity and stability and the role of aphidophagous predators in determining aphid abundance. Complementary measures such as augmentative or inundative release of reared predators, such as the two-spotted ladybird (*Adalia bipunctata*), have been evaluated (Wyss et al. 1999a, b; Kehrli and Wyss 2001) and shown potential to assist naturally occurring antagonists.

Conclusions

Experimental evidence shows that sown wildflower strips can increase the arthropod diversity and the density of natural enemies. The presence of a structurally complex vegetation, rich in species and structure, with a long flowering period and without great disturbance provides favourable conditions for natural enemies. Therefore, improved natural pest control in annual and perennial crops is possible.

In the future, manipulative experiments in the field are necessary to assess the net effect of natural enemies (including intraguild interference effects) on pest population and consecutive yield (plant performance). Finally, understanding the long-term effects of semi-natural habitats, including field margins, on natural enemies and pests on field and landscape level is important, if habitat manipulation is to be a viable and widely used tool for improving natural pest control.

Acknowledgements

We thank Dr H. Luka, C. Daniel and the two anonymous referees for commenting on earlier versions of this manuscript.

References

Aebischer, N.J. (1991). Twenty years of monitoring invertebrates and weeds in cereal fields in Sussex. In *The Ecology of Temperate Cereal Fields* (L. Firbank, N. Carter, J. Darbyshire and G. Potts, eds), pp. 305–331. Blackwell Scientific Publications, Oxford.

Alston, D.G. (1994). Effect of apple orchard floor vegetation on density and dispersal of phytophagous and predaceous mites in Utah. *Agriculture, Ecosystems and Environment* 50: 73–84.

Altieri, M.A. and Schmidt, L.L. (1985). Cover crop manipulation in Northern California orchards and vineyards: effects on arthropod communities. *Biological Agriculture and Horticulture* 3: 1–24.

Altieri, M.A. and Whitcomb, W.H. (1979). The potential use of weeds in the manipulation of beneficial insects. *Horticultural Science* 14: 12–18.

Andersen, A. (1997). Densities of overwintering carabids and staphylinids (Col., Carabidae and Staphylinidae) in cereal and grass fields and their boundaries. *Journal of Applied Entomology* 121: 77–80.

Anonymous (2003). Agricultural Report 2002. Swiss Federal Office for Agriculture. http:// www.blw.admin.ch/agrarbericht3/d/index.htm

Barone, M. and Frank, T. (2003). Habitat age increases reproduction and nutritional condition in a generalist predator. *Oecologia* 135: 78–83.

Baumann, L. (1996). The influence of field margins on populations of small mammals – A study of the population ecology of the common vole *Microtus arvalis* in sown weed strips. M.Sc. thesis. University of Bern, Switzerland. 27 pages.

Bohan, D.A., Bohan, A.C., Glen, D.M., Symondson, W.O.C., Wiltshire, C.W. and Hughes, L. (2000). Spatial dynamics of predation by carabid beetles on slugs. *Journal of Animal Ecology* 69: 367–379.

Boller, E.F., Gut, D. and Remund, U. (1997). Biodiversity in three trophic levels of the vineyard agro-ecosystem in northern Switzerland. In *Vertical Food Web Interactions. Ecological Studies* (Dettner et al., eds), vol. 130, pp. 299–318. Springer-Verlag, Berlin.

Bommarco, R. (1998). Reproduction and energy reserves of a predatory carabid beetle relative to agroecosystem complexity. *Ecological Applications* 8: 846–853.

Briner, T. (2002). Population dynamics, spatial and temporal patterns of the common vole (*Microtus arvalis*, Pall) in wildflower strip, using mark-recapture method and a new system for automatic radio tracking. Ph.D. thesis, University of Berne, Switzerland. 53 pages.

Broggi, M.F. and Schlegel, H. (1989). Mindestbedarf an naturnahen Flächen in der Kulturlandschaft – Dargestellt am Beispiel des Schweizerischen Mittellandes. Bericht 31 des Nationalen Forschungsprogrammes Boden. Liebefeld-Bern. 180 pages.

Brown, M.W. and Schmitt, J.J. (1996). Impact of ground cover plants on beneficial arthropods in an apple orchard in West Virginia, USA. *IOBC/WPRS Bulletin* 19 (4): 332–333.

Brown, M.W. and Welker, W.V. (1992). Development of the phytophagous arthropod community on apple as affected by orchard management. *Environmental Entomology* 21 (3): 485–492.

Bugg, R.L. and Dutcher, J.D. (1989). Warm-season cover crops for pecan orchards: horticultural and entomological implications. *Biology, Agriculture and Horticulture* 6: 123–148.

Bugg, R.L., Phatak, S.C. and Dutcher, J.D. (1990). Insects associated with cool-season cover crops in southern Georgia: implications for pest control in truck-farm and pecan agro-ecosystems. *Biology, Agriculture and Horticulture* 7: 17–45.

Bürki, H.M. and Hausammann, A. (1993). Überwinterung von Arthropoden im Boden und an Ackerkräutern künstlich angelegten Ackerkrautstreifen. Agrarökologie 7, Bern, Stuttgart, Wien, Hauptverlag. 158 pages.

Carroll, D.P. and Hoyt, S.C. (1984). Natural enemies and their effects on apple aphid, *Aphis pomi* DeGeer (Homoptera: Aphididae), colonies on young apple trees in Central Washington. *Environmental Entomology* 13: 469–481.

Clark, M.S. (1999). Ground beetle abundance and community composition in conventional and organic tomato systems of California's Central Valley. *Applied Soil Ecology* 11: 199–206.

Coli, W.M., Ciurlino, R.A. and Hosmer, T. (1994). Effect of understore and border vegetation composition on phytophagous and predatory mites in Massachusetts commercial apple orchards. *Agriculture, Ecosystems and Environment* 50: 49–60.

Collins, K.L., Boatman, N.D., Wilcox, A., Holland, J.M. and Chaney, K. (2002). Influence of beetle banks on cereal aphid predation in winter wheat. *Agriculture, Ecosystems and Environment* 93: 337–350.

Denys, C. and Tscharntke, T. (2002). Plant insects communities and predator-prey ratios in field margin strips, adjacent crop fields, and fallows. *Oecologia* 130: 315–324.

Dritschilo, W. and Wanner, D. (1980). Ground beetle abundance in organic and conventional corn field. *Environmental Entomology* 9: 629–631.

Dyer, L.E. and Landis, D. (1996). Effects of habitat, temperature, and sugar availability on longevity of *Eriborus terebrans* (Hymenoptera: Ichneumonidae). *Environmental Entomology* 25: 1192–1201.

Fahrig, L. (1997). Relative effects of habitat loss and fragmentation on population extinction. *Journal of Wildlife Management* 61: 603–610.

Flexner, J.L., Westigard, P.H., Gonzalves, P. and Hilton, R. (1991). The effect of groundcover and herbicide treatment on two-spotted spider mite density and dispersal in southern Oregon pear orchards. *Entomologia Experimentalis et Applicata* 60: 111–123.

Frank, T. (1997). Slug damage and numbers of the slug pests, Arion lusitanicus and *Deroceras reticulatum*, in oilseed rape grown beside sown wildflower strips. *Agriculture, Ecosystems and Environment* 67: 67–78.

Frank, T. (1998). Slug damage and number of slugs (Gastropoda: Pulmonata) in winter wheat in fields with sown wildflower strips. *Journal of Molluscudal Studies* 64: 319–328.

Frank, T. (1999). Density of adult hoverflies (Dipt., Syrphidae) in sown weed strips and adjacent fields. *Journal of Applied Entomology* 123 (6): 351–355.

Frank, T. (2003). Influence of slug herbivory on the vegetation development in an experimental wildflower strip. *Basic Applied Ecology* 4: 139–147.

Frank, T. and Nentwig, W. (1995). Ground dwelling spiders (Araneae) in sown weed strips and adjacent fields. *Acta Oecologica* 16: 179–193.

Frei, G. and Manhart, C. (1992). Nützlinge und Schädlinge an künstlich angelegten Ackerkrautstreifen in Getreidefeldern. Bern; Stuttgart; Wien: Haupt. 140 pages.

Friedli, J. and Frank, T. (1998). Reduced applications of metaldehyde pellets for reliable control of the slug pests *Arion lusitanicus* and *Deroceras reticulatum* in oilseed rape adjacent to sown wildflower strips. *Journal of Applied Ecology* 35: 504–513.

Fry, G.A. (1994). The role of field margins in the landscape. In *Field margins: Integrating agriculture and conservation, British Crop Protection Council Mono* (N. Boatman, ed.) 58: 31–40.

Fye, R.E. (1983). Cover crop manipulation for building pear psylla (Homoptera: Psyllidae) predator populations in pear orchards. *Journal of Economic Entomology* 76: 306–310.

Giller, K.E., Beare, M.H., Lavelle, P., Izaac, A.-M.N. and Swift, M.J. (1997). Agricultural intensification, soil biodiversity and agroecosystem function. *Applied Soil Ecology* 6: 3–16.

Glück, E. and Ingrisch, S. (1990). The effect of bio-dynamic and conventional agriculture management on Erigoninae and Lycosidae spiders. *Journal of Applied Entomology* 110: 136–148.

Günter, M. (2000). Anlage und Pflege von mehrjährigen Buntbrachen unter den Rahmenbedingungen des Schweizerischen Ackerbaugebietes. Ph.D. thesis, University of Berne, Switzerland, Agrarökologie 37. 159 pages.

Gurr, G.M. and Nicol, H.I. (2000). Effect of food on longevity of adults of *Trichogramma carverae* Oatman and Pinto and *Trichogramma brassicae Bezdenko* (Hymenoptera : Trichogrammatidae). *Australian Journal of Entomology* 39: 185–187.

Gurr, G.M., van Emden, H.F. and Wratten, S.D. (1998). Habitat manipulation and natural enemy efficiency: implications for the control of pests. In *Conservation Biological Control* (P. Barbosa, ed.), pp. 155–183. Academic Press, San Diego.

Hagley, E.A.C. and Allen, W.R. (1990). The green apple aphid, *Aphis pomi* DeGeer (Homoptera: Aphididae), as prey of polyphagous arthropod predators in Ontario. *Canadian Entomologist* 122: 1221–1228.

Hajek, A.E. (2002). Biological control of insects and mites. In *Encyclopedia of Pest Management* (D. Pimentel, ed.), pp. 57–59. Marcel Dekker, New York, Basel.

Haughton, A.J., Bell, J.R., Gates, S., Johnson, P.J., Macdonald, D.W., Tattersall, F.H. and Hart, B.H. (1999). Methods of increasing invertebrate abundance within field margins. *Aspects of Applied Biology* 54: 163–110.

Hausammann, A. (1996a). The effects of weed strip-management on pests and beneficial arthropods in winter wheat fields. *Journal of Plant Diseases and Protection* 103: 70–81.

Hausammann, A. (1996b). Strip management in rape crop: is winter rape endangered by negative impacts of sown weed strips. *Journal of Applied Entomology* 120: 505–512.

Heitzmann, A. (1995). Angesäte Ackerkrautstreifen – Veränderungen des Pflanzenbestandes während der natürlichen Sukzession. Agrarökologie 13, Bern, Stuttgart, Wien, Hauptverlag. 152 pages.

Hickmann, J.M. and Wratten, S.D. (1996). Use of Phacelia tanacetifolia strips to enhance biological control of aphids by hoverfly larvae in cereal fields. *Journal of Economic Entomology* 89: 832–840.

Hokkanen, H. and Holopainen, J.K. (1986). Carabid species and activity densities in biologically and conventionally managed cabbage fields. *Journal of Applied Entomology* 102: 353–363.

Jervis, M.A., Kidd, N.A.C., Fitton, M.G., Huddleston, T. and Dawah, H.A. (1993). Flower-visiting by hymenopteran parasitoids. *Journal of Natural History* 7: 67–105.

Jervis, M.A., Kidd, N.A.C. and Heimpel, G.E. (1996). Parasitoid adult feeding behaviour and biocontrol – a review. *Biocontrol News and Information* 17: 11–26.

Jmhasly, P. and Nentwig, W. (1995). Habitat managment in winter wheat and evaluation of subsequent spider predation on insect pests. *Acta Oecologica* 16: 389–403.

Kehrli, P. and Wyss, E. (2001). Effects of augmentative releases of the coccinellid, *Adalia bipunctata*, and of insecticide treatments in autumn on the spring population of aphids of the genus *Dysaphis* in apple orchards. *Entomologia Experimentalis et Applicata* 99: 245–252.

Keller, M. (1999). The importance of seed source in programmes to increase species diversity in arable systems. Ph. D. thesis. Swiss Federal Institute of Technology (ETH), Zürich. 90 pages.

Kielty, J.P., Allen, L.J. and Underwood, N. (1999). Prey preferences of six species of Carabidae (Coleoptera) and one Lycosidae (Araneae) commonly found in UK arable crop fields. *Journal of Applied Entomology* 123: 193–200.

Kienegger, M. and Kromp, B. (2001). The effect of strips of flowers on selected groups of beneficial insects in adjacent broccoli plots. *Mitteilungen der deutschen Gesellschaft für allgemeine und angewandte Entomologie* 13: 583–586.

Kienzle, J., Zebitz, C.P.W. and Brass, S. (1997). Floral and faunal species diversity and abundance of aphid predators in ecological apple orchards. *Biological Agriculture and Horticulture* 15: 233–240.

Koricheva, J., Mulder, C.P.H., Schmid, B., Joshi, J. and Huss-Danell, K. (2000). Numerical responses of different trophic groups of invertebrates to manipulations of plant diversity in grasslands. *Oecologia* 125: 271–282.

Kromp, B. (1989). Carabid beetle communities (Carabidae, Coleoptera) in biologically and conventionally farmed agroecosystems. *Agriculture, Ecosystems and Environment* 27: 241–251.

Kruess, A. and Tscharntke, T. (1994). Habitat fragmentation, species loss, and biological control. *Science* 264: 1581–1584.

Landis, D.A., Wratten, S.D. and Gurr, G.M. (2000). Habitat management to conserve natural enemies of arthropod pests in agriculture. *Annual Review of Entomology* 45: 175–201.

Lang, A. (2003). Intraguild interference and biocontrol effects of generalist predators in a winter wheat field. *Oecologia* 134: 144–153.

Leius, K. (1967). Influence of wild flowers on parasitism of tent caterpillar and codling moth. *Canadian Entomologist* 99: 444–446.

Lemke, A. and Poehling, H.M. (2002). Sown weed strips in cereal fields: overwintering site and 'source' habitat for *Oedothorax apicatus* (Blackwall) and *Erigone atra* (Blackwall) (Araneae: Erigonidae). *Agriculture, Ecosystems and Environment* 90: 67–80.

Lethmayer, C. (1995). Effects of sown weed strips on pest insects. Ph.D. thesis, University of Bern, Switzerland. 59 pages.

Lethmayer, C., Nentwig, W. and Frank, T. (1997). Effects of weed strips on the occurrence of noxious coleopteran species (Nitidulidae, Chrysomelidae, Curculinonidae). *Journal of Plant Diseases and Protection* 104: 75–92.

Liang, W. and Huang, M. (1994). Influence of citrus orchard ground cover plants on arthropod communities in China: a review. *Agriculture, Ecosystems and Environment* 50: 29–37.

Lund, R.D. and Turpin, F.T. (1977). Carabid damage to weed seeds found in Indiana cornfields. *Environmental Entomology* 6: 695–698.

Lys, J.A. and Nentwig, W. (1992). Augmentation of beneficial arthropods by strip-management. 4. Surface activity, movements and activity density of abundant carabid beetles in a cereal field. *Oecologia* 92: 373–382.

Lys, J.A. and Nentwig, W. (1994). Improvement of the overwintering sites for Carabidae, Staphylinidae and Araneae by strip-management in a cereal field. *Pedobiologia* 38: 238–242.

Lys, J.A., Zimmermann, M. and Nentwig, W. (1994). Increase in activity density and species number of carabid beetles in cereals as a result of strip-management. *Entomologia Experimentalis et Applicata* 73: 1–9.

Marshall, E.J.P. and Moonen, A.C. (2002). Field margins in northern Europe: their functions and interactions with agriculture. *Agriculture, Ecosystems and Environment* 89: 5–21.

Marshall, E.J.P., West, T.M. and Winstone, L. (1994). Extending field boundary habitats to enhance farmland wildlife and improve crop and environmental protection. *Aspects of Applied Biology* 40: 387–391.

Morales-Ramos, J.A., Rojas, M.G. and King, E.G. (1996). Significance of adult nutrition and oviposition experience on longevity and attainment of full fecundity of Catolaccus grandis (Hymenoptera: Pteromalidae). *Annals of the Entomological Society of America* 89 (4): 555–563.

Moreby, S.J., Aebischer, N.J., Southway, S.E. and Sotherton, N.W. (1994). A comparison of the flora and arthropod fauna of organically and conventionally grown winter wheat in southern England. *Annals of Applied Biology* 125: 13–27.

Nentwig, W. (1989). Augmentation of beneficial arthropods by strip-management. 2. Successional strips in a winter wheat field. *Journal of Plant Diseases and Protection* 96: 89–99.

Nentwig, W. (1998). Weedy plant species and their beneficial arthropods: potential for manipulation in field crops. In *Enhancing Biological Control. Habitat Management to Promote Natural Enemies of Agricultural Pests* (C.H.B. Pickett and R.L. Bugg, eds), pp. 49–71. University of California Press, Berkeley.

Nentwig, W., Frank, T. and Lethmayer, C. (1998). Sown weed strips: artificial ecological compensation areas as an important tool in conservation biological control. In *Conservation Biological Control* (P. Barbosa, ed.), pp. 133–153. Academic Press, San Diego.

Paoletti, M.G., Schweigl, U. and Favretto, M.R. (1995). Soil macroinvertebrates, heavy metals and organochlorines in low and high input apple orchards and a coppiced woodland. *Pedobiologia* 39: 20–33.

Pfiffner, L. (2000). Significance of organic farming for invertebrate diversity – enhancing beneficial organisms with field margins in combination with organic farming. In *The Relationship Between Nature Conservation, Biodiversity and Organic Agriculture. Proceedings of an International Workshop* (S. Stolton, B. Geier and J.A. McNeely, eds), pp. 52–66. Vignola, Italy, 1999. IFOAM, Tholey-Theley, Germany.

Pfiffner, L. and Luka, H. (2000). Overwintering of arthropods in soils of arable fields and adjacent semi-natural habitats. *Agriculture, Ecosystems and Environment* 78: 215–222.

Pfiffner, L. and Luka, H. (2003). Effects of low-input farming systems on carabids and epigeal spiders in cereal crops – a paired farm approach in NW-Switzerland. *Basic and Applied Ecology* 4: 117–127.

Pfiffner, L. and Niggli, U. (1996). Effects of bio-dynamic, organic and conventional farming on ground beetles (Col. Carabidae) and other epigaeic arhtropods in winter wheat. *Biological Agriculture and Horticulture* 12: 353–364.

Pfiffner, L. and Schaffner, D. (2000). Anlage und Pflege von Ackerkrautstreifen. In *Streifenförmige ökologische Ausgleichsflächen in der Kulturlandschaft* (W. Nentwig, ed.), pp. 41–53. Verlag Agrarökologie Bern – Hannover.

Pfiffner, L., Luka, H., Jeanneret, P. and Schüpbach, B. (2000a). Evaluation Ökomassnahmen: Effekte ökologischer Ausgleichsflächen auf die Laufkäferfauna. *Agrarforschung* 7: 212–227.

Pfiffner, L., Luka, H. and Lutz, M. (2000b). Significance of field margins for enhancing beneficial arthropods in arable and vegetable crops (Grosses Moos, Switzerland). In *13th International Congress of IFOAM* (T. Alföldi, W. Lockeretz and U. Niggli, eds), p. 464. Basel.

Pfiffner, L., Merkelbach, L. and Luka, H. (2003). Do sown wildflower strips enhance the parasitism of lepidopteran pests in cabbage crops? *IOBC/WPRS Bulletin* 26 (4): 111–116.

Prokopy, R.J. (1994). Integration in orchard pest and habitat management: A review. *Agriculture, Ecosystems and Environment* 50: 1–10.

Remund, U., Gut, D. and Boller, E.F. (1992). Beziehungen zwischen Begleitflora und Arthropodenfauna in Ostschweizer Rebbergen. *Schweizerische Zeitschrift für Obst- und Weinbau* 128: 527–540.

Salveter, R. (1998). The influence of sown herb strips and spontaneous weeds on the larval stages of aphidophagous hoverflies (Dipt., Syrphidae). *Journal of Applied Entomology* 122: 103–114.

Solomon, M., Fitzgerald, J. and Jolly, R. (1999). Artificial refuges and flowering plants to enhance predator populations in orchards. *IOBC/WPRS Bulletin* 22 (7): 31–38.

Sotherton, N.W. (1985). The distribution and abundance of predatory arthropods overwintering in field boundaries. *Annals of Applied Biology* 106: 17–21.

Stapel, J.O., Cortesero, A.M., Moraes, C.M., Tumlinson, J.H. and Lewis, W.J. (1997). Extrafloral nectar, honeydew, and sucrose effects on searching behavior and efficiency of *Microplitis croceipes* (Hymenoptera: Braconidae) in cotton. *Environmental Entomology* 26: 617–623.

Sunderland, K.D. (1999). Mechanisms underlying the effects of spiders on pest populations. *Journal of Arachnology* 27: 308–316.

Sunderland, K.D. (2002). Invertebrate pest control by carabids. In *The Agroecology of Carabid Beetles* (J.M. Holland, ed.), pp. 165–214. Intercept, Andover.

Sunderland, K.D. and Samu, F. (2000). Effects of agricultural diversification on the abundance, distribution, and pest control potential of spiders: a review. *Entomologia Experimentalis et Applicata* 95: 1–13.

Sunderland, K.D., Axelsen, J.A., Dromph, K., Freier, B., Hemptinne, J.-L., Holst, N.H., Mols, P.J.M., Petersen, M.K., Powell, W., Ruggle, P., Triltsch, H. and Winder, L. (1997). Pest control by a community of natural enemies. In *Arthropod Natural Enemies in Arable Land. 72. Acta Jutlandica III* (W. Powell, ed.), pp. 271–326. Aarhus University Press.

Symondson, W.O.C., Sunderland, K.D. and Greenstone, M.H. (2002). Can generalist predators be effective biocontrol agents? *Annual Review of Entomology* 47: 561–594.

Thiele, H.U. (1977). *Carabid Beetles in their Environments*. Springer-Verlag, Heidelberg.

Thistlewood, H.M.A., Borden, J.H. and McMullen, R.D. (1990). Seasonal abundance of the mullen bug, *Campylomma verbasci*, on apple and mullen in the Okanagen Valley. *Canadian Entomologist* 122: 1045–1058.

Thomas, C.F.G. and Marshall, E.J.P. (1999). Arthropod abundance and diversity in differently vegetated margins of arable fields. *Agriculture, Ecosystems and Environment* 72: 131–144.

Tourneur, J.C., Bouchard, D. and Pilon, J.G. (1991). Le complexe des enemies naturels des pucerons en pommeraie au Québec. In *Lutte Biologique* (V. Charles and D. Coderre, eds), pp. 179–193. G. Morrin, Québec.

Ullrich, K. (2001). The influence of wildflower strips on plant and insect (Hetroptera) diversity in an arable landscape. Ph.D. thesis, Swiss Federal Institute of Technology (ETH), Zürich, Switzerland. 127 pages.

van Emden, H.F. (1990). Plant diversity and natural enemy efficiency in agroecosystems. In *Critical Issues in Biological Control* (M. Mackauer, L.E. Ehler and J. Roland, eds), pp. 63–80. Intercept, Andover.

Vogt, H. and Weigel, A. (1999). Is it possible to enhance biological control of aphids in an apple orchard with flowering strips? *IOBC/WPRS Bulletin* 22 (7): 39–46.

Vogt, H., Weigel, A. and Wyss, E. (1998). Aspects of indirect plant protection strategies in orchards: are flowering strips an adequate measure to control apple aphids? In *Proceedings of the 6th European Congress of Entomology*, pp. 625–626. Ceske Budejovice, Czech Republic, August 1998.

Wäckers, F. and Stepphuhn, A. (2003). Characterising nutritional state and food source use of parasitoids collected in fields with high and low nectar availability. *IOBC/WPRS Bulletin* 26 (4): 203–208.

Weiss, E. and Stettmer, C. (1991). Unkräuter in der Agrarlandschaft locken blütenbesuchende Nutzinsekten an. Agrarökologie 1, Bern, Stuttgart, Wien, Hauptverlag. 104 pages.

Wise, D.H. (1994). *Spider in Ecological Webs*. Cambridge University Press, Cambridge.

Wyss, E. (1994). Biocontrol of apple aphids by weed strip-management in apple orchards. Ph.D. thesis, University of Berne, Switzerland. 84 pages.

Wyss, E. (1995). The effects of weed strips on aphids and aphidophagous predators in an apple orchard. *Entomologia Experimentalis et Applicata* 75: 43–49.

Wyss, E. (1996). The effects of artificial weed strips on diversity and abundance of the arthropod fauna in a Swiss experimental apple orchard. *Agriculture, Ecosystems and Environment* 60: 47–59.

Wyss, E., Niggli, U. and Nentwig, W. (1995). The impact of spiders on aphid populations in a strip-managed apple orchard. *Journal of Applied Entomology* 119: 473–478.

Wyss, E., Villiger, M., Hemptinne, J.L. and Müller-Schärer, H. (1999a). Effects of augmentative releases of eggs and larvae of the ladybird beetle, *Adalia bipunctata*, on the abundance of the rosy apple aphid, *Dysaphis plantaginea*, in organic apple orchards. *Entomologia Experimentalis et Applicata* 90: 167–173.

Wyss, E., Villiger, M. and Müller-Schärer, H. (1999b). The potential of three native insect predators to control the rosy apple aphid, Dysaphis plantaginea. *BioControl* 44: 171–182.

Zangger, A., Lys, J.A. and Nentwig, W. (1994). Increasing the availability of food and the reproduction of *Poecilus cupreus* in a cereal field by strip-management. *Entomologia Experimentalis et Applicata* 71: 111–120.

Chapter 12

Habitat manipulation for insect pest management in cotton cropping systems

Robert K. Mensah and Richard V. Sequeira

Introduction

There is abundant evidence of a link between monocultures, or a lack of crop diversity at the field level, and pest problems in agricultural production systems (Pimentel 1971; van Emden and Williams 1974; Vandermeer 1989; Andow 1991). Ecologically diverse cropping systems tend to support a greater abundance and diversity of beneficial arthropods such as predators and parasitic wasps (Herzog and Funderburk 1986; van Lenteren 1987; Russell 1989; Conlong 1990; van Emden 1990; Andow 1991; Corbett et al. 1991; Conlong 1994; Mensah and Madden 1994; Way and Heong 1994; Conlong 1995; Mensah and Khan 1997; Landis et al. 2000). In Australia, the adoption of within-field monocultures is believed to be a major factor underlying the nature and severity of pest problems in crop production systems such as cotton (Mensah 1999).

Natural enemies, including generalist arthropod predators and specialist parasitoids, are important agents of insect pest regulation (Beirne 1967; Murdoch et al. 1985; Clark et al. 1994; Dent 1995; Van Driesche and Bellows 1996). The role of natural enemies in insect pest regulation in annually disrupted, multi-pest agroecosystems such as cotton has been frequently overshadowed by the use of other means of pest management such as chemical insecticides. Generalist predators are a particularly underrated group of natural enemies (Luff 1983; Symondson et al. 2002).

Many generalist predators and specialist parasitoids have been recorded in Australian cotton cropping systems (Room 1979). Despite their occurrence, the potential of generalist predators has not been fully or widely exploited in cotton pest management (Mensah 1999). This may be due to several reasons, including a lack of information on the pest-management potential of such mortality agents and a lack of techniques to maximise their abundance and effectiveness. Lack of ecological diversity can limit the proliferation of some groups of natural enemies by the absence of suitable alternative food sources, particularly at times of scarcity, overwintering sites and shelter during unfavourable weather conditions or situations (Mensah 1999; Landis et al. 2000).

The seminal reviews of Andow (1991) and Altieri (1994) were instrumental in highlighting the potential role of ecological diversification at the field level in integrated pest management (IPM). These reviews clearly showed that, in many instances, growing several crops in the same space reduces pest problems relative to monocultures of the same species. The examples listed in Table 12.1 exemplify the wide range of cropping systems and crops where interplanting of secondary crops within primary (main) crops can be useful in managing crop pests.

Table 12.1: Examples of cropping systems where increased biodiversity through intercropping or multiple cropping has contributed to a reduction in pest outbreaks on the primary crops.

Main crop	Interplanted crop	Pest(s) controlled
Oats	Beans	Aphids
Cotton	Alfalfa or lucerne Corn (maize) Sorghum Cowpea	Lygus bug, green mirids Cotton bollworm Cotton bollworm Boll weevil
Corn (maize)	Beans Clover, soybeans Squash	Fall armyworm, leafhoppers European cornborer Aphids, spider mites
Melons	Wheat	Aphids, whiteflies
Peaches	Strawberries	Oriental fruit moth
Cassava	Cowpeas	Whiteflies
Beans	Winter wheat	Potato leafhopper, bean aphid
Peanuts	Beans	Aphids
Tomato	Cole crops	Flea beetles
Cole crops	Beans Clover	Cabbage flea beetle, cabbage aphid, Imported cabbage worm

Altieri and Letourneau (1982), Andow (1991) and Mensah and Khan (1997).

In Australia, cotton is grown largely as a monoculture. Within these cotton monocultures, the highly migratory pests *Helicoverpa* spp. and *Creontiades dilutus* can move in and out of crops, and rapidly achieve high population densities (Mensah and Khan 1997). Unless natural enemies are well-established before these pests arrive, they cannot respond rapidly enough to suppress the pest population before damage is sustained (Fitt 1989; Gregg et al. 1995; Mensah 1997, 2002).

In monocultures, manipulating the availability of host plants by intercropping with various alternative crops which may provide refuge for beneficial insects could enhance natural enemy conservation, biological control of crop pests and diversion of pests from target crops (Thomas et al. 1991; Tonhasca 1993; Alderweireldt 1994; Mensah 1999, 2002; Sequeira 2001).

As part of the Australian cotton industry's effort to reduce the use of chemical insecticides, an extensive research program has been established to develop an IPM framework for cotton that places greater emphasis on the role of beneficial insects. A key focus of this research is in-field diversification of the traditional monoculture cotton cropping system by intercropping. Alternative crops are deployed to conserve and augment populations of beneficial insects, thereby simultaneously augmenting biological control of the main pests and diverting pest pressure from cotton to the refuge crop.

In this chapter we review recent Australian research and commercial-scale attempts at intercropping where refuge or trap crops have been used to diversify cotton production systems. In the present context, the alternative (companion) crop is defined as a secondary or minor crop that is grown within or adjacent to the major crop (cotton, in this case) within a predefined cropping unit such as a field or paddock. We refer to the alternative crops as 'refuge' or 'trap' crops in accordance with their primary usage as sources of beneficial insects or sinks for pest insects, respectively. Implicit in this use of terminology is the acknowledgement that any crop may serve as both a source and sink for pests and beneficial insects, depending on growth stage, crop management practices and various other circumstances.

Refuges for beneficial insect conservation

By virtue of monoculture cropping, Australian cotton production systems do not offer long-term refuge sites to natural enemies. When the cotton crop matures and is harvested, natural enemies have to disperse to new locations in search of food, mates and overwintering or hibernating sites (Mensah 1999). These new locations may be far away from cotton-growing sites, resulting in late recolonisation of cotton crops by natural enemies following reinfestation by the pest (Mensah 1999). The lag time in population response to pest infestation is one explanatory factor that may underlie the perceived inability of natural enemies to effectively suppress cotton pests such as *Helicoverpa* spp. in high pest pressure situations.

Attributes and choice of beneficial insect refuges

Cotton intercropping can provide resources such as food and shelter, thereby increasing the residence time and enhancing the abundance and effectiveness of the natural enemies (Mensah 1999). The intercrop can also be managed to provide refuge and resources to natural enemies during the period when the cotton crop is not available (Mensah 1999). Mensah (1999) demonstrated the potential for using intercrops to conserve natural enemy populations. Mensah and Khan (1997) used intercrops such as lucerne (*Medicago sativa*) as strips within commercial cotton crops (Figure 12.1) to divert *Helicoverpa* spp. and other pests such as green mirids away from cotton crops.

Various other intercrop options have been evaluated in other studies (Sequeira 2001; Sequeira et al. 2001; Annells and Strickland 2002). Based on this research, an IPM program that incorporates intercropping and natural enemies as primary components has been developed for the cotton industry (Mensah 2002).

Mensah (1999) evaluated the utility of intercropping alternative crops such as sunflower, lucerne, safflower, sorghum and tomato for management of insect pests in commercial cotton fields. Higher densities of predatory insects were found on lucerne followed by sunflower,

Figure 12.1: Lucerne strips (centre) interplanted in commercial cotton crops at Warren in New South Wales, Australia.

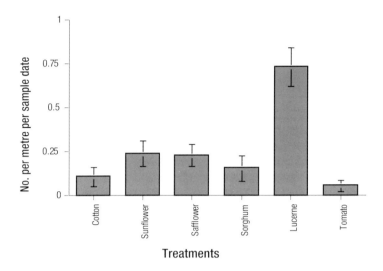

Figure 12.2: Number of predators recorded on refuge crops interplanted in commercial cotton crops in Narrabri in New South Wales, Australia (Mensah 1999).

safflower and sorghum (Figure 12.2). The lowest numbers of predators were recorded on cotton and tomato. There were significant differences between various categories of predatory insects, including beetles, bugs, lacewings and spiders per metre row recorded in the minor crops tested (see Mensah 1999 for details).

In later work on the potential of intercrops to conserve and increase the abundance of natural enemy populations in Australian cotton fields, Owens and Mensah (2000) evaluated the utility of crops such as lucerne, saltbush, faba beans, wheat, canola, chickpeas, vetch and lablab for intercropping in cotton to conserve natural enemies. Predatory insects identified in the above studies are listed in Table 12.2.

Owen and Mensah's (2000) study confirmed the earlier finding that greater densities of predators were recorded in lucerne crops than in other crops, though relatively high densities of

Table 12.2: Major predators of cotton pests identified from alternative winter crops in Narrabri, New South Wales, Australia.

Order	Family	Species	Group
Coleoptera	Coccinellidae	*Coccinella transversalis*	Predatory beetles
		Diomus notescens	
	Melyridae	*Dicranolaius bellulus*	
Hemiptera	Nabidae	*Nabis kinbergii*	Predatory bugs
	Lygaeidae	*Geocoris lubra*	
	Pentatomidae	*Cermatulus nasalis*	
		Oechalia schellembergii	
	Reduviidae	*Coranus triabeatus*	
Neuroptera	Chrysopidae	*Chrysopa* spp.	Predatory lacewings
	Hemerobiidae	*Micromus tasmaniae*	
Araneida	Lycosidae	*Lycosa* spp.	Spiders
	Oxyopidae	*Oxyopes* spp	
	Araneidae	*Araneus* spp.	

Source: After Mensah (1999).

Table 12.3: Number of predators per metre per sample date recorded from refuge crops interplanted in cotton crops at Narrabri in New South Wales, Australia, May–July 2000.

Crop	Predatory beetles	Predatory bugs	Predatory lacewings	Spiders
Lucerne	0.025 a	0.022 ab	0.043 b	0.177a
Saltbush	0.005 ab	0 c	0 c	0.114 ab
Faba beans	0.006 ab	0.048 a	0.033 bc	0.045 bc
Wheat	0.002 b	0 c	0.006 c	0.007 c
Canola	0 b	0 c	0.002 c	0.004 c
Mung bean	0 b	0 c	0 c	0.018 bc
Vetch	0 b	0 c	0 c	0 c
Lablab	0.031 a	0.007 bc	0.090 a	0.033 bc

Means within a column followed by the same letter are not significantly different ($p > 0.05$), Tukey-Kramer multiple comparisons test.
Source: Owens and Mensah (2000).

predatory bugs were found in faba beans, and lablab contained many lacewings (Table 12.3). Overall, these studies identified lucerne as a good intercropping candidate for the conservation of natural enemies in cotton . The high beneficial insect densities of lucerne may be attributed to the abundance of resources provided by this crop, including floral nectar, alternate prey, shelter, mating and oviposition sites (Mensah 1999). These resources, in addition to the fact that lucerne is a perennial species and can be managed by cutting to provide a suitable long-term habitat for natural enemies, makes it an ideal candidate for intercropping with cotton.

Deployment and efficacy of beneficial insect refuges

To be an effective natural enemy refuge, an intercrop planted in strips must be wide enough to permit independent cultivation or management and narrow enough for the crops or vegetation to interact agronomically (Vandermeer 1989; Parajulee et al. 1997). Mensah (1999) reported that the numbers of predators per metre recorded in lucerne strips were higher than in the cotton crop. Predator numbers declined with increasing distance from the lucerne strip, reaching their lowest level 300 m into the cotton crop (Figure 12.3). As a result, it was suggested that

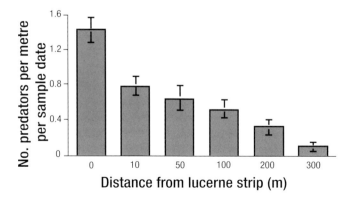

Figure 12.3: Effect of lucerne strips on densities of predators in an adjacent commercial cotton crop at Auscott near Narrabri, New South Wales, Australia (after Mensah 1999).

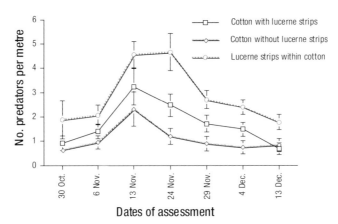

Figure 12.4: Comparison of predatory insect densities in lucerne strips, and cotton with or without lucerne strips in New South Wales, Australia (Mensah 1999).

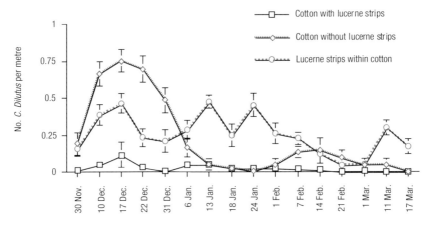

Figure 12.5: Comparison of *Creontiades dilutus* in cotton with and without lucerne strips at Norwood near Moree in New South Wales, Australia (Mensah and Khan 1997).

lucerne strips interplanted in cotton be 300 m apart and 8–12 m wide, and run the whole length of the field. Though lucerne strips tended to attract and retain beneficial insects, giving only limited dispersal to the adjacent cotton, cotton interplanted with lucerne tended to have more natural enemies than did cotton without lucerne (Figure 12.4).

The behaviour of lucerne strips as sinks or sources of beneficial insects is likely to depend on the timing of their establishment and colonisation in relation to the establishment of the cotton crop. The intercrop may act as a source of natural enemies if it is established and colonised before the establishment of the major crop (Corbett and Plant 1993). Alternatively, the intercrop may act as a sink, drawing beneficial insects out of the major crop when both crops are established at the same time (Mensah 1999). Despite such effects, the movement of natural enemies into and out of lucerne intercrops can be manipulated through the application of yeast-based natural enemy food attractant sprays such as Envirofeast® (Bayer, Australia) to the cotton crop or by agronomic management practices such as alternately slashing each half of the lucerne strips (Mensah and Harris 1995; Mensah 1999, 2002).

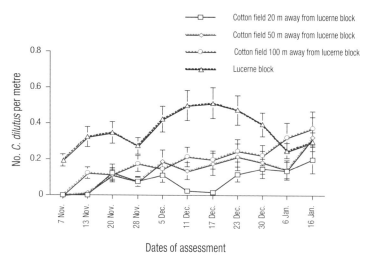

Figure 12.6: Effect of a centrally located lucerne block on *Creontiades dilutus* densities in cotton crops located at different distances from the lucerne block, at Norwood near Moree in New South Wales, Australia (Mensah and Singleton 1998).

Intercrops as traps for manipulation of pest populations

The use of intercropping to divert pest pressure away from the main crop has shown considerable promise in commercial cotton-cropping systems. In this situation the intercrop, by virtue of its greater attractiveness, functions as a sink to either prevent the target pest from reaching the main crop or concentrate it in a certain part of the field where the population can be controlled (Mensah 1999; Sequeira et al. 2001; Annells and Strickland 2002).

Studies by Mensah and Khan (1997) showed that strip-planting lucerne can reduce green mirid infestation within commercial cotton fields (see Figure 12.5). Mensah and Singleton (1998) planted lucerne in a centrally located block within commercial cotton fields and showed that the lucerne block can influence the density of green mirids in adjacent cotton at least 50 m distant (Figure 12.6).

Movement of mirids from the lucerne strips back into the adjacent cotton crop can be minimised by maintaining the attractiveness of strips through the cotton-growing season, especially the early squaring and flowering period. Once the lucerne begins to flower and set seeds, vegetative growth is limited and it becomes less attractive to green mirids (Mensah and Khan 1997). Additionally, lucerne in the seed-setting stage or facing drought situations may 'dry off', resulting in the movement of mirids from the lucerne into the adjacent cotton. Lucerne may then become a source of mirids rather than a sink for the adjacent cotton.

The attractiveness of the lucerne crop can be maintained by slashing or mowing half of each strip. The first cut should coincide with the appearance of the first square in cotton. Subsequent cuts should occur just as the other half of the lucerne strip starts to flower. Using narrow lucerne strips that are less than 12 m wide, or slashing all the lucerne strips if the density of mirids is low (less than 1 per metre) (Mensah and Khan 1997), can improve the movement of natural enemies from the lucerne strips to the adjacent cotton.

Sequeira (2001) documented the development, implementation and impact of the first area-wide program aimed at *Helicoverpa* spp. management in commercial cotton-cropping systems. Trap crops of chickpea and pigeon pea planted in cotton fields at the beginning (spring) and end of the cotton season (summer), respectively, were used to concentrate *Helicoverpa* spp. into small management units where they can be controlled. The program recommends 1% of the

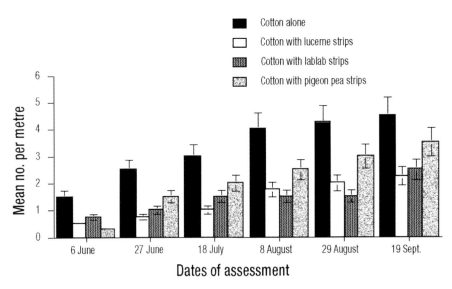

Figure 12.7: Cumulative number of *Helicoverpa* spp. larvae per metre on cotton interplanted with different companion crops in the Kimberley region in Western Australia, 2000.

Reproduced with permission from Annells, A.J. and Strickland, G.R. *Proceedings of the 11th Australian Cotton Conference.* Brisbane, 2002.

total cropping area or a minimum of 2 ha be set aside for trap cropping each growing season. The spring and summer trap crops are now an integral part of the cotton-growing protocol in central Queensland and other parts of Australia characterised by mild winter weather conditions (Sequeira and Playford 2001).

Cotton systems research and development underway in the Ord River irrigation area of Western Australia exemplifies the use of cotton intercropping for pest management. Annells and Strickland (2002) reported significant reductions in the abundance of *Helicoverpa* spp. larvae and insecticide applications in cotton interplanted with lucerne, lablab and pigeon pea (Figure 12.7).

Sequeira and Playford (2001) reported significant reductions in the abundance of *Helicoverpa* spp. larvae and reduced insecticide applications in cotton interplanted with trap crops.

Conclusions

In the studies reported here, it is evident that in-field diversification can play a major role in pest management in monoculture-cropping systems such as cotton. This can be achieved by using refuge crops to increase the abundance and effectiveness of natural enemies, and trap crops to divert pest pressure from the main crop. In Australia and many parts of the world, cotton is grown as a short-lived annual crop in areas that are sometimes dry and with sparse or non-existent surrounding natural vegetation. During the growing season, the irrigated cotton crops may be the only source of shelter and food for many insects, particularly *Helicoverpa* spp. and sucking pests such as green mirids. In Australia, these pests colonise the main crop before their natural enemies, thereby making it impossible for the latter to exert significant pest control. As a result, growers resort to insecticide-based control when pest populations reach damaging levels.

The development of a natural enemy-based IPM program for the cotton-cropping systems in Australia involves interplanting cotton with refuge crops such as lucerne (Mensah 1999, 2002), chickpea, pigeon pea and others (Sequeira 2001; Annells and Strickland 2002). This technique has artificially increased the diversity of cotton-cropping systems and enhanced the availability and diversity of prey, sheltering sites and mates for natural enemies.

Intercrops have also been used successfully as trap crop systems to divert pest infestations away from cotton. This has allowed cotton pests to be managed culturally rather than by the use of synthetic insecticides, thereby reducing the amount of insecticides used against these pests in cotton. The use of trap crops in cotton-cropping systems exemplifies the 'push–pull' strategy (Pyke et al. 1987), where the pest is pulled in by the trap crops, often with the assistance of tactics designed to push it off the main crop (see Khan and Pickett, ch. 10 this volume).

An unrelated but significant use of an in-field crop diversification strategy in Australia is associated with the deployment of transgenic cotton technology. Small areas of secondary crops are grown as 'pest refuges' for the pre-emptive management of resistance in genetically engineered (Bt) cotton crops. These Bt cottons express Cry1Ac and Cry2Ab insecticidal toxins that are toxic to lepidopteran species including *Helicoverpa* spp. (Fitt et al. 1994). It is anticipated that *Helicoverpa* spp., particularly *H. armigera*, will eventually develop resistance to the toxins. To slow the rate of resistance development, a management plan using refuge crops has been developed and adopted by the whole cotton industry since the first commercial release of Bt cottons in Australia in 1996. Each Bt cotton (Ingard® or Bollgard II®) grower is required to grow a refuge crop capable of producing sufficient Bt-susceptible *Helicoverpa* spp. moths to mate with any survivors from the Bt cotton crops (McGaughey and Whalon 1992; Roush 1997). Refuge crop options for Bt cottons include conventional (i.e. non-genetically modified) cotton, sorghum, corn and pigeon pea (Environmental Protection Agency 2001; Sequeira and Playford 2001).

In monocultural cropping systems in Australia and elsewhere, when the crop matures and is harvested the land is left bare at least until a winter crop or the next 'back-to-back' cotton crop is planted. Soils on these lands are subject to erosion, but the presence of permanent refuge crops has addressed some of these problems. In systems where lucerne strips have been used to provide refuge for beneficial insects or as traps for mirids, soil fertility has been increased. In addition, the lucerne strips have provided additional income to the farmers, particularly in times of drought when animal feed is scarce.

Despite the many advantages of refuge and trap crops in terms of habitat diversification and sustainable pest management in crop production, intercropping for pest management is not without problems. Intercrops can act as sinks for beneficial insects as well as for pest species. This aspect of intercropping needs to be resolved, to allow growers to develop greater confidence in this type of strategy. Until this is done the use of refuge and trap crops in pest management will be constrained.

Acknowledgements

We thank Wendy Harris, Ray Morphew and Ammie Forster (NSW Agriculture, Narrabri), for their technical assistance. Special thanks to Dave Anthony, Stefan Henggeller and Merrill Johnson (Auscott, Narrabri), Peter Glennie and Kylie May (Norwood, Moree), David Coulton and Iain Macpherson (Alcheringa, Goondiwindi), John O'Brien and Ross O'Brien (Bellevue, Warren), Chris Hogendyk, Shane Bodiam and Harvey Gaynor (Auscott, Warren) for their co-operation. We also thank Emeritus Professor John Madden (University of Tasmania, Hobart) and Dallas Gibb (NSW Agriculture, Narrabri) for reviewing the manuscript. The Australian Cotton Research and Development Corporation provided funding for this project (grant DAN 89C, 98C, 119C and 151C), for which we are grateful.

References

Alderweireldt, M. (1994). Habitat manipulations increasing spider densities in agroecosystems: possibilities for biological control? *Journal of Applied Entomology* 118: 10–16.

Altieri, M.A. (1994). *Biodiversity and Pest Management in Agroecosystems*. Food Products Press, New York. 185 pages.

Altieri, M.A. and Letourneau, D.K. (1982). Vegetation management and biological control in agroecosystems. *Crop Protection* 1: 405–430.

Andow, D.A. (1991). Vegetation diversity and arthropod population response. *Annual Review of Entomology* 36: 561–586.

Annells, A.J. and Strickland, G.R. (2002). Evaluation of companion crops as part of an integrated pest management package for genetic modified cotton in the Kimberley. *Proceedings of the 11th Australian Cotton Conference*. Brisbane. 874 pages.

Beirne, B.P. (1967). *Pest Management*. CRC Press, Cleveland, Ohio. 123 pages.

Bugg, R.L., Ehler, L.E. and Wilson, L.T. (1987). Effect of common knotweed (*Polygonum aviculare*) on abundance and efficiency of insect predators of crop pests. *Hilgardia* 55: 1–51.

Clark, M.S., Luna, J.M., Stone, N.D. and Youngman, R.R. (1994). Generalist predator consumption of army worm and effect of predator removal on damage in no-till corn. *Environmental Entomology* 23: 617–622.

Conlong, D.E. (1990). A study of pest–parasitoid relationships in natural habitats: an aid towards the biological control of *Eldana saccharina* in sugarcane. *Proceedings of the South African Sugar Technologists Association* 64: 111–115.

Conlong, D.E. (1994). A review and perspectives for the biological control of the African sugarcane stalk-borer *Eldana saccharina*. *Agriculture, Ecosystems and Environment* 48: 9–17.

Conlong, D.E. (1995). Results of preliminary pitfall trapping trials for potential arthropod predators of *Eldana saccharina*. *Proceedings of the South African Sugar Technologists Association* 69: 79–82.

Corbett, A. and Plant, R.E. (1993). Role of movement in the response of natural enemies to agroecosystem diversification: a theoretical evaluation. *Environmental Entomology* 22: 519–531.

Corbett, A., Leigh, T.F. and Wilson, L.T. (1991). Interplanting alfalfa as a source of *Metaseiulus occidentalis* for managing spider mites in cotton. *Biological Control* 1: 188–196.

Dent, D. (1995). *Integrated Pest Management*. Chapman & Hall, London.

Environmental Protection Authority (2001). Bt cotton refuge requirements for 2001 cotton growing season in Australia. www.epa.gov/pesticides/biopesticides/othersdocs/bt_cotton_refuge_2001.htm

Fitt, G.P. (1989). The ecology of Heliothis in relation to agroecosystems. *Annual Review of Entomology* 34: 17–52.

Fitt, G.P., Mares, C.L. and Llewellyn, D.J. (1994). Field evaluation and potential ecological impact of transgenic cottons in Australia. *Biocontrol Science and Technology* 4: 535–548.

Gregg, P.C., Fitt, G.P., Zalucki, M.P. and Murray, D.A.H. (1995). Insect migration in an arid continent. II. *Helicoverpa* spp. in eastern Australia. In *Insect Migration: Tracking Resources Through Space and Time* (V.A. Drake and A.G. Gatehouse, eds), pp. 151–172. Cambridge University Press, Cambridge.

Herzog, D.C. and Funderburk, J.E. (1986). Ecological basis for habitat management and pest control. In *Ecological Theory and Integrated Pest Management Practice* (M. Kogan, ed.), pp. 217–250. Wiley, New York.

Landis, D.A., Wratten, S.D. and Gurr, G.M. (2000). Habitat management to conserve natural enemies of arthropod pests in agriculture. *Annual Review of Entomology* 45: 175–201.

Luff, M.L. (1983). The potential predators for pest control. *Agriculture, Ecosystems and Environment* 10: 159–181.

McGaughey, W.H. and Whalon, M.E. (1992). Managing insect resistance to *Bacillus thuringiensis* toxins. *Science* 258: 1451–1455.

Mensah, R.K. (1997). Local density responses of predatory insects of *Helicoverpa* spp. to a newly developed food supplement 'Envirofeast' in commercial cotton in Australia. *International Journal of Pest Management* 43: 221–225.

Mensah, R.K. (1999). Habitat diversity: implications for the conservation and use of predatory insects of Helicoverpa spp. in cotton systems in Australia. *International Journal of Pest Management* 45: 91–100.

Mensah, R.K. (2002). Development of an integrated pest management programme for cotton. Part 1: establishing and utilizing natural enemies. *International Journal of Pest Management* 48: 87–94.

Mensah, R.K. and Harris, W.E. (1995). Using Envirofeast spray and refugia technology for cotton pest control. *Australian Cotton Grower* 16: 30–33.

Mensah, R.K. and Khan, M. (1997). Use of *Medicago sativa* (L.) interplantings/trap crops in the management of the green mirid, *Creontiades dilutus* (Stål) in commercial cotton in Australia. *International Journal of Pest Management* 43: 197–202.

Mensah, R.K. and Madden, J.L. (1994). Conservation of two predator species for biological control of *Chrysophtharta bimaculata* in Tasmanian forests. *Entomophaga* 39: 71–83.

Mensah, R.K. and Singleton, A. (1998). Integrated pest management (IPM) in cotton based on Envirofeast and lucerne technologies: where are we? *Proceedings 9th Australian Cotton Conference,* pp. 363–377. Broadbeach, Queensland.

Murdoch, M.M., Cheeson, J. and Cheeson, P.L. (1985). Biological control in theory and practice. *American Naturalist* 125: 344–366.

Owens, J. and Mensah, R.K. (2000). *Evaluating Refuge Crops for Predators in Cotton.* Technology Transfer Centre Press, Australian Cotton Research Institute, Narrabri, Australia. 20 pages.

Parajulee, M.N., Montandon, R. and Slosser, J.E. (1997). Relay intercropping to enhance abundance of insect predators of cotton aphid in Texas cotton. *International Journal of Pest Management* 43: 227–232.

Pimentel, D. (1971). Population control in crop systems: monocultures and plant spatial patterns. *Proceedings of Tall Timbers Conference on Ecology: Animal Control and Habitat Management* 2: 209–221.

Pyke, B., Rice, M., Sabine, B. and Zalucki, M. (1987). The push-pull strategy-behavioural control of Heliothis. *Australian Cotton Grower* May–July: 7–9.

Room, P.M. (1979). Seasonal occurrence of insects other than *Helicoverpa* spp. feeding on cotton in the Namoi valley of New South Wales. *Journal of the Australian Entomological Society* 16: 165–174.

Roush, R.T. (1997). Bt-transgenic crops: just another pretty insecticide or a chance for a new start in resistance management? *Pesticide Science* 51: 328–334.

Russell, E.P. (1989). Enemies hypothesis: a review of the effect of vegetational diversity on predatory insects and parasitoids. *Environmental Entomology* 12: 625–629.

Sequeira, R. (2001). Inter-seasonal population dynamics and cultural management of *Helicoverpa* spp. in a central Queensland cropping system. *Australian Journal of Experimental Agriculture* 41: 249–259.

Sequeira, R.V. and Playford, C.L. (2001). Abundance of *Helicoverpa* (Lepidoptera: Noctuidae) pupae under cotton and other crops in central Queensland: implications for resistance management. *Australian Journal of Entomology* 40: 264–269.

Sequeira, R.V., McDonald, J.L., Moore, A.D., Wright, G.A. and Wright, L.C. (2001). Host plant selection by *Helicoverpa* spp. in chickpea-companion cropping systems. *Entomologia Experimentalis et Applicata* 101: 1–7.

Symondson, K.D., Sunderland, K.D. and Greenstone, M.H. (2002). Can generalist predators be effective biocontrol agents? *Annual Review of Entomology* 47: 561–594.

Thomas, M.B., Wratten, S.D. and Sotherton, N.W. (1991). Creation of 'island' habitats in farmland to manipulate populations of beneficial arthropods: predator densities and emigration. *Journal of Applied Ecology* 38: 906–917.

Tonhasca, A. Jr (1993). Effects of agroecosystem diversification on natural enemies of soybean herbivores. *Entomologia Experimentalis et Applicata* 69: 83–90.

Vandermeer, J. (1989). *The Ecology of Intercropping.* Cambridge University Press, Cambridge. 237 pages.

Van Driesche, R.G. and Bellows, J.T.S. (1996). *Biological Control.* Chapman & Hall, New York.

van Emden, H.F. (1990). Plant diversity and natural enemy efficiency in agroecosystems. In *Critical Issues in Biological Control* (M. Mackauer, L.E. Ehler and J. Roland, eds). Intercept Press, Andover.

van Emden, H.F. and Williams, G.F. (1974). Insect stability and diversity in agroecosystems. *Annual Review of Entomology* 19: 455–475.

van Leteren, J.C. (1987). Environmental manipulation advantageous to natural enemies of pests. In *Integrated Pest Management, Quo Vadis? An International Perspective* (V. Delucchi, ed.). Parasitis, Geneva.

Way, M.J. and Heong, K.L. (1994). The role of biodiversity in the dynamics and management of insect pests of tropical irrigated rice – a review. *Bulletin of Entomological Research* 84: 567–587.

Chapter 13

Pest management and wildlife conservation: compatible goals for ecological engineering?

Cilla Kinross, Steve D. Wratten and Geoff M. Gurr

Introduction

On the assumption that primary producers will implement some of the habitat modifications outlined in this book with the aim of improving biological pest control, are there likely to be benefits for wildlife? Less than 5% of the earth's terrestrial area remains in an unmanaged state (Pimentel et al. 1992). The dominant land use is agriculture and this has had severe ramifications on the conservation of wildlife, particularly since intensification after the Second World War (Gall and Orians 1992; Stoate 1998). Loss of wildlife habitat and changing agricultural practices have been the main causes, leading to long-term declines in wildlife populations (Paoletti et al. 1992). Most species of wildlife (and by this we include all flora and fauna, vertebrate and invertebrate) live outside parks (Pimentel et al. 1992), so even if the reserve system were representative of the full range of terrestrial habitats, it cannot be relied upon for adequate protection for all native species. As agricultural activities continue to intensify, the role of conserving what is left of the original wildlife in agricultural landscapes will increasingly fall on the shoulders of the major landowners – the primary producers. Failure to do this, given the large areas of land devoted to agriculture, will have major detrimental effects on the ecosystem services provided by biodiversity.

Some farmers are aware of the benefits of farm biodiversity. Quite apart from those using the habitat manipulation methods for pest control featured in this book, many (46–63% in the US, depending on geographical region) are implementing practices to conserve wildlife for hunting, for example (Conover 1998).

Henry et al. (1999) use the term 'conservation corridors' to describe linear patches of created or modified habitat that differ from the surrounding matrix. While acknowledging that not all these patches will function as corridors, nor that they were necessarily created with the primary aim of conservation, the term is useful to describe the array of ecologically modified or created habitats found in agricultural landscapes. They are usually linear and have at least some potential for conservation of biodiversity. Typically, these features are 1–50 m wide and hundreds of metres long – small in scale compared with the typical scale of conservation in forests, but they can constitute valuable off-reserve resources within the broader matrix of habitat (Lindenmayer and Franklin 2002). As explored below, such features may be established as part of a network (albeit fragmented) and may link with remnant, natural or semi-natural vegetation fragments and so have effects on a larger, landscape scale. 'Conservation corridor' is, therefore, a useful

Figure 13.1: A British 'composite landscape feature' comprising (left to right) winter game cover crop, sown mixed-species plant strip and well-established beetle bank (Photograph: G.M. Gurr).

term for a wide range of features. These include shelterbelts, windbreaks, hedges, fencerows and riparian buffer zones, all of which generally contain woody vegetation; as well as herbaceous features such as set-aside field margins, uncropped wildlife strips, conservation headlands, beetle banks (Wratten 1992), game cover crops and 'composite landscape features' (Figure 13.1). The habitat manipulation features pictured in Gurr et al. (ch. 1 this volume) and described by various contributors to this book are also often linear and may be considered conservation corridors, but there is a paucity of studies in which the pest management and conservation utility of such features are simultaneously considered.

The questions which this chapter tries to address are as follows: are there some indirect benefits to wildlife that will accrue by putting into practice some of the habitat manipulation recommended in this book? Or will these conservation corridors merely attract opportunistic and/or undesirable species due to the suboptimal size, habitat and large edge–interior ratio of these largely linear features? Is it possible that wildlife habitat can be significantly improved by quite modest modifications to the conservation corridors? And what is the best scale at which to approach the problem: site or landscape?

To address these questions, the chapter reviews relevant literature to identify the most significant ecological functions of conservation corridors, using examples from different taxa and various parts of the world. The benefits and disadvantages of conservation corridors are discussed, and ways that they might be improved for wildlife without compromising their pest-control function are outlined.

The role of habitat

To what extent can conservation corridors extend habitat for wildlife? Habitat includes resources such as food and foraging substrates, shelter from the weather and predators, dispersal corridors and breeding habitat. Interest in the habitat function of the most widely studied of corridors, hedgerows, goes back many years. In the early part of the twentieth century, Richards (1928) described the role of hedgerows in conserving bryophytes (mosses) in England. Later, Petrides (1942) investigated the habitat function for birds and mammals of hedges in New York. In a milestone review of the ecological role of hedges in Britain, Pollard et al. (1974) concluded that they provided habitat for a few woodland plants, many species of invertebrates, several species of amphibians, all six species of reptiles and most British mammals. Their habitat value is probably most significant in countries where landscape changes have been most dramatic. In Ireland, the European country which has the least wooded area in the European Community, hedges are considered important wildlife refuges, particularly for woodland fauna (Webb 1988). Bickmore (2001) studied the function of hedges in England and Wales and, though historically the principal driver for their establishment was enclosure, aesthetics and wildlife now dominate as the reasons for establishment. Interest in the ecology of conservation corridors has grown considerably, although the literature remains almost silent on the less charismatic taxa of reptiles and amphibians (but see Maisonneuve and Rioux (2001) for the value of riparian strips to herpetofauna in Canada, for example). Moreover, the research is very biased towards North America and Europe.

Very little research appears to have been done in Australasia, Asia, South America or Africa, by comparison. Considerable work has been done on the benefits of reafforested areas such as the coffee plantations of Jamaica (Johnson 2000); rehabilitated mine sites in South Africa (Lubke et al. 1996; van Aarde et al. 1996) and Indonesia (Passell 2000); regenerating abandoned farmland (Chapman and Chapman 1999) in East Africa; and forestry plantations in Hong Kong (Kwok and Corlett 2000) and Chile (Estades and Temple 1999); conserving owls in woodlots in India (Rishi and Sandu 1999) and woodmouse habitat in Latvia (Tattersall 1998). In addition, a review of the use of revegetated areas by vertebrate fauna in Australia has been presented by Ryan (1999).

Food resources and foraging substrates

Probably the most important aspect of habitat that can be provided by conservation corridors is food, as well as substrates for foraging. In Western Australia, revegetated areas along roadsides in agricultural areas provide significant amounts of foods to both invertebrates and birds, and contain higher densities of invertebrates than does remnant woodland (Majer et al. 2001). Windbreaks and hedges also provide foraging opportunities and/or flightpaths for bats (e.g. Robinson and Stebbings 1997) and habitat for other small mammals (e.g. Yahner 1982a). Conservation corridors are particularly important in areas where food is limited in winter, such as northern Europe and America (Hill et al. 1995; Vickery et al. 2002). They are also important habitats for butterflies. As many as 26 species of butterflies can be found in field margins, including hedgerows, in the UK (Dover and Sparks 2000).

Though trees are essential for some animals, field margins and beetle banks created in field centres constitute valuable habitat and provide food resources such as insects and spiders for other wildlife species (Hill et al. 1995). An example is the harvest mouse, *Micromys minutus*, a species vulnerable to extinction in the UK. Beetle banks can be a particularly useful habitat for this species (Boatman 1999). Moreover, the availability of certain items, such as plant bugs and caterpillars during the breeding season, can be critical for some game bird species (Boatman 1999). Game crops can also be planted. These will not only aid game birds (Stoate 1998) but will provide food for non-game birds, particularly seed-eaters (Hill et al. 1995). These landscape enhancements (Figure 13.1) also have higher trophic level effects, and can provide useful hunting grounds for owls and raptors (Vickery et al. 2002).

Accordingly, at least some forms of ecological engineering for pest management offer food resources to wildlife species and, as a result of the insectivorous nature of many birds and small mammals, pest arthropods are likely to be represented in their respective diets. Their impact on pests may extend over considerable areas, especially in the case of birds, but work remains to be done on the extent to which vertebrates may regulate pest populations.

Breeding habitat

Many species reproduce successfully in linear habitats. Although most of the evidence of breeding comes from birds, some mammals are also clearly benefiting from habitat manipulation. Harvest mice, for example, make their nests in the long grass of beetle banks in southern England (Boatman 1998). Birds have been particularly well studied. For example, almost two-thirds of the bird species in Ireland nest in hedges (Webb 1988) and field margins in England provide nesting habitat for ground-nesting species such as yellowhammers *Emberiza citrinella* and whitethroats *Sylvia communis* (Boatman 1999). Nest density can be remarkably high. In a North American study on fencerows, Shalaway (1985) reported 43.5 nests/ha in shelterbelts, an extraordinary 10 times higher than in natural shrub habitats. Yahner (1982b) found that densities ranged from 28.8 nests/ha to 186.4 nests/ha, although >95% of the nests belonged to only five species. In eastern Australia, evidence of breeding (from opportunistic sightings only) was observed for 25 bird species in windbreaks (Kinross 2000); these included several species identified by Reid (1999) as declining in the landscape.

Enhanced field margins can elevate reproductive success in other ways apart from nest sites, particularly in relation to territorial behaviour. For example, hedgerows provide a territory boundary for male pheasants (*Phasianus colchicus*) (Boatman 1999) and windbreaks provide useful singing posts for other birds (Cassel and Wiehe 1980; O'Connor 1984).

Not all sites are suitable for breeding, of course. Windbreaks in the Czech Republic do not provide nest sites for migratory bird species (Balát 1986) and immature sites cannot provide satisfactory conditions for specialists such as obligate hollow-nesters (Cassel and Wiehe 1980). Some sites may be too narrow to provide breeding areas safe from predators (Yosef 1994) and the actual evidence relating to nest predation and edges is not conclusive. Rodewald (2002) provides a review of this debate and a contrary position: his findings indicate that nest success is more closely linked to landscape integrity than to edge effect.

Shelter from predation

Reproductive success may also depend on the number and species of predators in the locality. It was found by Shalaway (1985) that the absence of arboreal snakes, chipmunks and ground squirrels was part of the reason for a high success rate of nests in fencerows in Michigan.

The ability to thwart detection from predators in dense vegetation may also be an important factor in successful dispersal and safe foraging for birds (Hinsley and Bellamy 2000). Johnson and Adkisson (1985) demonstrated that North American blue jays, *Cyanocitta cristata,* closely followed fencerows while taking food to their winter caches. Windbreaks in Hungary (Legány 1991) and Australia (CK, personal observation) also provide secure places to which food can be brought and consumed after foraging in adjacent fields.

Physical shelter

Field margins and other habitat modifications provide shelter not only from predation, but from the weather and from activities associated with agriculture. For example, field boundaries act as refuge for animals during fertiliser application (Vickery et al. 2002) and pesticide spraying (Joyce et al. 1999). Unsprayed conservation headlands favour flowers, beneficial insects and game birds as well as wood mice *Apodemus sylvaticus* (Tew et al. 1992).

In the UK, hedgerows provide shelter for many species of arthropods and bird species (Hill et al. 1995) and even uncultivated strips can provide some shelter to game birds (Stoate 1998). In the snow-prone parts of Europe and North America, however, where shelterbelts often contain deciduous trees, there is little winter habitat (Cassel and Wiehe 1980). Where shelterbelts include evergreen trees, however, bird densities can rise in winter and such field margins may provide critical shelter from the elements (Balát 1986). Australian windbreaks have the advantage of providing year-round shelter if native tree species are used. In Western Australia, revegetated areas provide shelter from the weather to both vertebrates and invertebrates (Majer et al. 2001).

Dispersal corridors

Conservation corridors contain not only food, shelter and breeding habitat, but increase landscape connectivity (see Figure 13.2). Corridors therefore provide additional habitat for wide-ranging species (Harris 1984), mechanisms for dispersal of juveniles (Machtans et al. 1996) and migration and links between habitat patches (Wegner and Merriam 1979). This movement increases genetic interchange (Harris 1984) and reduces patch isolation by 'rescuing' populations from potential local extinction (Brown and Kodric-Brown 1977).

The importance of isolation will vary according to a species' vagility. Each species will perceive the degree of connectivity of a landscape differently and it will be hard to provide a

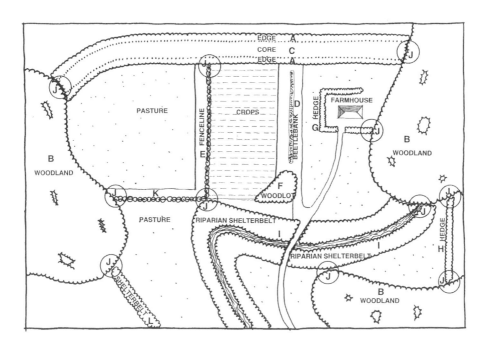

Figure 13.2: Conservation corridors on the farm scale and their habitat value. (A) Habitat for edge-tolerant species; (B) Source of core habitat for woodland species (edge-sensitive); (C) Shelterbelt/corridor, 30–60 m wide, for dispersal, foraging, nesting and shelter; (D) Overwintering refuge for predatory invertebrates and summer nest site for mammals and birds. High-activity hunting site for predatory birds, e.g. the barn owl (*Tyto alba*) in Europe; (E) Foraging and dispersal habitat for edge-tolerant species; (F) Corner planted for forest bird and mammal species; (G) Habitat for farmland bird and mammal species; (H) Corridors for invertebrates (e.g. butterflies), birds and mammals; (I) Increases connectivity and extends optimal riparian habitat; (J) Node, intersections increase core and provide mosaic for woodland species; (K) Non-treed margin, e.g. set-aside, wildlife strips, conservation headlands, game cover crops. Nest site for ground-nesting birds and food for invertebrates, birds and small mammals; (L) Extended habitat for woodland species.

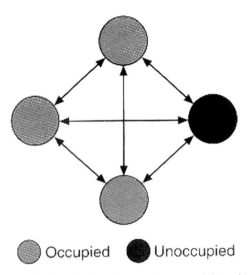

Occupied   Unoccupied

Figure 13.3: Schematic representation of the metapopulation model in which a species is present in only some of the potentially suitable patches of habitat between which individuals are able to move (Pullin 2002).

solution for the whole community (With and Crist 1995). Even within an ecological group such as epigeic forest fauna, a large difference in dispersal ability was found between species: some invaded new hedges quickly and became established, but others were much slower or had populations that fluctuated strongly over time (Gruttke and Kornacker 1995).

For species that can move readily between habitat patches in a highly disturbed environment, survival may be possible even if the size of individual patches is such that the chances of extinction within each is high. The term 'metapopulation' was coined by Levins (1969) to describe a number of populations that utilise such an arrangement, i.e. that exchange individuals through emigration and immigration. The persistence of such systems demands that the factors that may cause extinction not apply synchronously to all patches (Hanski 1997), as this would not allow recolonisation from occupied patches (Figure 13.3). Metapopulations are acknowledged as important in the dynamics of natural enemy (Bonsall and Hassell 2000) and pest systems (e.g. Jervis 1997).

Identifying whether source-sink and metapopulation processes are operating would not be easy in such a modified landscape as an agricultural matrix with 'islands' of habitats. It would be difficult to distinguish between the extinction in a small patch and shifts in freely moving or seasonally nomadic populations (Andrén 1994; Simberloff 1995). Consequently, there are few studies that empirically demonstrate either the rescue effect or corridor function, presumably due to the practical difficulties encountered in such investigations (Nicholls and Margules 1991; Inglis and Underwood 1992). Hence, most of the literature is rather speculative in this respect.

Generally, the presence of corridors is seen as positive. One of the conclusions of the van Dorp and Opdam (1987) comparative study of woodlot connectivity and birds in the Netherlands was that extinction appeared to be highest in the smallest and most remote woodlots, with recolonisation being highest in better-connected ones. They concluded that an extensive hedgerow network aided the species richness of an area. It has also been suggested that hedgerows can act as corridors for butterflies, but the evidence is not conclusive (Dover and Sparks 2000). Study of hedgerows in central New York State indicated that they provided a corridor for forest herbs as the richness of the flora and the similarity to the forest plant assemblage declined with distance from the forest source. Some of these hedges, however, were remnants themselves and may have contained relic plants, thus diluting the evidence of corridor effect (Corbit et al. 1999). Even fragmented corridors probably act as 'stepping-stones' for woodland birds, particularly migratory species (Yahner 1983; Shafer 1997) and bats (Bennett 1990) and may also aid woodland invertebrate species to traverse the landscape and utilise croplands (Maudsley 2000).

Studies based on experimental use of corridors have focused on small mammals and birds in North America and Europe. For example, Wegner and Merriam (1979) demonstrated that well-vegetated fencerows could overcome the dispersal barriers for both mammals and birds created by fields, although the evidence for bird dispersal came from observation only. Haas (1995) reported some bird species (individually marked) dispersing preferentially between patches of woodland connected by corridors. There is evidence of man-made corridor use by small mammals in newly planted shelterbelts in Poland (Ryszkowski et al. 2003) and by arboreal dormice *Muscardinus avellanarius* in hedgerows in southern UK (Bright 1998).

The extent of benefits to wildlife

It is clear from the above that wildlife – both flora and fauna – can benefit from conservation corridors, including the types that may be established either expressly or at least partly to aid pest management. Such corridors may provide resources for foraging, reproduction and dispersal of wildlife species. However, a note of caution should be sounded at this point due to the possibility that the corridors may benefit undesirable or already common species and either not benefit or, worse, disadvantage desirable or rare species. It may also be the case that they cannot create a fully functioning community or ecosystem. Indeed, as reflected in the studies reviewed above, much research has focused on species- rather than community-level responses. Accordingly, conservation corridors may act as landscape sinks for organisms with low reproductive success. These are not trivial issues and need to be investigated prior to drawing conclusions about conservation corridor value.

Species that can take advantage of conservation corridors include only those that have adaptive traits that facilitate dispersal or that can utilise habitats with a high edge–area ratio, i.e. narrow islands in a matrix that might be quite hostile to migration. These may be generalist species commonly associated with disturbed landscapes, that do not require any particular conservation strategy. They may indeed be undesirable species, such as introduced pests. Sedentary species will clearly not benefit; those that are less vagile or cannot detect landscape patterns at this scale may take years or centuries to colonise the habitat and/or utilise the corridor successfully (Burel and Baudry 1990), as reflected in a study of colonisation of remediated mine sites by different taxa (Nichols and Nichols 2003). Furthermore, good habitat is likely to be colonised quickly by offspring and this could reduce genetic diversity as colonisation attempts by outsiders are blocked by already-occupied habitat (Demers et al. 1995).

It is clear that shelterbelts and windbreaks do contain a large proportion of the generalist-opportunistic guild (Ryan 1999; Pierce et al. 2001), particularly in respect to breeding bird populations (Yahner 1982b). However, conservation corridors of sufficient width, complexity and maturity do also cater for some forest-interior species, at least for birds, including those with regionally declining populations (Capel 1988; Green et al. 1994; Kinross 2000; Taws 2000) although care is needed in interpreting these results, due to small samples.

Conservation corridors containing woody vegetation may also act as a barrier or have negative effects. Grassland birds may be negatively affected by the adoption of shelterbelts in North American plains habitat (Pierce et al. 2001), the reafforested parts of South Africa (Allan et al. 1997) or hedgerows in British fields (Hinsley and Bellamy 2000). Hedgerows can also be a barrier to some species of butterflies (Dover and Sparks 2000) and other invertebrates (Maudsley 2000; Wratten et al. 2003). Species such as these clearly cannot be helped by planting trees in narrow strips; they require alternative conservation strategies such as reservations of large, pristine areas of forest or grassland.

Perhaps of greater concern is the extent to which a fully functioning ecosystem can be created by provision of conservation corridors. This would be discerned by a high species and guild diversity, including the presence of specialists, an adequate structural diversity of vegetation and groundcover, including a litter layer and the recycling of water and nutrients (Majer 1990), and soil formation (Lubke et al. 1996). Evidence from mine-site rehabilitation is encouraging, at least in respect of species diversity (van Aarde et al. 1996) and soil formation (Lubke et al. 1996), but comparisons may not be valid. Mine sites are often revegetated with the aim of providing habitat for the original biota (as opposed to improving agricultural productivity); the restored areas are also larger and often closer to areas of natural bush that encourage recolonisation and succession (Ryan 1999). The time involved may also be very long. For example, nine years was not sufficient to create a functioning community of epigeic forest invertebrates in new hedges (Gruttke and Kornacker 1995) and animal taxa were still colonising mine sites eight years after rehabilitation (Nichols and Nichols 2003).

The other contentious issue is whether conservation corridors are actually acting as sinks in the landscape. The agricultural landscape has been seen as analogous to the 'source–sink' model of Pulliam (1988), with remnant bushland usually acting as 'source' habitats and windbreaks generally as 'sinks' of suboptimal breeding habitat, whose populations need to be replenished from 'sources' to avoid extinction in the landscape (Opdam 1990). If there is insufficient immigration to top up these populations, they may become locally extinct (Mader et al. 1990; Gruttke and Kornacker 1995); indeed, patches of remnant bushland may act as sinks rather than sources, especially if their size is small.

It is interesting to note that Burel and Baudry (1990) suggested that corridors may also act as sources. This situation could occur when agricultural land has been abandoned and the corridors are harbouring the flora and fauna available for recolonisation – a good example being that of earthworms (Lagerlof et al. 2002). Field margins may also act as a source of recolonisation after pesticide application. Should the colonisers be pest species this has clear negative implications for pest management, but natural enemies are also likely to be among colonising species (Joyce et al. 1999). Because natural enemies are often more sensitive to pesticide mortality than are pests (Nicholls and Altieri, ch. 3 this volume), the source function of conservation corridors for arthropods is likely to give a net positive effect for pest management.

The majority of the literature, however, describes conservation corridors as sinks. The evidence comes from the fact that some corridors contain elevated numbers of predators and brood parasites, leading to a lower reproductive rate (Yosef 1994; Brawn and Robinson 1996). They may also contain large numbers of invasive or aggressive species, particularly birds (Yahner 1983; Wilson and Lindenmayer 1996). Moreover, there is some evidence that corridors may be of disproportionate benefit to introduced species, with evidence coming from rodents in road corridors (Downes et al. 1997), birds in shelterbelts (Emmerich and Vohs 1982; Yahner 1983) and understorey plants in windbreaks (CK, personal observation). These undesirable species, as well as disease, may move along corridors, killing or preventing the dispersal of desirable species (Simberloff and Cox 1987).

Not all studies produce such negative results. For example, in a study of eastern Australian farm windbreaks, the bird species that were found in higher densities in windbreaks than in remnant vegetation were all small, native, insectivorous or nectarivorous species – opportunistic and generalist perhaps, but not large, aggressive or predatory (Kinross 2000). They are likely to contribute to suppression of arthropod pests. This finding may be partly due to the relative paucity of brood parasites in Australia compared with North America (Ford et al. 2001). It may also be due to the landscape which, despite being fragmented, retained a reasonably high level of connectivity and still contained areas of remnant vegetation within a reasonable distance. This contrasts with some regions in which windbreaks have been studied, where whole landscapes may be considered to be sinks (Brawn and Robinson 1996).

Other reasons that might explain these preliminary optimistic results from Australia lie in the characteristics of the windbreaks themselves. Although direct comparisons are difficult, many of the windbreaks in Australia appear to be wider and have higher structural and plant species diversity than those of North America and Europe. Even greater differences become obvious when comparing these sites with hedges. Perhaps the question of whether windbreaks and field margins provide suboptimal habitat could be rephrased: How can conservation corridors be designed so that they not only meet their production objectives, but also provide habitat which contributes to wildlife conservation aims? The following section addresses this question.

Management of conservation corridors

The literature suggests that the key attributes that can be modified to improve windbreaks and hedges for wildlife, particularly birds, are: patch size or width, vegetation diversity, maturity and composition, and overall landscape heterogeneity and connectivity (Osborne 1984; Green et al. 1994; Lakhani 1994; Parish et al. 1994). Site management may also be important, due to the effect on factors such as hedge bottom quality (Rands 1986; Hinsley and Bellamy 2000) and the availability of resources in field margins (Vickery et al. 2002).

Size

As a general rule, for small mammals, breeding birds and deer, the larger the windbreak the better (Hess and Bay 2000). As the conservation features being considered here are primarily linear in nature, the variable that reduces edge–area ratio is width. Several studies have shown that width is critical in the determination of species composition and diversity, both for birds (e.g. Shalaway 1985; Green et al. 1994) and invertebrates (Mader 1988).

It has been suggested that narrow windbreaks impede dispersal because of the increased potential for a corridor to act as a sink for predators (Merriam and Saunders 1993) (see discussion above). Animals such as land snails that avoid edges may also have their dispersal slowed due to the need to return frequently from edges to core habitat (Baur and Baur 1992). Conversely, wider corridors could encourage territorial species to take up residency and prevent dispersal through aggression (Arnold 1995) or competition (Demers et al. 1995). At this stage, however, there is little evidence that dispersal is being impeded by introduced species, predators or parasites (Schmeigelow et al. 1997). Until more evidence is available from field studies, it is likely that wider is better, at least for the majority of species.

The question remains: How wide is wide? Determining the thresholds for width is not easy and clearly will vary from taxon to taxon. Another impediment is that most of the work in this respect has been done on remnant forest strips, rather than man-made conservation corridors. For example, roadside corridors as narrow as 45 cm in Western Australia were sufficient to contain the bird species found in the (albeit degraded) patches that they were linking (Saunders and Rebeira 1991). In riparian strips, however, in eastern US, corridor widths of 75–175 m were required to support 90% of the local avifauna (Spackman and Hughes 1995) and some forest-interior species have been recorded as absent in forest edges 500 m wide (Kilgo et al. 1998).

Attempts at quantifying width thresholds in conservation corridors are even rarer. Morgan and Gates (1982) suggested that nesting birds need to disperse their nests in a 15–20 m-wide shrubby border to avoid spatial density-dependent mortality from predation and other mortalities. This is supported by preliminary findings from eastern Australian farm windbreaks, all less than 55 m wide. Modelling of the abundance of uncommon woodland bird species suggested that 15 m was the minimum width that provided useful native bird habitat and that a width of 27 m was better (Kinross 2000).

Even less is understood about the thresholds required for forest invertebrates and plants. Forest carabid beetles that need a mesic environment may require widths of 15 m or greater to provide the correct light and microclimate (Šustek 1994, cited in Joyce et al. 1999). For similar reasons, suitable habitat for forest species appears to be reduced in hedges less than 7 m wide (Corbit et al. 1999) and Burel and Baudry (1990) considered any hedge less than 4 m wide poor habitat for forest plant species.

In practice, hedges are often narrow (1–2 m wide) but even at this fine resolution, a difference in width can be important for some bird species, particularly those already adapted to agricultural landscapes such as blackbird *Turdus merula* and goldfinch *Carduelis carduelis* (Green et al. 1994). Little is known about the widths required for conservation corridors without trees, but Stoate et al. (2001) suggested that many field boundaries should be widened to provide habitat for uncommon British species such as the whitethroat *Sylvia communis*.

Vegetation structure and composition

Where width is not an issue, the most important characteristic determining bird community composition is that of vegetation structure, particularly the presence of trees (Hamel 2003). Tree height and site maturity have also been shown to be important. For example, the abundance of some small mammal species is closely related to windbreak or hedge maturity (Trnka et al. 1990; Bryja and Zukal 2000). Conservation corridors will therefore require time to allow the successional processes to take place before less opportunistic species will become established.

Understorey density, including the presence of a shrub layer, is also very important to help birds avoid predation (Berg and Pärt 1994), particularly during the breeding season (Lack 1988b). Shalaway (1985) reported that nest density and diversity in North American fencerows increased with the number of shrubs and that 75% of open nests were found in shrubs.

Field margins containing trees, shrubs and an understorey provide a range of foraging substrates, such as branches, leaves, flowers and grass stems. The structural diversity provided will attract a higher number of species than will a more simple vegetational structure such as a grassland for birds (Osborne 1984; MacDonald and Johnson 1995; Jobin et al. 2001) and invertebrates (Maudsley 2000). Even conservation corridors without trees will be more highly utilised by vertebrates and invertebrates if they contain a diverse mix of plants, including wildflowers (Clarke et al. 1997; Vickery et al. 2002).

Merely providing for a high density of species or even a diverse fauna may not be sufficient to aid wildlife conservation aims, if habitat for species with declining populations is not provided (Van Horne 1983). These species may have very specific requirements, such as a favoured food plant, and this may be more important than other factors such as plant diversity, width or tree height.

Similarly, there are undesirable species. In Australia there is evidence that planting exotic species attracts introduced birds, and that native plants attract native birds in a much higher proportion than could be expected by chance (Green 1986). It is also probably imprudent to provide species such as the European starling, *Sturnus vulgaris*, with a rich source of overwintering food such as berries in countries such as Australia where the species is considered a pest. Of course there are exceptions to this general principle, as exotic plants can provide useful resources in the form of food, shelter and nest-sites. Indeed, it has been suggested that conservation corridors could be designed to be 'functional mimics' (Lefroy and Hobbs 2000) that may perform a useful role as part of a functioning ecosystem and concomitantly serve the purpose of attracting native fauna, even when non-native plant species are used. They also act as refugia for predatory and parasitic arthropods that are important in pest management.

When selecting plant species, care is needed to provide a balance of food resources. Windbreaks in Australia that contain nectar-producing shrubs and trees provide for nectarivorous birds and insects and also attract insectivorous birds (Kinross 2000). Species that have extended or frequent flowering periods may be particularly important (Biddiscombe 1985). However, excess nectar availability can lead honeyeaters to overwinter in areas where there is insufficient insect food, leading to nutrient deficiencies (Paton et al. 1983). Also, a rich source of nectar can attract large honeyeaters that aggressively drive off smaller birds (Ford et al. 2001). The role of these nectar sources for beneficial invertebrates is likely to be significant, though very little researched. It is clear, however, from other chapters within this book (especially Jervis et al., ch. 5) that plant-derived foods such as nectar and pollen play a critical role in the life-history of many biological control agents.

Landscape complexity and connectivity

As many ecological functions operate at the landscape level, particularly those of metapopulations and corridors, then consideration of the landscape as a whole could be a useful approach when planning for wildlife conservation (Henry et al. 1999; Lindenmayer and Franklin 2003), just as it is in the case of arthropod pest management (Schmidt et al, ch. 4 this volume). Site characteristics are clearly important predictors of wildlife community composition, but may be less important than their spatial arrangement (Burel and Baudry 1990). A single site is likely to contain only a very small subset of the total flora and fauna with the potential to colonise them (Baudry et al. 2000). For example, tall hedges do not suit all species, so it is better to aim for structural diversity in the landscape rather than just within each conservation corridor (Hinsley and Bellamy 2000). Similarly, it is important to realise that diversity at a site scale may not automatically confer diversity at a regional scale (Van Horne 1983).

Improving connectivity and extending habitat

Once it has been established that improving connectivity is both desirable and feasible, and that the width and habitat quality provisions will be considered, the first step is to consider the location and pattern. It may be a simple task of linking two or more isolated patches to form a pattern that will facilitate the dispersal of offspring and provide foraging corridors around the farm. Where there is a choice of placement, it is recommended that corridors be placed along riparian zones in order to optimise conservation goals, as these sites are favoured foraging and dispersal habitats (Fisher and Goldney 1997).

Conversely, it may not be necessary to join two patches, instead extending existing ones to provide additional foraging and breeding habitat. Conservation corridors that adjoin existing remnants contain higher plant species diversity than isolated ones (Kaule and Krebs 1989; Harvey 2000) and can be seen as an extension of habitat for those species. This can also work in reverse and it has been shown that woodlands connected to hedgerows have richer bird communities than those without (Vanhinsbergh et al. 2002).

Finally, it is necessary to consider landscape complexity. A landscape is usually more than a network of corridors. It contains 'nodes' – extra-wide sections of the corridor, small remnant patches or even the intersection point between several corridors (Forman and Godron 1986). Hedgerows that contain nodes are more attractive to birds (Constant et al. 1976; Lack 1988a). Furthermore, when compared with mid-length sections, hedge intersections contain higher densities of birds (Lack 1988a) but also pest predators such as carabid beetles (Joyce et al. 1999) and butterflies (Dover and Sparks 2000). They also provide a higher plant species richness (Riffell and Gutzwiller 1996). Several reasons for this were offered by Lack (1988a): easier defence of territory, efficiency of obtaining food, more shelter, extra protection from predation, increased habitat diversity and less intensive field management in corners. Joyce (1999) added the suggestion that the microclimate in nodes may be similar to that of a forest and may provide suitable sites for reproduction.

Although not always possible for individual small-property owners, it might be useful to move up the scale and consider landscape planning at the level of several square kilometres. It would then be possible to account not only for hedge diversity and nodes but hedgerow network diversity (Baudry et al. 2000). At this level one can consider the location and quantity of source habitat such as remnant woodland and riparian strips that are already in the landscape. If the source habitat is less than 10–30%, as typical of many agricultural areas, it may be below the threshold considered of value to species sensitive to fragmentation (Andrén 1994). This is not to argue that such landscapes should be written off as consisting only of sink habitats, but to raise awareness of the difficulties of the task in such landscapes.

Management practices for conservation corridors

For advice on day-to-day management practices, as opposed to design principles, that will aid wildlife conservation, there is little empirical evidence from research that can help. Most of the recommendations made in research articles and in the 'grey' literature appear to be based on limited research and are often unsubstantiated. However, there are some notable exceptions, particularly in relation to field margins and hedges.

The relationship between different types of treeless field margins and bird foraging was investigated by Vickery et al. (2002). Margins investigated were grass margins (grass-only strips and grass/wildflower strips), naturally regenerated set-aside margins, uncropped wildlife strips, game cover crops and conservation headlands. The authors concluded that the most useful margins for winter food supplies were the game cover crops and the naturally regenerated rotational set-asides. In summer, a supply of invertebrates and seeds was provided by the other types of margins, although the grass-only strips were of lesser value. For sites with trees, floristic and structural diversity were shown to be important factors, although all unsprayed margins were of some benefit and resulted in an increased diversity and abundance of invertebrates (Vickery et al. 2002).

In the UK, the Game Conservancy Trust provides some guidance on how field margins and other farm habitats can be improved for wildlife. At least some of this, such as the structure of 'composite landscape feature' (Figure 13.1), is evidence-based. The Trust recommends sowing grassy field margins and beetle banks with tussock-forming perennial grasses, preferably close to hedgerows, and 2–8 m wide. It also suggests that ploughing of rotational set-asides and mowing of other areas should be left until after the breeding season. Field margins may be enhanced for game birds such as partridges (*Perdix perdix*) by allowing the grasses to grow tall or by providing grassy banks (Boatman 1999). Mowing of field margins can therefore be generally seen as detrimental and can remove nesting areas, but it may also provide opportunities for ground-foraging birds (Vickery et al. 2002).

It is suggested that not all hedgerows be cut at the same time, to provide landscape diversity (Maudsley 2000; Marshall et al. 2001). Similarly, the Game Conservancy Trust suggests that hedgerows should be trimmed every other year (during winter) and may have their density increased by coppicing. However, Maudsley (2000) has observed that coppicing appears to be generally detrimental to invertebrates. Hedge habitat may be enhanced for butterflies by ensuring that the base of hedges is not regularly cut (Dover and Sparks 2000). Hedge-bottom structural and floristic diversity can also be improved by resowing (Maudsley 2000).

Windbreaks and shelterbelts often lack forest animal species due to the paucity of woody debris, hollow logs and nest-hollows. These could be provided by translocation of woodpiles, which would otherwise be burnt, and erection of artificial nest-boxes (Majer et al. 2001). For invertebrate pollinators, nest boxes on farmland have proved useful (Barron et al. 2000). It has also been noticed that avenues of trees and shrubs leading to farmhouses in Australia have a higher proportion of ground-foraging birds such as robins and some thornbills than do windbreaks that are highly densely packed (CK, personal observation). This remains to be tested, but it is logical to assume that ground-foraging birds require clear ground space close to or within the windbreak, so they have both foraging substrate and shelter close by. Having a gap within the windbreak also aids access for maintenance and pest control, so is worth considering, providing that the ecological outcomes are monitored.

Conclusion

What can be concluded from the literature about the utility of conservation corridors as wildlife habitat? There is clearly concern that windbreaks could be composed entirely of edge habitat, thus acting as landscape sinks in which reproduction is insufficient to replace loss through predation and other causes. Most authors concede that conservation corridors, even though they provide useful habitat for generalist species and forest-edge species, provide secondary habitat compared with remnant woodland or forest and may even impede the conservation of some species, particularly grassland birds. Few species with specialised habitat requirements were recorded utilising revegetated areas in Australia (Ryan 1999) and even fewer have been observed breeding (Yahner 1982b), despite high densities of both birds and nests. Conservation corridors without trees are even less likely to be providing for a range of species that are regionally declining or threatened with extinction.

On the other hand, Webb (1988) suggested that windbreaks and similar vegetation can provide surrogate habitat for most species which occurred in the area prior to clearing, and the general conclusion is that they are good for wildlife (Schwilling 1982). In Britain, many birds have been able to adapt to different conditions over time and new habitats have been colonised, permitting a widening of the species' tolerance (Williamson 1967). Certainly, many species in the northern hemisphere are now considered residents of windbreaks and hedges.

Howe and Davis (1991) suggested that small sinks can be valuable to overall population survival, given good connectivity and excellent habitat quality, and providing that the source population is itself secure. The other point worth noting is that, even if windbreaks attract only edge specialists, some of these are now known to be in decline and require management (Keller et al. 1993). This is a very important point. In the past very little was known about habitat enhancement and so conservation corridors benefited only ubiquitous species (Hill et al. 1995). At the same time, once-common species are becoming scarce through mismanagement of the countryside. While there are still many gaps, particularly regarding invertebrates, there is a much greater understanding of how landscapes can be designed with a view to arresting the decline of wild plant and animal species (Hill et al. 1995).

There may also be substantial benefits for pest control. The relationship between provision of habitat for beneficial species and the suppression of pest species may be operating as a positive feedback loop. The literature indicates that most species of birds and bats utilising conservation corridors are at least partially insectivorous. It has been demonstrated in Canada that field margins do not contribute significantly as breeding habitats of bird species that may damage crops. Furthermore, they provide shelter to species that may contribute to biological pest control (Jobin et al. 2001).

The design of landscapes that provide habitat diversity, species diversity and ecosystem services (as well as being self-reproducing) is not an exact science. Conservation biologists have a fairly good understanding of factors such as the proportion of conservation land versus production land at the catchment scale for goals such as hydrogeological balance and vertebrate species viability (Bennett 1999; Platt 2002). At the farm scale, however, much still needs to be learnt. Although linear plantings such as hedges, shelterbelts and windbreaks are not currently optimal habitat, they are providing a useful resource and making a positive contribution to the abundance and diversity of vertebrate and invertebrate wildlife in agricultural areas. Future agricultural landscapes may more frequently reflect the types of features and level of connectivity shown in Figure 13.2. While some of these features have provided suboptimal habitat in the past, it is becoming increasingly clear that the quality of on-farm habitats, and thereby the long-term viability of wildlife populations, can be improved with only minimal modifications or cost to the primary producer. As knowledge of this form of ecological engineering advances it will contribute to the ecological sustainability of agricultural landscapes through enhanced wildlife conservation and, often, reduced densities of agricultural pests and the environmental impacts associated with conventional pest control measures.

References

Allan, D.G., Harrison, J.A., Navarro, R.A., van Wilgen, B.W. and Thompson, M.W. (1997). The impact of commercial afforestation on bird populations in Mpumalanga Province, South Africa – insights from Bird-Atlas data. *Biological Conservation* 79: 173–185.

Andrén, H. (1994). Effects of habitat fragmentation on birds and mammals in landscapes with different proportions of suitable habitat: a review. *Oikos* 71: 355–366.

Arnold, G. (1995). Incorporating landscape pattern into conservation programs. In *Mosaic Landscapes and Ecological Processes* (L. Hansson, L. Fahrig and G. Merriam, eds), pp. 293–308. Chapman & Hall, London.

Balát, F. (1986). The avian component of a well-established windbreak in the Breclav area of Czechoslovakia. *Folia Zoologica* 35: 229–238.

Barron, M.C., Wratten, S.D. and Donovan, B.J. (2000). A four-year investigation into the efficacy of domiciles for enhancement of bumble bee populations. *Agricultural and Forest Entomology* 2: 141–146.

Baudry, J., Bunce, R.G.H. and Burel, F. (2000). Hedgerows: an international perspective on their origin, function and management. *Journal of Environmental Management* 60: 7–22.

Baur, A. and Baur, B. (1992). Effect of corridor width on animal dispersal: a simulation study. *Global Ecology and Biogeography Letters* 2: 52–56.

Bennett, A.F. (1990). *Habitat Corridors: Their Role in Wildlife Management and Conservation*. Department of Conservation and Environment, Melbourne, Australia.

Bennett, A.F. (1999). *Linkages in the Landscape: The Role of Corridors and Connectivity in Wildlife Conservation*. International Union for the Conservation of Nature, World Conservation Union, Cambridge.

Berg, Å. and Pärt, T. (1994). Abundance of breeding farmland birds on arable and set-aside fields at forest edges. *Ecography* 17: 147–152.

Bickmore, C.J. (2001). The function of hedges in England and Wales – reasons for their establishment. In *Hedgerows of the World: Their Ecological Functions in Different Landscapes* (C. Barr and S. Petit, eds), pp. 329–338. IALE, Guleph.

Biddiscombe, E.F. (1985). Bird populations of farm plantations in Hotham River Valley, W.A. *Western Australian Naturalist* 16: 32–39.

Boatman, N. (1998). *The Allerton Project. Game Conservancy Trust Review 1997*, pp. 61–65. Game Conservancy Ltd, Fordingbridge, UK.

Boatman, N. (1999). *Marginal benefits? How field edges and beetle banks contribute to game and wildlife conservation. Game Conservancy Trust Review 1998*. Game Conservancy Ltd, Fordingbridge, UK.

Bonsall, M.B. and Hassell, M.P. (2000). The effects of metapopulation structure on indirect interactions in host–parasitoid assemblages. *Proceedings of the Royal Society of London B. Biological Sciences* 267: 2207–2212.

Brawn, J.D. and Robinson, S.K. (1996). Source-sink population dynamics may complicate interpretation of long-term census data. *Ecology* 77: 3–12.

Bright, P.W. (1998). Behaviour of specialist species in habitat corridors: arboreal dormice avoid corridor gaps. *Animal Behaviour* 56: 1485–1490.

Brown, J.H. and Kodric-Brown, A. (1977). Turnover rates in insular biogeography: effect of immigration on extinction. *Ecology* 58: 445–449.

Bryja, J. and Zukal, J. (2000). Small mammal communities in newly planted biocorridors and their surroundings in southern Moravia (Czech Republic). *Folia Zoologica* 49: 191–197.

Burel, F. and Baudry, J. (1990). Hedgerow networks as habitats for forest species: implications for colonising abandoned agricultural land. In *Species Dispersal in Agricultural Land* (R.G.H. Bunce and D.C. Howard, eds), pp. 238–255. Belhaven Press, London.

Capel, S.W. (1988). Design of windbreaks for wildlife in the Great Plains of North America. *Agriculture, Ecosystems and Environment* 22/23: 337–347.

Cassel, J.F. and Wiehe, J.M. (1980). Uses of shelterbelts by birds. In *Management of Western Forests and Grasslands for Nongame Birds* (R.M. DeGraaf, ed.), pp. 78–87. USDA Forest Service.

Chapman, C.A. and Chapman, L.J. (1999). Forest restoration in abandoned agricultural land: a case study from East Africa. *Conservation Biology* 13: 1301–1311.

Clarke, J.H., Jones, N.E., Hill, D.A. and Tucker, G.M. (1997). The management of set-aside within a farm and its impact on birds. *Proceedings of the 1997 International Brighton Crop Protection Conference: Weeds*. 17–20 November, Brighton, UK. British Crop Protection Council, Farnham.

Conover, M.R. (1998). Perceptions of American agricultural producers about wildlife on their farms and ranches. *Wildlife Society Magazine* 26: 597–604.

Constant, P.M., Eybert, C. and Mahed, R. (1976). Avifaune reproductrice du bocage de l'Ouest. In *Les Bocages: Histoire, Ecologie, Economie* (M.J. Missonnier, ed.), pp. 327–332. University of Rennes, France.

Corbit, M., Marks, P.L. and Gardescu, S. (1999). Hedgerows as habitat corridors for forest herbs in central New York, USA. *Journal of Ecology* 87: 220–232.

Crabb, J., Firbank, L., Winter, M., Parham, C. and Dauven, A. (1998). Set-aside landscapes: farmer perceptions and practices in England. *Landscape Research* 23: 237–254.

Demers, M.N., Simpson, J.W., Boerner, R.E.J., Silva, A., Berns, L. and Artigas, F. (1995). Fencerows, edges and implications of changing connectivity illustrated by two contiguous Ohio landscapes. *Conservation Biology* 9: 1159–1168.

Dover, J. and Sparks, T. (2000). A review of the ecology of butterflies in British hedgerows. *Journal of Environmental Management* 60: 51–63.

Downes, S.J., Handasyde, K.A. and Elgar, M.A. (1997). Variation in the use of corridors by introduced and native rodents in south-eastern Australia. *Biological Conservation* 82: 379–383.

Emmerich, J.M. and Vohs, P.A. (1982). Comparative use of four woodland habitats by birds. *Journal of Wildlife Management* 46: 43–49.

Estades, C.F. and Temple, S.A. (1999). Deciduous-forest bird communities in a fragmented landscape dominated by exotic pine plantations. *Ecological Applications* 9: 573–585.

Fisher, A.M. and Goldney, D.C. (1997). Use of birds of riparian vegetation in an extensively fragmented landscape. *Pacific Conservation Biology* 3: 275–288.

Ford, H.A., Barrett, G.W., Saunders, D.A. and Recher, H.F. (2001). Why have birds in the woodlands of Southern Australia declined? *Biological Conservation* 97: 71–88.

Forman, R.T.T. and Godron, M. (1986). *Landscape Ecology*. John Wiley & Sons, New York.

Gall, G.A.E. and Orians, G.H. (1992). Agriculture and biological conservation. *Agriculture, Ecosystems and Environment* 42: 1–8.

Green, R.E., Osborne, P.E. and Sears, E.J. (1994). The distribution of passerine birds in hedgerows during the breeding season in relation to characteristics of the hedgerow and adjacent farmland. *Journal of Applied Ecology* 31: 677–692.

Green, R.J. (1986). Native and exotic birds in the suburban habitat. In *The Dynamic Partnership: Birds and Plants in Southern Australia* (H.A. Ford and D.C. Paton, eds). Government Printer, South Australia.

Gruttke, H. and Kornacker, P.M. (1995). The development of epigeic fauna in new hedges – a comparison of spatial and temporal trends. *Landsape and Urban Planning* 31: 217–231.

Haas, C.A. (1995). Dispersal and use of corridors by birds in wooded patches on an agricultural landscape. *Conservation Biology* 9: 845–854.

Hamel, P.B. (2003). Winter bird community differences among methods of bottomland hardwood forest restoration: results after seven growing seasons. *Forestry* 76: 189–197.

Hanski, I. (1997). Metapopulation dynamics: from concepts to observations and models. In *Metapopulation Biology: Ecology Genetics and Evolution* (I.A. Hanski and M.E. Gilpin, eds), pp. 69–91. Academic Press, San Diego.

Harris, L.D. (1984). *The Fragmented Forest: Island Biogeographic Theory and the Preservation of Biotic Diversity*. University of Chicago Press, Chicago.

Harvey, C.A. (2000). Colonization of agricultural windbreaks by forest trees: effects of connectivity and remnant trees. *Ecological Applications* 10: 1762–73.

Henry, A.C., Hosack, D.A., Johnson, C.W., Rol, D. and Bentrup, G. (1999). Conservation corridors in the United States: benefits and planning guidelines. *Journal of Soil and Water Conservation* 54: 645–650.

Hess, G.R. and Bay, J.M. (2000). A regional assessment of windbreak habitat suitability. *Environmental Monitoring and Assessment* 61: 237–254.

Hill, D.A., Andrews, J., Sotherton, N. and Hawkins, J. (1995). Farmland. In *Managing Habitats for Conservation* (W.J. Sutherland and D.A. Hill, eds), pp. 230–266. Cambridge University Press, Cambridge.

Hinsley, S.A. and Bellamy, P.E. (2000). The influence of hedge structure, management and landscape context on the value of hedgerows to birds: a review. *Journal of Environmental Management* 60: 33–49.

Howe, R.W. and Davis, G.J. (1991). The demographic significance of 'sink' populations. *Biological Conservation* 57: 239–255.

Inglis, G. and Underwood, A.J. (1992). Comments on some designs proposed for experiments on the biological importance of corridors. *Conservation Biology* 6: 581–586.

Jervis, M.A. (1997). Metapopulation dynamics and the control of mobile agricultural pests: fresh insights. *International Journal of Pest Management* 43: 251–252.

Jobin, B., Choinière, L. and Bélanger, L. (2001). Bird use of three types of field margins in relation to intensive agriculture in Québec, Canada. *Agriculture, Ecosystems and Environment* 84: 131–143.

Johnson, M.D. (2000). Effects of shade-tree species and crop structure on the winter arthropod and bird communities in a Jamaican shade coffee plantation. *Biotropica* 32: 133–145.

Johnson, W.C. and Adkission, C.S. (1985). Dispersal of beech nuts by Blue Jays in fragmented landscapes. *American Midland Naturalist* 113: 319–324.

Joyce, K.A., Holland, J.M. and Doncaster, C.P. (1999). Influences of hedgerow intersections and gaps on the movement of carabid beetles. *Bulletin of Entomological Research* 89 (6): 523–531.

Kaule, G. and Krebs, S. (1989). Creating new habitats in intensively used farmland. In *Biological Habitat Reconstruction* (G.P. Buckley, ed.). Belhaven Press, London.

Keller, C.M.E., Robbins, C.S. and Hatfield, J.S. (1993). Avian communities in riparian forests of different widths in Maryland and Delaware. *Wetlands* 13: 137–144.

Kilgo, J.C., Sargent, R.A., Chapman, B.R. and Miller, K.V. (1998). Effect of stand width and adjacent habitat on breeding bird communities in bottomland hardwoods. *Journal of Wildlife Management* 62: 72–83.

Kinross, C.M. (2000). The ecology of bird communities in windbreaks and other avian habitats on farms. Ph.D. thesis. Charles Sturt University, Bathurst, New South Wales.

Kučera, J.C.N. (1992). Contribution of knowledge of small terrestrial mammals in southern Moravia agroecoenoses. *Folia Zoologica* 41: 221–231.

Kwok, H.K. and Corlett, R.T. (2000). The bird communities of a natural secondary forest and a Lophostemon confertus plantation in Hong Kong, South China. *Forest Ecology and Management* 130: 227–234.

Lack, P.C. (1988a). Hedge intersections and breeding bird distributions in farmland. *Bird Study* 35: 133–136.

Lack, P.C. (1988b). Nesting success of birds in trees and shrubs in farmland hedges. *Ecological Bulletin* 39: 191–193.

Lagerlof, J., Goffre, B. and Vincent, C. (2002). The importance of field boundaries for earthworms (Lumbricidae) in the Swedish agricultural landscape. *Agriculture, Ecosystems and Environment* 89: 91–103.

Lakhani, K.H. (1994). The importance of field margin attributes to birds. In *Field Margins: Integrating Agriculture and Conservation* (N. Boatman, ed.), pp. 77–84. British Crop Protection Council, University of Warwick.

Lefroy, T. and Hobbs, R. (2000). Functional mimics of natural ecosystems as a basis for sustainable agriculture. In *Nature Conservation 5: Nature Conservation in Production Environments* (J.L. Craig, N. Mitchell and D.A. Saunders, eds), pp. 179–187. Surrey Beatty & Sons, Sydney.

Legány, D.A. (1991). Significance of shelter-belts and rows of trees in respect of ornithology and nature conservancy. *Aquila* 98: 169–180.

Levins, R. (1969). Some demographic and genetic consequences of environmental heterogeneity for biological control. *Bulletin of the Entomological Society of America* 15: 237–240.

Lindenmayer, D.B. and Franklin, J.F. (2002). *Conserving Forest Biodiversity: A Comprehensive Multi-scaled Approach.* Island Press, Washington.

Lubke, R.A., Avis, A.M. and Moll, J.B. (1996). Post-mining rehabilitation of coastal sand dunes in Zululand, South Africa. *Landscape and Urban Planning* 34: 335–345.

MacDonald, D.W. and Johnson, P.J. (1995). The relationship between bird distribution and the botanical and structural characteristics of hedges. *Journal of Applied Ecology* 32: 492–505.

Machtans, C.S., Villard, M.-A. and Hannon, S.J. (1996). Use of riparian buffer strips as movement corridors by forest birds. *Conservation Biology* 10: 1366–1379.

Mader, H.J. (1988). Effects of increased spatial heterogeneity on the biocenosis in rural landscapes. *Ecological Bulletin* 39: 169–179.

Mader, H.J., Schell, C. and Kornacker, P. (1990). Linear barriers to arthropod movements in the landscape. *Biological Conservation* 54: 209–222.

Maisonneuve, C. and Rioux, S. (2001). Importance of riparian habitats for small mammal and herpetofaunal communities in agricultural landscapes of southern Quebec. *Agriculture, Ecosystems and Environment* 83: 165–175.

Majer, J.D. (1990). Rehabilitation of disturbed land: long-term prospects for the recolonization of fauna. In *Australian Ecosystems: 200 Years of Utilization, Degradation and Reconstruction* (D.A. Saunders, A.J.M. Hopkins and R.A. How, eds). Surrey Beatty & Sons, Sydney.

Majer, J.D., Recher, H.F., Graham, R. and Watson, A. (2001). The potential of revegetation programs to encourage invertebrates and insectivorous birds. *Bulletin 20*, School of Environmental Biology, Curtin University of Technology.

Marshall, E.J.P., Maudsley, M.J., West, T.M. and Rowcliffe, H.R. (2001). Effects of management on the biodiversity of English hedgerows. In *Hedgerows of the World: Their Ecological Functions in Different Landscapes. Proceedings of the European IALE Congress* (C. Barr and S. Petit, eds), pp. 361–365. University of Birmingham, UK.

Maudsley, M.J. (2000). A review of the ecology and conservation of hedgerow invertebrates in Britain. *Journal of Environmental Management* 60: 65–76.

Merriam, G. and Saunders, D.A. (1993). Corridors in restoration of fragmented landscapes. In *Nature Conservation 3: Reconstruction of Fragmented Ecosystems: Global and Regional Perspectives* (D.A. Saunders, R.J. Hobbs and P.R. Ehrlich, eds), pp. 71–87. Surrey Beatty & Sons, Sydney.

Morgan, K.A. and Gates, J.E. (1982). Bird population patterns in forest edge and strip vegetation at Remington Farms, Maryland. *Journal of Wildlife Management* 46: 933–944.

Nicholls, A.O. and Margules, C.R. (1991). The design of studies to demonstrate the biological importance of corridors. In *Nature Conservation 2: The Role of Corridors* (D.A. Saunders and R.J. Hobbs, eds). Surrey Beatty & Sons, Sydney.

Nichols, O.G. and Nichols, F.M. (2003). Long-term trends in faunal colonisation after bauxite mining in the jarrah forest of southwestern Australia. *Restoration Ecology* 11: 261–272.

O'Connor, R.J. (1984). The importance of hedges to songbirds. In *Agriculture and the Environment: Proceedings of ITE Symposium No. 13* (D. Jenkins, ed.), pp. 117–123. Institute of Terrestrial Ecology, Cambridge, Monks Wood Experimental Station.

Opdam, P. (1990). Dispersal in fragmented populations: the key to survival. In *Species Dispersal in Agricultural Habitats* (R.G.H. Bunce and D.C. Howard, eds). Belhaven Press, London.

Osborne, P. (1984). Bird numbers and habitat characteristics in farmland hedgerows. *Journal of Applied Ecology* 21: 63–82.

Paoletti, M.G., Pimentel, D., Stinner, B.R. and Stinner, D. (1992). Agroecosystem biodiversity: matching production and conservation biology. *Agriculture, Ecosystems and Environment* 40: 3–23.

Parish, T., Lakhani, K.H. and Sparks, T.H. (1994). Modelling the relationship between bird population variables and hedgerow and other field margin attributes. I. Species richness of winter, summer and breeding birds. *Journal of Applied Ecology* 31: 764–775.

Passell, H.D. (2000). Recovery of bird species in minimally restored Indonesian tin strip mines. *Restoration Ecology* 8: 112–118.

Paton, D.C., Dorward, D.F. and Fell, P. (1983). Thiamine deficiency and winter mortality in Red Wattlebirds, Anthochaera carunculata (Aves: Meliphagidae) in suburban Melbourne. *Australian Journal of Zoology* 31: 147–154.

Petrides, G. (1942). Relation of hedgerows in winter to wildlife in central New York. *Journal of Wildlife Management* 6: 261–280.

Pierce, R.A., Farrand, D.T. and Kurtz, W.B. (2001). Projecting the bird community response resulting from the adoption of shelterbelt agroforestry practices in Eastern Nebraska. *Agroforestry Systems* 53: 333–350.

Pimentel, D., Stachow, U., Takacs, D.A., Brubaker, H.W., Dumas, A.R., Meaney, J.J., O'Neil, J.A.S., Onsi, D.E. and Corzilius, D.B. (1992). Conserving biological diversity in agricultural/forestry systems. *Bioscience* 42: 354–362.

Pimentel, D., Wilson, C., McCullum, C., Huang, R., Dwen, P., Flack, J., Tran, Q., Saltman, T. and Cliff, P. (1997). Economic and environmental benefit of biodiversity. *Bioscience* 47: 747–757.

Platt, S.J. (2002). *How to Plan Wildlife Landscapes: A Guide for Community Organisations.* Department of Natural Resources and Environment, Melbourne.

Pollard, E., Hooper, M.D. and Moore, N.W. (1974). *Hedges.* Collins, London.

Pulliam, H.R. (1988). Sources, sinks and population regulation. *American Naturalist* 132: 652–661.

Pullin, A.S. (2002). *Conservation Biology.* Cambridge University Press, Cambridge.

Rands, M.R.W. (1986). Effect of hedgerow characteristics on partridge breeding densities. *Journal of Applied Ecology* 23: 479–487.

Reid, J.R.W. (1999). *Threatened and Declining Birds in the New South Wales Sheep-Wheat Belt: 1). Diagnosis, Characteristics and Management.* Consultancy Report to NSW National Parks and Wildlife Service. CSIRO Wildlife and Ecology, Canberra.

Richards, P.W.N. (1928). Ecological notes on the bryophytes of Middlesex. *Journal of Ecology* 16: 267–300.

Riffell, S.K. and Gutzwiller, K.J. (1996). Plant-species richness in corridor intersections – is intersection shape influential? *Landscape Ecology* 11: 157–168.

Rishi, L.B. and Sandhu, J.S. (1999). Role of eucalyptus woodlots in owl conservation in Punjab. *Tigerpaper* 26: 15–17.

Robinson, M.F. and Stebbings, R.E. (1997). Home range and habitat use by the serotine bat, Eptesicus serotinus, in England. *Journal of Zoology* 243: 117–136.

Rodewald, A.D. (2002). Nest predation in forested regions: landscape and edge effects. *Journal of Wildlife Management* 66: 634–640.

Ryan, P.A. (1999). The use of revegetated areas by vertebrate fauna in Australia: a review. In *Temperate Eucalypt Woodlands in Australia: Biology, Conservation, Management and Restoration* (R.J. Hobbs and C.J. Yates, eds), pp. 318–335. Surrey Beatty, Sydney.

Ryszkowski, L., Karg, J. and Bernacki, Z. (2003). Biocenotic function of the mid-field woodlots in West Poland: study area and research assumptions. *Polish Journal of Ecology* 51: 269–281.

Saunders, D.A. and Rebeira, C.P. (1991). Values of corridors to avian populations in a fragmented landscape. In *Nature Conservation 2: The Role of Corridors* (D.A. Saunders and R.J. Hobbs, eds), pp. 221–240. Surrey Beatty & Sons, Sydney.

Schmeigelow, F.K.A., Machtans, C.S. and Hannon, S.J. (1997). Are boreal birds resilient to forest fragmentation? An experimental study of short-term community responses. *Ecology* 78: 1914–1932.

Schwilling, M.D. (1982). Nongame wildlife in windbreaks. In Forestry Committee of the Great Plains Agricultural Council.

Shafer, C.L. (1997). Terrestrial nature reserve design at the urban/rural interface. In *Conservation in Highly Fragmented Landscapes* (M.W. Schwartz, ed.), pp. 345–378. Chapman & Hall, New York.

Shalaway, S.D. (1985). Fencerow management for nesting birds in Michigan. *Wildlife Society Bulletin* 13: 302–306.

Simberloff, D. (1995). Habitat fragmentation and population extinction of birds. *Ibis* 137: S105–S111.

Simberloff, D.S. and Cox, J. (1987). Consequences and costs of conservation corridors. *Conservation Biology* 1: 63–71.

Spackman, S.C. and Hughes, J.W. (1995). Assessment of minimum stream corridor width for biological conservation: species richness and distribution along mid-order streams in Vermont, USA. *Biological Conservation* 71: 325–332.

Stoate, C. (1998). The role of game management in songbird conservation. *Game Conservancy Trust Review* 1998: 68–73.

Stoate, C., Morris, R.M. and Wilson, J.D. (2001). Cultural ecology of Whitethroat (Sylvia communis) habitat management by farmers: field-boundary vegetation in lowland England. *Journal of Environmental Management* 62: 329–341.

Stoate, C., Brockless, M.H. and Boatman, N.D. (2002). A multifunction approach to bird conservation on farmland: a ten-year appraisal. *Aspects of Applied Biology* 67: 191–196.

Šustek, Z. (1994). Windbreaks as migration corridors for carabids in an agricultural landscape. In *Carabid Beetles: Ecology and Evolution* (K. Desender, ed.), pp. 377–382. Kluwer Academic, Dordrecht.

Tattersall, F.H., Tew, T.E. and Macdonald, D.W. (1998). Habitat selection by arable woodmice: a review of work carried out by the wildlife conservation research unit. *Proceedings of the Latvian Academy of Science Section* B 52: 31–36.

Taws, N. (2000). The Greening Australia Birdwatch project. *Canberra Bird Notes* 25: 89–93.

Tew, T.E., MacDonald, D.W. and Rands, M.R.W. (1992). Herbicide application affects microhabitat use by arable wood mice. *Journal of Applied Ecology* 29: 532–539.

The Wildlife Trust (2003). Practical biodiversity – article 1. On-line at url: http://www.ukagriculture.com/conservation/farming matters/.

Trnka, P., Rozkosny, R., Gaisler, J. and Houskova, L. (1990). Importance of windbreaks for ecological diversity in agricultural landscape. *Ecology* (CSSR) 9: 241–258.

van Aarde, R.J., Ferreira, S.M., Kritzinger, J.J., Vandyk, P.J., Vogt, M. and Wassenaar, T.D. (1996). An evaluation of habitat rehabilitation on coastal dune forests in northern Kwazulu-Natal, South Africa. *Restoration Ecology* 4: 334–345.

van Dorp, D. and Opdam, P.F.M. (1987). Effects of patch size isolation and regional abundance on forest bird communities. *Landscape Ecology* 1: 59–73.

Vanhinsbergh, D., Gough, S., Fuller, R.J. and Brierley, E.D.R. (2002). Summer and winter bird communities in recently established farm woodlands in lowland England. *Agriculture Ecosystems and Environment* 92: 123–136.

Van Horne, B. (1983). Density as a misleading indicator of habitat quality. *Journal of Wildlife Management* 47: 893–901.

Vickery, J., Carter, N. and Fuller, R.J. (2002). The potential value of managed cereal field margins as foraging habitats for farmland birds in the UK. *Agriculture, Ecosystems and Environment* 89: 41–52.

Webb, R. (1988). The status of hedgerow field margins in Ireland. In *Environmental Management in Agriculture: European Perspectives* (J.R. Park, ed.), pp. 125–131. Belhaven Press, London.

Wegner, J.F. and Merriam, G. (1979). Movements of birds and small mammals between a wood and adjoining farmland habitats. *Journal of Applied Ecology* 16: 349–357.

Williamson, K. (1967). Buntings on a barley farm. *Bird Study* 15: 34–37.

Wilson, A.-M. and Lindenmayer, D. (1996). Wildlife corridors – their potential role in the conservation of biodiversity in rural Australia. *Australian Journal of Soil and Water Conservation* 9: 22–28.

With, K.A. and Crist, T.O. (1995). Critical thresholds in species' responses to landscape structure. *Ecology* 76: 2446–2459.

Wratten, S.D. (1992). Weeding out the cereal killers. *New Scientist* August: 33–37.

Wratten, S.D., Bowie, M.H., Hickman, J.M. Evans, Sedcole, J.R. and Tylianakis, J.M. (2003). Field boundaries as barriers to movement of hover flies (Diptera: Syrphidae) in cultivated land. *Oecologia* 134: 605–611.

Yahner, R.H. (1982a). Microhabitat use by small mammals in farmstead shelterbelts. *Journal of Mammalogy* 63: 440–445.

Yahner, R.H. (1982b). Avian nest densities and nest site selection in farmstead shelterbelts. *Wilson Bulletin* 94: 156–175.

Yahner, R.H. (1983). Seasonal dynamics, habitat relationships and management of avifauna in farmstead shelterbelts. *Journal of Wildlife Management* 47: 85–104.

Yosef, R. (1994). The effects of fencelines on the reproductive success of Loggerhead Shrikes. *Conservation Biology* 8: 281–285.

Chapter 14

Ecological engineering for enhanced pest management: towards a rigorous science

Geoff M. Gurr, Steve D. Wratten and Miguel A. Altieri

Introduction

In the context of ecological engineering for arthropod pest management, the contributions to this book provide comprehensive evidence that diversity is a powerful tool for pest suppression in agricultural systems. Indeed, attempts to quantify the value of biodiversity in terms of ecosystem services generate large dollar values (e.g. Costanza et al. 1997; Pimentel 1997; see Table 14.1), including pest suppression in crops worth $US100 billion worldwide per annum.

Farmers are increasingly aware of the ecosystem services performed by agricultural biodiversity; indigenous and peasant farmers in the developing world have always relied on biodiversity for agroecosystem function. Others have been influenced by broader regulatory schemes such as 'set-aside' (Crabb et al. 1998), the conservation reserve program (Frawley and Walters 1996), 'LEAF' (Linking Environment and Farming) (Drummond 2002) and various payment systems that have been proposed to reward farmers for 'producing nature' (e.g. Slangen 1997; Musters et al. 1999). The 2001 foot-and-mouth epidemic in Britain raised awareness, among farmers, the broader public and policy-makers, of the direct economic value of farmlands for various amenity uses including farm-based tourism.

Organic farming, for instance, can benefit wildlife such as birds (Beecher et al. 2002; Mader et al. 2002) but, as pointed out by Vickery (2002), such whole-farm approaches are in the minority. Alternative strategies that can readily be incorporated into conventional farming systems are important. These do exist, and examples of farming practices and landscape features that favour biodiversity have been comprehensively reviewed by Paoletti (1999). They include the following:

- hedgerows;
- polycultures;
- agroforestry;
- herbacious strips within crops;
- appropriate field margins;
- small fields surrounded by hedgerows.

It is no coincidence that all the practices listed have been mentioned by contributors to this volume in the context of ecological engineering for pest management. In fact, many farmers and

Table 14.1: Estimated annual value of farm-related benefits of ecosystem services and related resources from biodiversity. Estimates are from a broader inventory with a total value of 2928×10^9.

Ecosystem service/activity	US value ($US $\times 10^9$)	Worldwide value ($US $\times 10^9$)
Waste disposal	62	760
Soil formation	5	25
Nitrogen fixation	8	90
Bioremediation of chemicals	22.5	121
Crop breeding (genetics)	20	115
Livestock breeding (genetics)	20	40
Biotechnology	2.5	6
Biocontrol of crop pests	12	100
Host plant resistance in crops	8	80
Pollination	40	200
Fishing	29	60
Hunting	12	25
'Wild' foods	0.5	180
Wood products	8	84
Ecotourism	18	500
Pharmaceuticals from plants	20	84
Forest sequestration of carbon dioxide	6	135

Source: Pimentel et al. (1997).

practitioners have observed that the following features linked to agroecosystem biodiversity tend to enhance indigenous natural enemies and suppress pests in crops:

- organically managed or subjected to minimal inputs;
- diversified cropping systems;
- presence of flowering plants within or around the field;
- soils covered with mulch or vegetation in the off-season;
- presence of perennial plants;
- small fields surrounded by natural vegetation.

Biodiversity is undeniably a powerful tool for pest management, but it is not consistently beneficial and the discipline of ecological engineering is in the process of moving from a 'first approximation' – the simplistic assumption that diversity per se is beneficial. In a landmark review of the relevant literature, Andow (1991) determined that pests tended to be less abundant in polycultures than in monocultures in 48.5% of cases in annual crops systems and 60.5% of perennial crop cases. In the remaining cases (close to half of those in the literature) pest population densities were either unchanged, responded variably or were increased in polycultures. Further, as would be expected, a suppressive effect of polycultures on polyphagous pest species was far less commonly reported than was the case for monophagous pests. Combining statistics for annual and perennial crop systems, lowered pest densities were apparent in 63% of monophagous pests species but in only 23% of polyphagous pests.

Clearly, vegetational diversity is no guarantee of lowered pest densities, either by invoking resource concentration or enemies hypotheses. Indeed, recent studies have illustrated some ways

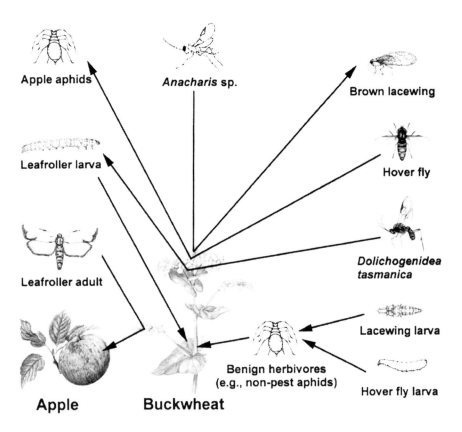

Figure 14.1: Summary of potential effects on arthropods of adding a buckwheat groundcover to an apple orchard (see text for discussion) (Wratten et al. 2000).

in which adding plant diversity to agroecosystems can have negative consequences. Figure 14.1 illustrates some of the possible interactions in ecological engineering of the New Zealand apple agroecosystem. The planting of a cover crop species such as buckwheat provides pollen and/or nectar resources for *Dolichogenidea tasmanica*, the key parasitoid of an important leafroller pest, *Epiphyas postvittana*. This parasitoid oviposits in early-instar larvae of the leafroller and its efficacy can be enhanced by providing adult foods. However, without laboratory or field analysis, many questions would remain unanswered. These questions include:

- whether buckwheat is the most beneficial food plant species for *D. tasmanica*;
- whether buckwheat provides other benefits, such as a moderated microclimate or supports alternative hosts, for *D. tasmanica*;
- whether agonists of beneficial insects – such as *Anacharis*, a parasitoid of the brown lacewing – derive a benefit from buckwheat (as found by Stephens et al. 1998);
- whether the target pest derives a benefit from buckwheat, either as larvae or from adult feeding on nectar (as found for potato tuber moth, *Phthorimaea operculella*, by Baggen and Gurr 1998).

The points listed above can contribute to the outcomes of vegetational diversification and account for the lack of 'success' in many cases (Andow 1991). Further complications arise when

Figure 14.2: 'Chocolate-box ecology'? Contrasting farmscapes and approaches to pest management in British wheat crops. (top) Conventional large monoculture, little non-crop vegetation and reliance on pesticidal control of pests (Photograph: G.M. Gurr). (bottom) Experimental use of field margin vegetation (*Phacelia tanacetifolia*) for enhanced biocontrol of pests (Hickman and Wratten 1996) (Photograph: S.D. Wratten).

considering the effects of buckwheat on other pest species and their natural enemies (such as apple aphids and hoverflies, respectively, in Figure 14.1). Clearly, trophic interactions of these types are important in the selection of plants in ecological engineering.

'Chocolate-box ecology'?

Habitat manipulation for enhanced pest control has been referred to as 'chocolate-box ecology' because of the picturesque nature of some of the tools used, for example strips of flowers (Figure 14.2). In some cases, floristically diverse vegetation is added without prior testing and ranking of the candidate plants, but this crude approach is not universal and habitat manipulation researchers now more commonly screen plant species to determine optimal species or use a range of selection criteria to determine appropriate botanical composition (Gurr et al. 1998; Pfiffner and Wyss, ch. 11 this volume). These approaches reflect the notion that the quality, not the quantity, of diversity is important (Polasezek et al. 1999). This requires the selection of the 'right kind' of diversity (Intachat 1998). This is illustrated by the work of Tooker and Hanks (2000), that showed that parasitic Hymenoptera tended to visit only a limited number of food plants – a mean of 2.9 plant species per parasitoid species. Therefore, providing nectar to a parasitoid of a key pest could fail unless an appropriate food plant species is identified by appropriate research.

'Directed' approaches to habitat manipulation

As argued in detail by Gurr et al. (2004), for habitat manipulation to evolve into a rigorous science and become a branch of ecological engineering, it needs to employ 'directed' approaches rather than rely on the 'shotgun' approach of simply increasing vegetational diversity that is implied as desirable, for example in the existence of commercial seed mixes containing many untested plant species. Directed approaches include experimental studies to identify optimal forms of diversity and to better understand the ways in which natural enemies respond to vegetational diversity at local and landscape levels. Jervis et al. (ch. 5 this volume)and Schmidt et al. (ch. 4 this volume) provide examples of such work, while Khan and Pickett (ch. 10 this volume) show how the contribution of bottom-up (i.e. direct effects of plants on herbivores) effects can be optimised by rigorous research. Such studies are supported by a range of methodological approaches, including molecular biology (Menalled et al., ch. 6 this volume), marking and tracking methods (Lavandero et al., ch. 7 this volume) and remote sensing (Coll, ch. 8 this volume).

Aspects of theory and basic biology are also important, and the ecological bases of habitat manipulation and role of life-history studies (Nicholls and Altieri, ch. 3 this volume and Jervis et al., ch. 5 this volume, respectively) merit greater attention by future workers. A recently opened frontier related to this is the use of modelling to direct conservation biological control (Bianchi 2003; Kean et al. 2003). Of course, the production of appropriate models requires a knowledge of the biology of the species involved as well as an understanding of the more general ecological mechanisms by which species interact within food webs. Combining modelling with theory and empiricism offers scope to select the most appropriate strategies to use in manipulating key natural enemy attributes, such as search rate. Workers who focus on particular types of production system, such as agroforestry (Altieri and Nicholls, ch. 9 this volume) and cotton (Mensah and Sequeira, ch. 12 this volume) are increasingly likely to use modelling and other directed approaches in the future.

Genetically engineered crops are likely to have profound effects on agriculture as they become still more widely used, especially in developed countries. The net effect may or may not be beneficial, and whether the risks of proceeding outweigh the potential benefits is actively debated. As explored by Altieri et al. (ch. 2 this volume), genetically engineered crops do offer at least some scope to work synergistically with ecological engineering techniques, but negative effects may outweigh the benefits. The risk of negative consequences for farm biodiversity from genetically engineered crops increases the need for farm landscapes to incorporate features that will favour wildlife. Often, such ideas as conservation corridors can fulfil this function, as well as providing pest-management benefits. As argued by Kinross et al. (ch. 13 this volume), optimal outcomes for wildlife species require a better understanding of the ways in which animals respond to manipulation approaches and of the implementation of optimal strategies. This supports the notion of directed approaches for pest management referred to above.

Conclusion

For those involved in habitat manipulation research and development, this is an exciting time. As the discipline moves beyond the first approximation that diversity per se is the route to success, rapid progress is being made in the power and range of methods available to researchers, as well as in the development of theory. As methods and theory are integrated and more widely used, ecological engineering will evolve into a rigorous branch of ecology. Whether or not genetic engineering and ecological engineering achieve synergies or become entrenched as alternative paradigms for pest management, the development of the latter discipline into a more rigorous branch of ecology will allow it to contribute to the challenge of meeting our needs for agricultural products in a sustainable fashion. This trend will follow the history, over recent decades, of increasing levels of research activity in ecological engineering for pest management (Gurr et al., ch. 1 this volume) and make this approach a still more attractive research field. We hope this book will encourage scholars to build on the advances made by its contributing authors.

References

Andow, D.A. (1991). Vegetational diversity and arthropod population response. *Annual Review of Entomology* 36: 561–586.

Baggen, L.R. and Gurr, G.M. (1998). The influence of food on *Copidosoma koehleri*, and the use of flowering plants as a habitat management tool to enhance biological control of potato moth, *Phthorimaea operculella*. *Biological Control* 11 (1): 9–17.

Beecher, N.A., Johnson, R.J., Brandle, J.R., Case, R.M. and Yong, L.J. (2002). Agroecology of birds in organic and nonorganic farmland. *Conservation Biology* 16: 1620–1631.

Bianchi, F.J.J.A. (2003). Usefulness of spatially explicit population models in conservation biological control: an example. *International Organisation for Biological Control WPRS Bulletin* 26: 13–18.

Costanza, R., D'Arge, R., de Groot, R., Farber, S., Grasso, M., Hannon, B., Limburg, K., Naeem, S., O'Neill, R.V., Paruelo, J., Raskin, R.G., Sutton, P. and van den Belt, M. (1997). The value of the world's ecosystem services and natural capital. *Nature* 387: 6630, 253–260.

Crabb, J., Firbank, L., Winter, M., Parham, C. and Dauven, A. (1998). Set-aside landscapes: farmer perceptions and practices in England. *Landscape Research* 23: 237–254.

Drummond, C. (2002). Integrated farm management: balancing profit with environmental care. In *Proceedings of the Joint British Grassland Society/British Ecological Society Conference*. 15–17 April, University of Lancaster.

Frawley, B.J. and Walters, S. (1996). Reuse of annual set aside lands: implications for wildlife. *Wildlife Society Bulletin* 24: 655–659.

Gurr, G.M., van Emden, H.F. and Wratten, S.D. (1998). Habitat manipulation and natural enemy efficiency: implications for the control of pests. In *Conservation Biological Control* (P. Barbosa, ed.), pp. 155–183. Academic Press, San Diego.

Gurr, G.M., Wratten, S.D., Tylianakis, J., Kean, J. and Keller, M. (2004). Providing plant foods for insect natural enemies in farming systems: balancing practicalities and theory. In *Plant-derived Food and Plant–carnivore Mutualism* (F.L. Wäckers, P.C.J. van Rijn and J. Bruin, eds), Cambridge University Press, Cambridge. In press.

Hickman, J.M. and Wratten, S.D. (1996). Use of *Phacelia tanacetifolia* strips to enhance biological control of aphids by hoverfly larvae in cereal fields. *Journal of Economic Entomology* 89: 832–840.

Intachat, J. (2002). The role of biodiversity in insect pest management in tropical forest plantations. FORSPA Publication. FAO Regional Office for Asia and the Pacific, Bangkok, Thailand. No. 30: 67–74.

Kean, J., Wratten, S., Tylaniakis, J. and Barlow, N. (2003). The population consequences of natural enemy enhancement. *Ecology Letters* 6: 604–612.

Musters, C.J.M., Kruk, M., De Graff, H.J. and Ter Keurs, W.J. (1999). Breeding birds as a farm product. *Conservation Biology* 15: 363–369.

Paoletti, M.G. (1999). Using bioindicators based on biodiversity to assess landscape sustainability. *Agriculture, Ecosystems and Environment* 74: 1–18.

Pimentel, D., Wilson, C., McCullum, C., Huang, R., Dwen, P., Flack, J., Tran, Q., Saltman, T. and Cliff, B. (1997). Economic and environmental benefits of biodiversity. *BioScience* 47: 747–757.

Polaszek, A., Riches, C. and Lenne, J.M. (1999). *The Effects of Pest Management Strategies on Biodiversity in Agroecosystems. Agrobiodiversity: Characterization, Utilization and Management*, pp. 273–303. CAB International, Wallingford, UK.

Slangen, L.H.G. (1997). How to organise nature production by farmers. *European Review of Agricultural Economics* 24: 508–529.

Stevens, M.J., France, C.M., Wratten, S.D. and Frampton, C. (1998). Enhancing biological control of leafrollers (Lepidoptera: Tortricidae) by sowing buckwheat (*Fagopyron esculentum*) in an orchard. *Biocontrol Science and Technology* 8: 547–558.

Tooker, J.F. and Hanks, L.M. (2000). Influence of plant community structure on natural enemies of pine needle scale (Homoptera: Diaspididae) in urban landscapes. *Environmental Entomology* 29: 1305–1311.

Vickery, J., Carter, N. and Fuller, R.J. (2002). The potential value of managed cereal field margins as foraging habitats for farmland birds in the UK. *Agriculture, Ecosystems and Environment* 89: 41–52.

Wratten, S.D., Gurr, G.M., Landis, D.A, Irvin, N.A. and Berndt, L.A. (2000). Conservation biological control of pests: multi-trophic-level effects. In *Proceedings of California Conference on Biological Control II*, pp. 73–80. July 2000, Riverside, California.

Index